For Dr. An Wang

with kind regards.

Walter Freiberger

AF723156

Statistical Computer Performance Evaluation

Statistical Computer Performance Evaluation

edited by **Walter Freiberger**
Division of Applied Mathematics
Brown University
Providence, Rhode Island

with the collaboration of

Ulf Grenander
Division of Applied Mathematics
Brown University
Providence, Rhode Island

Barry H. Margolin
Department of Statistics
Yale University
New Haven, Connecticut

Rhett F. Tsao
Thomas J. Watson Research Center, IBM
Yorktown Heights, New York

Proceedings of a Conference held at Brown University, Providence, Rhode Island, November 22-23, 1971 under the Auspices of the Division of Applied Mathematics and the Center for Computer and Information Sciences, and supported by the Office of Naval Research

Academic Press New York and London 1972

ACADEMIC PRESS, INC.
111 Fifth Avenue, New York, New York 10003

United Kingdom Edition published by
ACADEMIC PRESS, INC. (LONDON) LTD.
24/28 Oval Road, London NW1

LIBRARY OF CONGRESS CATALOG CARD NUMBER: 72-77727

PRINTED IN THE UNITED STATES OF AMERICA

CONTENTS

SESSION III: DESIGN OF EXPERIMENTS

Chairman: F. J. Anscombe

SESSION IV: INPUT ANALYSIS

Chairman: W. Timlake

SESSION V: REGRESSION ANALYSIS

Chairman: Cuthbert Daniel

AUTHORS, DISCUSSANTS, AND SESSION CHAIRMEN

Harold A. Anderson, Jr., Thomas J. Watson Research Center, IBM, Yorktown Heights, New York 10598

F. J. Anscombe, Department of Statistics, Yale University, New Haven, Connecticut 06520

Sant R. Arora, Sperry Rand Corporation, Univac Data Processing Division, Roseville, Minnesota 55113

J. L. Baer, Computer Science Group, University of Washington, Seattle, Washington 98195

Y. Bard, IBM Scientific Center, Cambridge, Massachusetts 02139

H. Beilner, Department of Computer Sciences, University of Stuttgart, Stuttgart, Germany

L. A. Belady, Department of Computer Science, University of California, Berkeley, California 94720

L. W. Comeau, IBM Federal Systems Division, Gaithersburg, Maryland 20760

Cuthbert Daniel, Box 150, R.D. 2, Rhinebeck, New York 12572

Marvin Denicoff, Office of Naval Research, Arlington, Virginia 22217

John M. Feeley, Computer Sciences Corporation, Pasadena, California 91107

Walter Freiberger, Division of Applied Mathematics, Brown University, Providence, Rhode Island 02912

H. P. Friedman, IBM Systems Research Institute, New York, New York 10017

J. Gerald, University of Chicago, Chicago, Illinois 61822

Ulf Grenander, Division of Applied Mathematics, Brown University, Providence, Rhode Island 02912

Jerrold M. Grochow, Office of the Director of Information Processing Services, Massachusetts Institute of Technology, Cambridge, Massachusetts 02139

D. Hatfield, IBM Scientific Center, Cambridge, Massachusetts 02139

H. Hellerman, State University of New York, Binghamton, New York 13901

Z. Jelinski, McDonnell Douglas Astronautics Co., Huntington Beach, California 92647

Edwin R. Lassettre, IBM Systems Development Division, Poughkeepsie, New York 12602

M. M. Lehman, Thomas J. Watson Research Center, IBM, Yorktown Heights, New York 10598

P. A. W. Lewis, Naval Postgraduate School, Monterey, California 93940

Barry H. Margolin, Department of Statistics, Yale University, New Haven, Connecticut 06520

P. Moranda, McDonnell Douglas Astronautics Co., Huntington Beach, California 92647

J. F. Mount, National Cash Register Company, Dayton, Ohio 45401

R. G. Munck, Computing Laboratory, Brown University, Providence, Rhode Island 02912

R. P. Parmelee, IBM Scientific Center, Cambridge, Massachusetts 02139

Emanuel Parzen, Department of Statistics, State University of New York, Buffalo, New York 14214

M. P. Racite, IBM Systems Development Division, Poughkeepsie, New York 12002

N. Rasmussen, IBM Scientific Center Complex, White Plains, New York 10601

G. R. Sager, Computer Science Group, University of Washington, Seattle, Washington 98195

Robert G. Sargent, Department of Industrial Engineering, Syracuse University, Syracuse, New York 13210

M. Schatzoff, IBM Scientific Center, Cambridge, Massachusetts 02139

Allan L. Scherr, IBM Systems Development Division, Poughkeepsie, New York 12002

Akira Sekino, Project MAC, Cambridge, Massachusetts 02139

Martin L. Shooman, Department of Electrical Engineering, Massachusetts Institute of Technology, Cambridge, Massachusetts 02139

K. V. Suryanarayana, Fayetteville State University, Fayetteville, North Carolina 28301

W. Timlake, IBM Scientific Center, Cambridge, Massachusetts 02139

Rhett F. Tsao, Thomas J. Watson Research Center, IBM, Yorktown Heights, New York 10598

G. Waldbaum, Thomas J. Watson Research Center, IBM, Yorktown Heights, New York 10598

Peter Wegner, Division of Applied Mathematics, Brown University, Providence, Rhode Island 02912

Frank L. Wu, Sperry Rand Corporation, Univac Data Processing Division, Roseville, Minnesota 55113

A. C. Yeh, IBM Systems Development Division, Poughkeepsie, New York, 12002

P. C. Yue, Thomas J. Watson Research Center, IBM, Yorktown Heights, New York 10598

PREFACE

It was the purpose of this conference to investigate a new and promising field in computer science: the application of quantitative, and particularly statistical, methods to the study of computer performance. The present state of the art of the field is somewhat paradoxical: there is, on the one hand, a wealth of data, gathered, usually, in rather haphazard fashion; and there are, on the other hand, theoretical models, mostly of queueing theory type, whose stringent assumptions are not examined and whose conclusions are not verified. We have tried to exclude both types of papers and to accept only those that dealt with real data in a reasonably sophisticated manner.

The organizing committee of the conference consisted of Professor Walter Freiberger (Chairman), Brown University, Professor Ulf Grenander, Brown University, Professor Barry H. Margolin, Yale University, and Dr. Rhett F. Tsao, Thomas J. Watson Research Center, IBM.

We are very grateful to the Office of Naval Research, particularly to Dr. Robert J. Lundegard, Director of its Mathematical and Information Sciences Division, and to Dr. Marvin Denicoff, Director of its Information Systems Program, for their advice and support which made the conference possible. Mrs. Katrina Avery edited and typed the entire manuscript with dedication, skill, and dispatch.

We are, of course, particularly indebted to the speakers, discussants, and chairmen who made the conference a success, and who helped open up a field which will not only enhance our understanding of computer systems but possibly also contribute to statistical methodology.

Walter Freiberger
Professor of Applied Mathematics
Director, Center for Computer and Information Sciences
Brown University

FOREWORD

We have lived through two decades in which challenges for measures and proofs of computer system efficiency were thrust aside, or nimbly avoided by relentless and even dazzling innovation. As response to those nasty questions about efficiency, we provided more speed, larger memories, 1st, 2nd, and 3rd generation hardware; paging, time-sharing, multiprogramming, multiprocessing, the progression from assembly through natural language. We performed. Certainly, we did perform. But, I would suggest that this conference symbolizes the fact that, at long last, we are about to get down to the serious business of measuring that performance. The time has come to raise questions, as this conference will, of how we go about defining not simply what the measures are, but what should be measured, and how the measuring should be done.

The legacy of the past is frightening complexity as well as breathtaking development. From whose standpoint do we evaluate efficiency—the programmer, the individual user, the manager of the computer installation, the functional organization? In what time-frame do we apply our measurement instruments—a single day, a month, a year of computation? Are we concerned about optimizing the throughput of one job, some unique combination of jobs, all of the jobs? Are these concerns and interests even compatible?

Additional to these global questions, there remains the practical matter of measuring the utilization of computing resources; this interest is tremendously complicated by the advent of the multiprogramming, multiprocessing environment. In opposition to the philosophy that inherent to these or any hardware or software innovation was the promise of greater efficiency—we are beginning to see doubts and even suspicions of inefficiency for these same developments.

The managers once again are breathing down our necks. The time has come to bring in the statisticians and the mathematicians. I am pleased to be in the company of such brave men.

Marvin Denicoff
Program Director
Informative Systems
Office of Naval Research

Statistical Computer Performance Evaluation

SESSION I

GENERAL APPROACHES

CHAIRMAN: WALTER FREIBERGER

QUANTITATIVE METHODS FOR EVALUATING COMPUTER SYSTEM PERFORMANCE: A REVIEW AND PROPOSALS

U. Grenander and R.F. Tsao

ABSTRACT: The purpose of this paper is to appraise the state of the art of computer system evaluation and to review some quantitative methods (i.e., analytical, simulation and empirical methods) which are applicable to the problem. The main theme of the paper is that statistical techniques will have to be used to reach satisfactory results.

I. INTRODUCTION: STATE OF THE ART

1.1 The problems: comparing, tuning and designing

The problem with which we are concerned is of more recent origin than the digital computer. In the early days of computer technology, the question of evaluating a computer reduced to finding some of the physical parameters of the machine, such as the speed of fundamental arithmetic operations, memory capacity and I/O limitations. As computer technology advanced and the architecture of the systems tended to become more complex, it gradually became clear that one or a few figures of merit like the ones mentioned would no longer suffice for evaluating computing systems. One then faced a formidable problem and, although it has received much attention in the literature, no definitive conclusions have yet been reached as to the solution. Much controversy surrounds this question and it is the purpose of this paper to appraise the state of the art and to evaluate the methods that can be used. The main theme of the paper is that modern statistical methodology offers us powerful tools enabling us to reach satisfactory results.

Let us first take a look at what constitutes the problem. We shall describe it on three different levels.

Say, to begin with, that we are given two systems and we simply want to compare them from the point of view of a potential buyer. Physical parameters measuring speed and capacity are certainly necessary but they are not sufficient, and the reason for this is simply that the value of the system to a particular owner depends to a high degree on what sort of workload he will put on it. It is not possible to base the comparison only on such parameters since the performance of the system cannot be evaluated in a vacuum without taking the workload into account. The difficulty, or at least a part of it, lies in characterizing workloads and relating them to the characteristics of the system.

On the next level we start from a given system and wish to tune it: i.e., to change algorithms or other hardware and/or software components in such a way that higher efficiency is obtained. The performance of a system is typically a multiparameter phenomenon and it becomes very complicated when the interrelationship among these system parameters has to be taken into account. Again, the workload appears as a crucial part of the problem. Therefore, it is far from easy to advance a single convincing numerical criterion as a measure of the efficiency. We need a set of rationally selected criteria for the performance with meaningful economic interpretation.

On the third, and highest, level of ambition, we start already with the design of the system. An additional complexity is introduced at this level by the fact that in the process of designing a system, there exist enormous numbers of alternative decisions. It is often difficult, if not impossible, to trace the effect of a particular design decision on the overall performance of a system after it is built. A thorough analysis should lead to a method of predicting changes in efficiency of a system due to changes in its design. Until now, the emphasis at this stage has been on making the system work with less attention paid to efficiency, but the attitude, both of the computer manufacturers and of the buyers, is clearly changing. There is no doubt that a more systematic and quantitative approach is needed.

It would be too optimistic to hope for a drastic breakthrough on all of the three levels via quantitative methodology alone. We shall, however, show how some significant improvements in methodology can help at each level (i.e., comparison, tuing and design) of system performance evaluation. Before turning to this, however, it will be useful to look at some ways in which the problem has been approached.

1.2 Current approach to evaluation: an appraisal

There are quite a few papers presenting a general survey and review of computer system performance evaluation; e.g., Calingaert [4] and Drummond [9]. The main purpose of this section, however, is to present our personal views on the appraisal of current approaches and to set the stage for discussion of some quantitative methods in the next section. Our appraisal of current approaches to evaluation can be summarized as follows:

(1) There appear to be two divergent efforts, which we shall refer to as analytical studies and system-oriented studies. Analytical studies include, in the most part, the probabilistic model building of queueing theory. System-oriented studies are mainly concerned with the actual measurement and simulation of a real computing system. There is no question that both efforts have been proven worthwhile, and we shall discuss them at length in the next section. There exists, however, a large gap between the complexity of the models theoretically analyzable and that of even the simplest real systems. It is questionable if the actual design of a real system has benefited from the analytical studies of oversimplified models. On the other hand, for some reasons which will become clear later, we doubt that the measurement and simulation activities of a particular system have improved significantly the general understanding of computer systems.

(2) As computer systems become more complex, the choice of criteria for evaluating systems becomes more difficult. Some effort has been made in the search for reasonable criteria either in tuning or comparing systems, e.g., Stimler [22] has proposed six criteria for time-sharing system performance, Schatzoff et al. [19] have applied another set of six measures for comparing time-sharing and batch-processing systems. It is clear from these studies that by its very nature the evaluation of system performance is a multidimensional problem. However, most efforts stop short of a quantitative investigation of the trade-off among these performance measures. The measures themselves are usually in terms of averages whose use is often necessitated by the primitive stage of the methodology applied. For example, one could argue that, from a time-sharing-user viewpoint, the variance of the response time is as important as the mean of the response time. One common characteristic which

is shared by all proposed criteria is that they all depend on what particular workload they are based on. It can be said, of course, that this is just as it should be, since the evaluation of systems must depend upon what sort of jobs they are called upon to execute. A more serious criticism is that they do not clearly express how they depend upon the workload. A method of evaluation should enable us not only to compare systems for a given workload but also to see how the evaluation is affected if the characteristics of the workload are changed.

(3) One major difficulty in the evaluation of computer systems seems to be the lack of quantitative understanding of the relation between workload and performance. Many techniques have been introduced to tackle this problem; e.g., instruction mix, kernel, benchmark, etc. Their success has been very limited. This is because the characterization of workload itself still leaves much to be desired. We believe that no real significant advance in the evaluation of systems can be expected until some breakthrough is made in the characterization of the workload. It is by no means an easy task. Some work has been done in this area under the broad heading of "program behavior" (e.g. [2], [7], [13, [25]).

(4) One area which has received much attention and has advanced impressively in recent years is the design of measurement tools, i.e. hardware or software devices for monitoring the events in a computer system. It is fair to state that today we can obtain almost any data we wish from a computer run if we are willing to pay the cost for measurement (especially in using software measurement tools). However, there are still problems remaining to be solved. In general, the hardware monitors lack selectivity of measurement (they must analyze the CPU at each instruction cycle of execution time) and the software monitors are slow and often cause considerable degradation. The results of the measurement process are sometimes suspicious because of this degradation. Nevertheless, the difficulty is not that we lack data; on the contrary, there is an abundance of data. Usually, little attention has been given to the design of data collection so that the result is a wealth of data which are difficult to interpret. Thus much of the measurement effort represents a waste of time and money. This, we believe, is due to too much attention being paid to how to collect data rather than what data to collect and what analysis to perform after the data have been collected.

(5) The discussion above naturally leads to, in the authors' view, the weakest point of the current approach to evaluation, namely the lack of quantitative methodology on which to base both evaluation of current systems and analysis of real systems data for model formulation and validation. It is because of this lack of methodology that most probability models have been presented without empirical validation and that masses of measurement data have been collected without proper analysis. We shall devote the rest of this paper to dealing with this subject.

II. THE QUANTITATIVE APPROACH

2.1 What is it?

In order to place this paper in its proper perspective, it is necessary to consider what quantitative methods are in the present context. By quantitative methods we mean those techniques applicable to problems arising from the following three phases of computer system evaluation:

(a) Collection of data: this phase involves the planning and conducting of the experiment for data collection as well as the techniques of getting measurement data from a computing system, i.e. measurement tools.

(b) Analysis of data: this phase consists of construction of models and estimation of parameters in the models as well as validation of models.

(c) Interpretation of data: this phase concerns the summarization of the results and new insights gained in the study as well as making decisions based on the results.

The emphasis of the application of quantitative methods is on the convergence of two aspects. The first aspect is the reliance on data either from experimentation on the real-life environment or from simulation. The second is the use of mathematical models. It is easy to collect massive amounts of confusing data unless one has some model (hypothesis) of a system in mind. On the other hand, a model without empirical validation is at best an intellectual exercise. A review of the state of the art of computing system evaluation reveals that there is a wealth of data gathered in rather haphazard fashion, and theoretical models, mostly of queueing type, whose assumptions and conclusions are not verified by real-life data.

The above unsatisfactory situation has existed in many

other fields and the way out has always been to focus attention on critical items such as selection of criteria and definition of goals as well as to apply powerful methodology to gain the insight we need. Some quantitative methods have already been applied successfully in other fields. To cite a few examples, the techniques of queueing theory and simulation have been used successfully to solve problems in telephone networks; experimental design has been demonstrated as a useful tool in the fields of agriculture and chemistry; regression methods are applied widely in economics and the behavioral sciences. What is needed in the area of system evaluation is to go beyond the domain of computing, and apply methods that have been proven useful in other fields.

Specifically, we shall classify quantitative methods into the following three categories:

a. Analytical methods
b. Simulation methods
c. Empirical methods

We shall discuss these methods in detail in the remainder of this paper.

2.2 Analytical methods

In the analytic approach, we try to describe the functioning of a computer system or one of its (hardware or software) components in rigorous mathematical terms; then we construct a mathematical model of it. Often the model will have randomness in it, and we shall have to appeal to the theory of probability for the solution. Modelling is not an easy task and it is often necessary to simplify the model to make it amenable to mathematical analysis, remembering that any mathematical model is only an approximation of real-life events.

What is typical of a computer operation is that the scarcity and cost of different resources force the requests for them to compete with each other. One branch of probability theory, queueing theory, has been developed to deal with such situations (originally for telephone traffic).

Anyone working with systems is aware of their high degree of complexity, and he is likely to be shocked when he sees the models suggested. He may react negatively when he notices how much of the real system has been left out and how restrictive the assumptions are that have been made in the analysis. He may even reject the whole idea of analytic study in this context. We shall try to evaluate how valid this skepticism is.

But first, let us examine the usual assumptions made in the analysis. The independence and the related assumptions on exponential distribution that crop up in different forms play a crucial role in queueing analysis: they help enormously in reducing the mathematical difficulties. The reason is that they ensure a "memoryless" property; e.g., the probability of getting a new request does not depend upon how long ago the last request was made. This leads to a drastic decrease in mathematical complexity.

Indeed, under these assumptions the equations that govern the performance of the system will consist of ordinary differential equations which can be solved analytically or at least numerically. Otherwise we would get some much more difficult functional equations. A further simplification will be achieved if we are only interested in the stationary (statistical equilibrium) behavior after it has settled down to steady state. With this additional assumption, the time derivatives vanish and we have simply to solve a system of linear equations.

Unfortunately, these assumptions are questionable for real systems since there is no empirical evidence available for them a priori. Intuitively speaking, it is clear that very short values of the interarrival times between requests, say, are not at all likely since the time scale of the human activity that leads to a request is of the order of magnitude of one tenth of a second. This does not agree with the exponential distribution which assigns the highest probability density to the smallest time interval of length zero. Another, less serious, objection is that the assumption does not take into account the variation of the intensity of computing activities over the day: in a typical time-shared setup the maximum would occur sometime in the middle of the day and much less intensity in the morning and evening. This could be taken into account, however, if we allow arrival intensity to vary slowly with the time of day, assuming, of course, that we know how it varies with time. The same remark could be made about time dependence in other parts of the computing system.

As for the stationarity assumption, it should be noted that this assumption is often misunderstood when discussing computer systems. It is often argued that the stationary solution is of no practical interest since it is never met in actual operation. To back up this argument reference is made to measurements of quantitities like CPU utilization, fraction problem state, I/O activities, or number of page faults per time unit. When such quantities are plotted against time they typically display intervals with heavy activity separated by other intervals with low activity. But this does not disprove stationarity. It is well known that there are stationary stochastic processes possessing exactly this type of "regime-like" behavior. Certainly, more empirical evidence is needed to support or disprove this assumption.

Even assuming that the queueing model allows analytic treatment, the problem still remains for what quantities to solve. For example, while studying the response time we might like to find the probability distribution of the response time, or actually prefer to compute the conditional probability distribution of the response time given that the request was of a certain type, say fixed CPU-time requirement, or belonged to a certain command type. It is often difficult to find the entire probability distribution and we have to compromise by just computing something less informative, such as the mean value or, in some cases, the standard deviation. There is no question that some information has been lost in the process.

With all the difficulties mentioned above, queueing modeling still remains one of the few analytic tools for studying computer systems. Since the concept of time-sharing and multiprogramming was introduced, queueing models have been appearing in the literature of computer science with increasing frequency. For a general survey of these works see, for example, McKinney [14] and Coffman [5].

In particular, it is interesting to note that in one of the earliest and best studies of its kind, Scherr [21] investigated a system (CTSS) analytically and supplemented his study both with simulations and measurements of the real system. Although his mathematical model was quite simple and included strong assumptions of the type mentioned above, it turned out that it described the actual functioning of the real system quite well. For example, the model predicted the average response time quite accurately under different operating conditions.

Another interesting paper is that of Lewis and Shedler [12]. These authors study a multiprogrammed system of two-level memory with demand paging. The main feature of their paper from the system viewpoint is the explicit representation of system overhead in the model. Mathematically, all fundamental stochastic variables but one have arbitrary probability distributions. The execution intervals, however, are assumed to be independent of each other and exponentially distributed. The model is more detailed than others of its kind and the authors manage to push through their mathematical analysis by the sophisticated use of advanced queueing theory. In spite of this impressive mathematical analysis, it is questionable whether this sort of approach would be practical if one wished to include more system features or to generalize the distribution even further and remove, say, the independence assumption for the execution intervals.

In addition to the stringent assumptions made in queueing analysis, another difficulty which has received too little attention to date is the empirical estimation of the parameters of queueing models and the calibration of the analytical result with real-life systems. It is because of these restrictions that we believe that the reliance on analytic methods will not be sufficient for the evaluation of computing systems. However, this does not mean that they should be discarded, for the following reasons.

First, the mathematical formalization and simplification is of value in itself since it forces us to be very precise in stating all the assumptions. It is not always admitted that we do make assumptions, perhaps more vague and fuzzy, in a more qualitative analysis. It must be better to bring out these hidden assumptions into the daylight for critical examination.

Second, when we go ahead to simulation we still need the mathematical model; the difference is that we solve it not analytically but by Monte Carlo techniques. Even the statistical approach employs the notion of the model, although these models are likely to be quite different and more phenomenological in nature.

Third, subsets and components of the whole system are more likely candidates for a successful treatment by analytic methods than are entire systems. We expect to see more of this in the future. An illuminating example of what can be achieved in this direction is the paper by Wang and Ghosh [26].

Summing up, we conclude that queueing models alone are not sufficient to study and evaluate computing systems, but that they help in understanding the behavior of the system. Other methods will have to be used in conjunction with them.

2.3 Simulation methods

If the system is too complicated to be amenable to complete analytical study, we can turn to simulation instead. Since this term is not always used in the same sense as in this paper, it should first be clearly defined. By simulation we mean here the process in which a model of the system is run with the job stream generated by a program written so that its output mirrors as closely as possible the statistical structure postulated for the sequence of requests, with their various attributes, as they occur in time.

This means that we still have to decide on a model, but what is different from the analytical approach is that we no longer look for closed-form solutions for the quantities of interest. Instead, we adopt the more limited goal of getting statistical estimates for them, based on the output of the simulation.

We gain in realism since we are no longer forced to impose more or less artificial assumptions for analytical convenience. This makes the simulation more lifelike. Indeed, this is the reason why simulation has been the most frequently used quantitative method in all three levels of computer evaluation, as mentioned in section 1 (e.g. [11], [17]).

However, we do not get this free. First, any practitioner of simulations knows that they tend to be expensive if a high degree of realism on a detailed level is required. This is not surprising when one takes into account all the processing of queues and lists that is needed. Hence, the accuracy of a simplified simulation model as a representation of the real-life system is often questioned. Moreover, since the results are statistical in nature, their precision can be measured by the standard deviations of the estimates obtained. Typically, such standard deviations are of the order of magnitude $1/\sqrt{N}$, where N is the number of iterations during the simulation. Therefore, the standard deviations decrease slowly as N increases and high precision may demand an economically impossible size of the simulation experiment. For reasons of precision, it is advisable to design the

experiment more carefully and not just rely on the brute force of the computer. There are various techniques for doing this and we cannot go into detail here. One is the so-called method of antithetic variables and is described by Page [18]. Another consists of relying on an analytic study of a similar but simpler system: both the real and the simplified system are simulated at the same time and the analytic results are exploited via a regression equation relating them to the ones sought (see Gaver and Shedler [10]).

Sometimes a different but similar procedure is used. The main difference is that while the input to a simulation is generated artificially by a specially written program, in the second approach, sometimes spoken of as emulation, one employs real input. Very often this is done on the system itself rather than on a simplified model of the system. The obvious disadvantage of an emulation compared to a simulation is that the results apply only to the situation represented by the actually recorded data that were used as input, so that we get either little or no insight into the performance under other operating conditions. For the simulation, on the other hand, we can vary significant parameters in order to get other job load characteristics that may be more typical for the environment that we aim for. Actually, the emulation may be more like the benchmark approach and suffers from the same limitations, while the simulation, if well done, can lead to more generally useful conclusions.

In spite of this, many systems experts have mixed feelings about simulation, and for a very good reason. They feel that there is often too much arbitrariness in the choice of input generations. It is true, of course, that we need not assume exponential distributions, independence, etc. as in the analytic models; nevertheless, we must define - in some way - how requests are generated in time and what their attributes will be. The simulation requires a model for input and with the rather sketchy knowledge we have at present about this demand mechanism there is considerable danger that we may commit serious mistakes while specifying the input generation. This points out the need for more thorough statistical studies of empirical data.

The same is true for smaller simulations intended to shed light on only a part of the system. To be concrete, let us say that we want to measure the performance of two or more paging algorithms. We then have to model program behavior: the statistical properties of the instruction sequence have to be specified as well as the statistics of the data used in conjunction with the program. This, however, is something we know little about at present. (For an excellent treatment of this subject, see Lewis and Yue [13].) It is tempting to use a Markov chain model for the references to the various pages that the program is made of, but it is not a very promising avenue. First, it is well known that the Markov specification does not give a close fit to actual programs. Second, even if it were true, we would have to specify all the entries in the probability transition matrix and it is not clear how we should determine all these parameters, perhaps hundreds of them.

The authors believe that the study of demand structure and especially program behavior deserves a penetrating study. It is likely that knowledge about this will be of value for future systems, whether paged or not, and independently of how computer architecture develops.

Now some more technical aspects of simulation. There has been much discussion in the computing literature, e.g. [23], about special languages for simulation, and many packages such as GPSS and SIMSCRIPT are now available. We are not convinced that much can be gained by using them. Instead, it may be just as easy to write the program in some general-purpose language since what is gained in convenience when using the special language is often lost through their relative lack of flexibility. If the simulation is large or moderate, in effect it is a good idea to write a scaled-down version to begin with, run it and try to increase the precision by devices such as those mentioned above. Such pilot experiments can offer valuable hints that can lead to great savings in the final run of the full-scale experiment. In this connection it should also be emphasized that a careful design of the statistical experiment will result in highly increased accuracy. We will have more to say about design of experiments in the next section.

To sum up: the main effort should go into what is the weakest link in simulations of this type, the specification of the demand mechanism. Only if this is done well can we expect meaningful results from the simulation. In order to do this we must turn to empirical methods and statistics gathering.

2.4 Empirical methods

Empirical methods are quantitative techniques for obtaining and analyzing data. These methods are mostly in the domain of statistics. However, the kind of empirical methods described in this paper are more in the spirit of indication through statistical analysis of data rather than formal statistical inference through rigorous distributional assumptions (see Mosteller and Tukey [16]). We believe that the latter course is not always open to us in the study of system evaluation problems. This may be due to one or more of the following reasons:

(a) A good assessment of uncertainty in this field often seems difficult,
(b) The existing formal inference procedures usually employ such stringent assumptions as to make their use unwise or misleading,
(c) No formal inference procedure has been developed for the problems under investigation.

What has been done in system evaluation is, so far, mostly a combination of intuition and experience, which should be transformed into a systematic body of knowledge with its own principles and working rules. Empirical methods appear to constitute a good starting point for such an undertaking.

As stated before, masses of measurement data for computers are gathered nowadays, particularly by monitoring devices. It is questionable that any meaningful result can be obtained by this "shotgun" approach. For any method to produce meaningful results it is necessary that the basic conditions for the realization of those results be inherent in the data. There is no point in collecting data for the mere satisfaction of accumulation. We should discipline ourselves not to collect data aimlessly; rather, we should collect data by planned experiments. The objective of any experimental investigation is to learn more about the system being investigated; i.e., to study the effect of variation of certain factors or the relation between certain factors in a system. The technique of experimental design is by no means foreign to scientific investigation in various fields ranging from the physical sciences to the social sciences. However, the authors know of only a few attempts to evaluate computer performance by careful design of experiments. In this connection we note that in one of the earliest studies of its kind,

Schatzoff et al. [19] conducted an experimental comparison of time-sharing and batch processing. Also, Morfield et al. [15] presented three experiments of response delay in a time-shared, on-line computing system. A comprehensive multi-factor paging experiment is reported by Tsao et al. [24]. It is fair to say that these are the exceptions rather than the rule in the study of computer performance, but that they indicate some promising possibilities which should be pursued more systematically.

One major difficulty in evaluating computer performance is the very large number of factors and their relationships involved in the problem under investigation. One is harassed by extraneous factors which influence the outcome of the evaluation, factors which one would like to eliminate but cannot control completely. Experimental design is concerned with these factors. The utility of a design is to eliminate these factors if possible or to minimize their effects by arranging the experiment so that the effects may be expected to cancel and partially cancel each other in the analysis of the resulting data from the experiment. Such a design may be nothing more than an obvious application of common sense, or it may be a highly elaborate arrangement based on the mathematical theory of finite geometries.

It is impossible for us in this paper to catalogue the literally hundreds of experimental designs that might be introduced. Instead, we shall point out some of the more important simple designs and indicate how they may contribute to the study of system performance. Detailed discussion of many experimental designs can be found in the statistical literature, e.g. Davis [6].

There are two general types of experimental designs, systematic and randomized. Systematic designs are those which do not have an element of randomness in them, while randomized designs imply randomization among the experimental units. The chief advantage of randomized design relative to systematic design is that it is subject to valid statistical analysis. We shall concentrate only on the randomized designs in this paper, since we believe that they will play a more important role in the evaluation problems.

The simplest of all randomized designs is the completely randomized design. The "treatments" are assigned completely at random over all experimental units. The term "treatment" in the design of experiments is used generally to denote any controlled variation in the conditions under investigation. It may be different computing systems, different algorithms, different workloads, and so on. The completely randomized design is suitable only if one is willing to assume a fairly constant experimental environment. This is rare in computer science.

Attempting to maintain a constant experimental environment while looking at the variable of interest, one can apply the principle of blocking. One typical design of this kind is known as the randomized block design. This design controls variation in one direction. Another is that of the Latin square design which controls source of variation in two directions. We shall illustrate the Latin square design by an oversimplified artificial example.

Suppose that we wish to compare three systems X, Y and Z using, say, CPU time as a basis for comparison. Further, suppose that each system will execute three of the following nine jobs J_i, i=1,2,...,9. The attributes of these nine jobs are:

Languages \ Usage	Scientific	Commercial	Mixed
FORTRAN	J_1	J_2	J_3
PL/I	J_4	J_5	J_6
ALGOL	J_7	J_8	J_9

A Latin square design would arrange the experiment in the following manner:

Y: J_1	X: J_2	Z: J_3
X: J_4	Z: J_5	Y: J_6
Z: J_7	Y: J_8	X: J_9

Note that the Latin square arrangement places all systems, X, Y and Z, in each row as well as each column with some degree of freedom. That is, there are a number of 3×3 arrangements that place all three treatments in each row and column. These arrangements will allow us to compare the differences among systems by blocking out the effect of languages and usages. We select one at random so as to permit some random assignment of experimental errors to the treatments.

Frequently in computer science problems arise in which the effects of a number of different factors on computing systems are to be evaluated. For example, in choosing CPU time as response (or dependent variable) the factors may be different memory sizes, different replacement algorithms or different jobs, etc. The different alternatives of a factor examined in an experiment are known as levels. Provided the plan has been correctly chosen, it is then possible to determine not only the effect of each individual factor but also the way in which each effect depends on the other factors; i.e., interaction. The concept of interaction between factors in an experiment is extremely important. If interactions are present, one cannot study the effect of one factor independently of the other. The individual effects alone are misleading and the response must be investigated at each combination of various levels of interacting factors. In this situation, the usual "one-factor-at-a-time" method is not suitable. In that method, each factor is investigated separately by observing various levels of a factor one at a time while holding all other factors constant. This method is not only inefficient but, more importantly, will not be able to detect factor interactions, and we may miss some critical points if we employ this method.

In evaluation of computer performance, we should aim at finding the interaction between factors. The factorial experiment is very useful in this environment. A factorial experiment is one that includes all combinations of the levels of the factors, taking one level from each factor. For example, if there are four factors and each factor has three levels, the complete factorial experiment would require 3^4 (81) runs. An application of factorial experiments in the evaluation of computer performance can be found in Tsao et al. [24]. It should be noted that the completely randomized, randomized block and Latin square designs will provide no information on factor interactions. They are more applicable to the comparing level of evaluation, while the factorial experiments are more applicable to the tuning and design levels.

In a well-designed experiment, consideration must be given to methods of analyzing data once they are obtained. The analysis of variance is a statistical technique for analyzing measurements from experimental designs, i.e., to decide which kinds of effects are important and to estimate the effects. There are three underlying assumptions made in the analysis of variance:

(a) normality of the experimental errors,
(b) equality of variance of the errors,
(c) statistical independence of the errors.

To date, there is not enough empirical evidence to support any one of the three assumptions in problems of system performance. The effects of departure from the underlying assumptions have been treated by Scheffé [20]. In general, as one would expect, the effect of departure can be very serious on formal inference such as significance tests.

Before leaving the subject of experimental design, it should be pointed out that experimentation is an iterative learning process. We must think in the overall context of gathering knowledge as well as of the knowledge gained by any one particular experiment. There is no single experiment that will solve the problem once and for all. Each experiment should be viewed as one of a series of experiments. The steps in this iterative learning process are: conjecture which leads to a design, design which leads to an experiment, experiment which provides data that are analyzed before forming a new conjecture. This process is repeated as many times as necessary, with new knowledge being gathered at each step. A well-designed experiment will speed up the learning process as much as possible.

We have stated that one of the basic empirical methods in collection of computer data is the use of experimental design techniques. However, there are circumstances in which it is rather difficult to experiment. The experimentation in computing systems often requires complete control of the system to execute every design point. It will usually demand a dedicated system and thus interrupts normal operation. Furthermore, there already exists a tremendous volume of data which are routinely compiled in a matter of every few seconds. What are we going to do with these data? Can we make any sense out of them? It seems that what we need are some techniques of extracting from masses of data the main features of the relationships hidden or implied in these figures.

In monitoring a computer system, there may be ten or twenty variables whose quantities are changed and recorded simultaneously, e.g. CPU time, I/O time, etc. It is of interest to examine the effects that some variables exert on others. There may be a simple functional relationship between variables. Or, more frequently, there may exist a complicated functional relationship which we may wish to approximate by some simple function, easy to grasp. Even where no physical relationships exist between variables, we still may be interested in summarizing a particular set of data by some simple function. The function may not have any physical meaning. But it may be extremely valuable for predicting the values of some variables by the values of other variables appearing in the function. Regression analysis is a collection of techniques for data analysis which are appropriate in obtaining such a mathematical relationship. In this paper "regression" will be used in its broad sense to include those techniques concerned with the fitting of straight lines, curves, or surfaces to sets of observations where the fit for one reason or another is imperfect. Extensive treatments of regression analysis can be found in nearly every standard statistical textbook, e.g. Draper and Smith [8].

Generally speaking, regression analysis deals with the following problem. Given a set of data containing the observations for several (independent) variables $X_1, X_2, \ldots, X_k$ and a (dependent) variable, Y, we want to fit these data by a function $Y = f(X_1, X_2, \ldots, X_k)$. For instance, we could choose a linear model, i.e.

$$Y = a_0 + \sum_{i=1}^{k} a_i X_i$$

where a_i, $i=0,1,\ldots,k$ are unknown parameters.

There are at least three questions that can be raised at this point:

(a) What criteria should be chosen to measure goodness of fit of the proposed model?

(b) How should the parameters a_i be estimated?

(c) Given (a), how well does the model fit the data?

Technical details involved in answering the above questions are omitted here, and for these topics the reader is advised to consult the above references. Instead, we shall review some applications of regression analysis to the computer performance problem.

It is interesting to note that although regression analysis is one of the most used and sometimes abused statistical methods in various fields, its application to the computer performance problem is relatively rare. In this connection we should mention one of the most ambitious attempts of its kind: the work of Bard [2].

Bard applied regression analysis to a set of data collected by monitoring an IBM System/360 model 67 computer running under the CP-67 time-sharing operating system at the IBM Cambridge Scientific Center. It is stated that "regression techniques were employed to estimate average CPU supervisor state time (CP time) for each of a number of significant system functions such as paging, spooling, and virtual I/O's; predict CP time, given measurements of significant system functions; estimate average percentage of CP time spent in each of the significant system functions; identify and investigate the source of aberrant observations; and estimate the effects of system hardware and software modifications." (Bard et al. [1]). The major variable of interest (or dependent variable) is CP time. The attempt is to relate CP time in a linear fashion to a set of potential explanatory variables (or independent variables).

Although it is somewhat doubtful, in our opinion, that Bard has achieved the objectives stated above, his work is definitely a pioneering effort in the application of regression techniques to the computer evaluation problem. It raised many interesting questions such as the physical interpretation of the fitted regression coefficients, the effect of somewhat arbitrary procedures of rejecting the so-called "outliers", etc. It also demonstrated the utility of regression techniques in handling masses of measurement data. There is no doubt in our minds that regression analysis will be a major tool in the evaluation of computer performance, especially at the <u>design</u> level. Its importance as a tool of learning consists in the fact that it permits us to relate to values of one variable the <u>expected</u> values of the other variables. Therefore, in this sense, regression analysis could be viewed as a tool of <u>scientific prediction</u>.

An important property of any predicting equation is its stability. When using regression prediction, there is an implicit assumption that conditions do not change as far as regression models are concerned. Numerous examples can be given for a change in circumstances which will make a relationship worthless for prediction. It is well to remember not only what independent variables entered the regression function but also under what conditions the regression function is applicable.

There are many other useful statistical techniques available to the study of computer performance problems, e.g. correlation analysis, time-series analysis, multivariate analysis, etc. Those we have described above are the ones which, in our opinion, consist of the most useful tools to the problems in the exploratory stage of the evaluation of computer performance. The use of more sophisticated statistical techniques usually requires more restrictive assumptions. At the present time we have little theoretical or empirical evidence to support these assumptions and justify the application of these methods.

To sum up: the empirical methods are the most promising tools for studying system performance problems. We should apply and develop statistical techniques (a) to strengthen the analytical methods by validation of the assumptions and estimation of the parameters in the analytical models, and (b) to provide a data-analytic tool for measurement or simulation of real systems.

III. CONCLUSION: THE ASSESSMENT

Our personal views can be summed up as follows. There is no doubt that the task of evaluating the performance of computer systems is of utmost importance, both to computer manufacturers and to users of computers. It deserves a more systematic study than it has received until now and we believe that modern statistical methodology offers powerful tools that have not yet been exploited to their full capability for this purpose.

While we do not deny the importance of the analytical methods, it seems that their usefulness to the problem of evaluation is severely restricted. Simulation methods are, no doubt, more widely applicable, but the main obstacle that has stood in the way is the lack of understanding of the statistical structures created in a computing system. It will not be easy to model these structures, but we believe that quite sophisticated and useful analyses will result in due time from a determined research effort in this direction. In particular, it seems imperative to learn about the input side, and model the demand behavior of users and of their programs. Program behavior will probably develop into an active area of computer science, although at present we can only see vaguely the outline of the theory. The internal statistical structures that arise as a consequence of any type of program behavior will also require more attention.

The main tools for learning about compumetrics, to use a newly coined term, will be statistical means of analysis: data analysis, design of experiments, regression analysis, to mention only a few of the most likely ones. We have advocated their use in the earlier sections of this paper, admittedly on rather loose grounds since they have not yet proven their full value; but it already seems clear that they promise considerable rewards. At the same time, it is likely that statistical theory will profit from its application to computers and that this will result in new models and methods of analysis.

Acknowledgment. The authors wish to thank L. Belady, H. Ernst and B. Margolin for reading the manuscript and making many valuable comments.

REFERENCES

1. Bard, Y., Margolin, B. Peterson, T. and Schatzoff, M. (1970). CP-67 measurement and analysis I: regression analysis. IBM Cambridge Scientific Center Technical Report G320-2061.
2. Bard, Y. (1970). CP-67 measurement and anlysis II: overhead and throughput. IBM Cambridge Scientific Center Technical Report G320-2062.
3. Belady, L.A. (1966). A study of replacement algorithms for a virtual storage computer. IBM Syst. J. 5,2.
4. Calingaert, P. (1967). System performance evaluation: survey and appraisal. Comm. ACM 10,2.
5. Coffman, E.G. (1967). Studying multiprogramming systems with the queueing theory. Datamation 13,6.
6. Davies, O.L. (1956). The design and analysis of industrial experiments. Hafner, N.Y.
7. Denning, P.J. (1968). The working set model for program behavior. Comm. ACM 11,5.
8. Draper, N.R. and Smith, H. (1967). Applied regression analysis. Wiley, N.Y.
9. Drummond, M.E., Jr. (1969). A perspective on system performance evaluation. IBM Syst. J. 8,4.
10. Gaver, D.P. and Shedler, G.S. (1971). Control variable methods in the simulation of a model of a multiprogrammed computer system, to appear in Naval Res. Logistic Quart.
11. Katz, J.H. (1966). Simulation of a multiprocessor computer system. Spring Joint Comp. Conf. Proc.

12. Lewis, P.A.W. and Shedler, G.S. (1970). A cyclic-queue model of system overhead in multiprogramming computer systems, IBM Res. Rpt. RC 2814.
13. Lewis, P.A.W. and Yue, P.C. (1971). Statistical analysis of program reference patterns in a paging environment. Presented at IEEE 1971 Computer Conf., Sept. 1971, Boston; also to appear as IBM Res. Rpt.
14. McKinney, J.M. (1969). A survey of analytical time-sharing models. Comp. Surveys 1,2.
15. Morfield, M.A., Wilsen, R.A., Grossberg, M. and Yntema, D.B. (1969). Initial experiments on the effects of system delay on on-line problem solving. Tech. Note 1969-5, Lincoln Lab., M.I.T.
16. Mosteller, F. and Tukey, J.W. Data analysis, including statistics. In Handbook of social psychology (G. Lindzey, ed.). Addison-Wesley, Reading, Mass.
17. Nielsen, N.R. (1967). Computer simulation of computer system performance. ACM National Meeting Proc.
18. Page, E.S. (1954). The Monte Carlo method solution of some integral equations. Proc. Camb. Phil. Soc. 50.
19. Schatzoff, M., Tsao, R., and Wiig, R. (1967). An experimental comparison of time-sharing and batch processing. Comm. ACM 10,5.
20. Scheffé, H. (1959). The analysis of variance. Wiley, N.Y.
21. Scherr, A.L. (1965). An analysis of time-shared computer systems. MAC-TR-18, M.I.T. Project MAC, Cambridge, Mass.
22. Stimler, S. (1969). Some criteria for time-sharing system performance. Comm. ACM 12,1.
23. Teichroew, D. and Lubin, J.F. (1966). Computer simulation - discussion of the technique and comparison of languages. Comm. ACM 9,10.
24. Tsao, R.F., Comeau, L.W. and Margolin, B.H. (1971). A multi-factor paging experiment: I. The experiment and the conclusion. These Proceedings.
25. Varian, L., and Coffman, E. (1968). Further experimental data on the behavior of programs in a paging environment. Comm. ACM 10,2.
26. Wang, C.P. and Ghosh, S.P. (1969). Analysis and modeling of a multi-module disk storage system with sector scheduling. IBM Res. Rpt. FJ 579.

A UTILITY THEORETIC APPROACH TO EVALUATION OF A TIME-SHARING SYSTEM

Jerrold M. Grochow

INTRODUCTION

One of the problems faced by the manager of a computer center in trying to provide a better time-sharing service is the determination of the relative worth to users of taking one action versus another. For instance, he must determine how his users might react to better reliability versus decreased response time or availability. He has to determine then how a particular investment (in a programmer, equipment, software, etc.) will affect reliability, availability and response time. Such determinations are made on an empirical basis every day. The purpose of this paper is to outline a possible procedure for the formalization of management decision-making in this area. The procedure draws heavily on the methods of statistical decision theory, particularly as they relate to assessment of preferences and multi-attributed utility functions.

THE METHOD OF ANALYSIS

In order to determine the system characteristics relevant to the decision-making process, we will construct a hierarchical structure which contains a collectively exhaustive (although not necessarily mutually exclusive) set of time-sharing system characteristics, $S = \{s_i\}$, which can be used to evaluate possible outcomes of any action. We will then attempt to determine a multi-dimensional utility function for each system user with values of these attributes as parameters, $U_i = U_i(S)$. User groups, G_j, will be defined on the basis of similarities in their utility functions and the utility function of the group will be taken to be an approximation of the user functions:

$$U_{G_j} = U_{G_j}(S) = f(U_1(S),U_2(S),\ldots,U_r(S)) \qquad (1)$$

where f is some approximating function. The utility value of the current system can then be determined for each user group by evaluation of all attributes and computation of the utility values:

$$U^0_{G_j} = U_{G_j}(s^0_1, s^0_2, \ldots, s^0_n) \qquad (2)$$

Similarly, a particular action can be analyzed for its probable contribution to changes in the attributes and new utility values can be calculated using new values of the attributes:

$$\text{Action: } s^0_i \rightarrow s^1_i$$

$$U^1_{G_j} = U^1_{G_j}(s^1_i, s^1_i, \ldots, s^1_n) \qquad (3)$$

The computer center management can now take these utility statements to represent the worth to a particular user group of the particular action. Management must also determine the importance of each user group with respect to the particular decision, and arrive at an overall utility value for the action:

$$U = g(U_{G_1}(S^1), \ldots, U_{G_m}(S^1)) \qquad (4)$$

The crucial step of picking an action should be done with the aid of the calculation of the expected utility of each action. To engage in the argument relating to the use of a single value (the utility) as the determinant of a managerial decision is beyond the scope of this investigation.

UTILITY THEORY

General information on the concepts of multi-dimensional utility analysis can be found in Raiffa [7]. The work of Keeney [3] deals with the particular subject of quasi-separable utility functions while an early paper by Yntema and Klem [12] deals with the problem of assessing a person's utility function under these assumptions.

Later work by Keeney in his doctoral dissertation [4] formulates the conditions for "utility independence" and shows the quasi-separability conditions to be those of mutual utility independence. For example, an attribute x is said to be utility-independent of an attribute y if the relative preferences for values of x are independent of the value of y:

$$U(x,y) = C_1(y) + C_2(y)\,U(x,y_0), \quad \text{all } y \qquad (5)$$

It should be noted that "x utility-independent of y" does not necessarily imply "y utility-independent of x". However, if both these conditions hold, then the utility function is separable and x and y are "mutually utility-independent". (The conditions of utility independence correspond to Raiffa's formulation of "strong" conditional utility independence [7].)

In dealing with utility functions of more than two attributes, it is possible that attribute x is utility-independent of y but not of z, or that there be any other combination of utility-independent or non-independent variables. Keeney shows that it is possible to make use of any information on utility independence that is available to simplify the utility functions. For instance, if we have a situation where x is utility-independent of (y,z) and (y,z) is utility-independent of x, then we can formulate a utility function of vector attributes as follows:

$$A = (x) \qquad\qquad B = (y,z) \qquad (6)$$

We see immediately that our three-attribute utility function, under the above conditions, reduces to a two-attribute quasi-separable utility function with vector arguments A and B.

Keeney's work goes on to show minimal combinations of preference curves, point utilities, or marginal utilities that can be combined with the various conditions of utility independence in order to evaluate the utility function over the entire space. For example, if we have a utility function of three attributes, U(x,y,z), such that attribute x is utility-independent of (y,z) and attribute y is utility-independent of (x,z), then the complete utility function is specified by assessing six one-attribute conditional utility functions: $U(x,y_0,z_0)$, $U(x_0,y,z_0)$, $U(x_0,y_0,z)$, $U(x_0,y_1,z)$, $U(x_1,y_0,z)$ and $U(x_1,y_1,z)$ for arbitrary values x_0,x_1,y_0,y_1 and z_0, subject to consistent scaling on the six. The proof of this statement follows directly from successive application of Eqs. (5) and (6) for different variables. The total utility function of x, y and z will then be a weighted sum and product form of the six conditional utility functions (see [4, ch. 3] for a more complete discussion).

Further statements by Keeney indicate the usefulness of

the above-mentioned techniques even under conditions of only approximate utility independence. He shows that when Eq. (5) is solved, the resulting equation is a utility function with five degrees of freedom:

$$U(x,y) = U(x_0,y)[1 - U(x,y_0)] + U(x_1,y)\ U(x,y_0)$$
$$U(x_0,y_0) = 0\ ; \qquad U(x_1,y_0) = 1 \tag{7}$$

Using the techniques shown in Eq. (6), we can then use this utility function as an approximation to one with more than two scalar attributes. This is viewed as being of major importance in many operational situations, as shown below.

EXPERIMENTAL DETERMINATION OF A MULTI-DIMENSIONAL UTILITY FUNCTION

An experiment was performed to show the feasibility of the determination of a general utility function for one group of users of a general-purpose time-sharing system. The experiment consisted of a series of interviews with users to determine their overall utility function for the following time-sharing characteristics: availability, response time to trivial requests, response time to compute-bound requests. These three characteristics were chosen to illustrate various degrees of utility independence and because earlier research showed them to be important components of an overall time-sharing evaluation scheme [2].

The experiment began with a series of interviews with users to determine the minimum amount of information necessary to formulate their utility function. The determining factor as to what information was necessary was the utility independence relationships that existed among the system characteristics (see the discussion in the previous section). The data were then combined according to the mathematical formulation of the utility function and tested for general correctness by further interviews (no statistical tests of accuracy were made as the usefulness of these procedures does not rely upon extreme precision of utility values).

The interview procedure consisted of several parts.

1. General determination of users' usage patterns of the time-sharing system.
2. Introduction to basic theory of utility assessment.
3. Focusing of attention on the three attributes of

the time-sharing system to be discussed: availability (A), response time to trivial requests (RT), and response time to compute-bound requests (RC). Another factor, reliability, was rejected for this example as being potentially dominating of all other factors and thus masking illustrative utility function behavior.
4. Determination of utility independence relations.
5. Determination of conditional utility functions and point utilities necessary for the calculation of the total (three-dimensional) utility function. (The determination of which conditional and point utilities were needed was performed by application of mathematical utility theory.)

Focus was directed to the three attributes to be considered by presenting a scenario which established their presence as the major variables. For instance, the question of reliability was removed from the picture by insuring that on the "experimental" time-sharing system reliability was at a uniformly high level (as long as this situation was not too far removed from the user's experiences it was felt that he could and would accept the assumption and thus ignore reliability as influencing his preferences). Another example is that the features necessary for him to perform his task adequately were assumed to be fully operable (again, it was necessary that this condition not be too far removed from actuality). In essence, we are, therefore, computing utility functions conditional on high levels of all characteristics that are not explicitly considered.

Independence relations were determined by a series of questions regarding the shape of various conditional utility curves. For example, if the user asserts that his conditional utility function for availability has one shape when response time is at a favorable level and a fairly different shape when response time is at an unfavorable level, then we can say that availability is not utility-independent of response time. If, on the other hand, the user asserts that the shape of his conditional utility curve for one variable is independent of the value of the second variable, then we can state that a utility independence relation exists. As shown later, utility independence relations can be used to determine a mathematical form for a complete utility function involving a relatively small number of conditional utility functions and point utilities.

Each user was then asked to assess conditional utility curves as necessary by questioning along the lines of a standard fractile assessment procedure [4]. Once the user had shown facility in the assessment of one-dimensional utility functions, he was asked to reassess them for specific values of the other variables suggested by the experimenter. For example, in assessing the conditional utility function for response time to trivial requests (RT), the user may have been assuming an availability of approximately A_1 and a response to compute-bound requests of RC_1. By having the user assess the conditional utility for RT with suggested values of A and RC we not only tested the validity of the utility independence relations but also provided data for the determination of the overall utility function.

Finally, the user was asked to assess his utility for the eight possible "end points" generated by the combination of the highest and lowest values of availability, response time to trivial requests and response time to compute-bound requests over which conditions of utility independence were believed to hold (the method used for this assessment was very similar to that of Yntema and Klem[12] mentioned above). This produced further information for the determination of the total utility function and for the scaling and fixing of origins.

The experimental procedure was very iterative in nature: for instance, if determination of a conditional utility function showed a possibly incorrect utility independence relation, then this part of the procedure was redone. A number of interviews were held with selected users to insure the validity of these results. Much general discussion was interspersed with the assessment procedure to give the users a chance to "live with" their statements, understand their implications on other assessments, etc. Although time-consuming, this was felt to be a very worthwhile exercise and an important learning experience.

AN EXAMPLE

The users interviewed were computer system programmers whose main tasks were the input and editing of programs (trivial requests) coupled with the compilation and testing thereof (compute-bound requests). The ratio of editing sessions to compile and testing "sessions" was fairly high (about five to one), indicating that a lot of "desk debugging" was being done (irrelevant to the problem, but interesting nonetheless). The three attributes were measured as follows:

Availability: percentage of successful logins (i.e., probability of an accepted login at any attempt).
Response time to trivial requests: average number of seconds to satisfy majority of requests.
Response time to compute-bound requests: average number of seconds to satisfy majority of requests.

Initial testing of the subjects indicated the following utility independence relationships between two variables when the third variable was in the relevant range:

1. RT ui A - Relative preferences for trivial request response time was independent of availability.

2. RT ui RC - Relative preference for response to trivial requests was independent of response time to compute-bound requests.

(These both seem to make sense since response time to editing requests was the major determinant of how much work could be accomplished. It is suspected that these may not hold for other classes of users - especially when there is a more even balance between trivial and compute-bound requests.)

3. RC ui A - Relative preferences for compute-bound response time was independent of availability.

(This also seems reasonable, as the subjects, in general, tended to separate preferences for the various values of system work conditions from anything else. This may also not hold for a less demanding user type.)

This, of course, leaves the following three pairs to be not utility-independent:

A. A not ui RT
B. A not ui RC
C. RC not ui RT.

(Again, these seem to make logical sense, as indicated in the following statements. If either response has an unfavorable value, then the relative preferences for A may very well change as the programmer will spend most of his time having to contend with the unfavorable response and may not wish to log in as badly. Also, in any particular session the programmer may set his relative expectations for RC in terms of the RT he is experiencing - a not uncommon practice of changing sights in view of concurrent conditions.)

It was now possible, using methods similar to those developed by Keeney [4], to determine exactly which conditional utility functions and point utilities had to be determined completely to specify this class of users' complete utility function. The following was proven.

Given the above conditions of utility independence of the three attributes A, RT, RC, the following four utility points and seven one-attribute conditional utility functions are sufficient completely to specify U(A,RT,RC):

Point utility for (A_0,RT_1,RC_0) and (A_1,RT_0,RC_0), (A_1,RT_0,RC_1), (A_1,RT_1,RC_0), as well as scaling points for minimum and maximum utility.

Conditional utility functions: $U(A,RT_0,RC_0)$, $U(A,RT_0,RC_1)$, $U(A,RT_1,RC_0)$, $U(A,RT_1,RC_1)$, $U(A_0,RT,RC_0)$, $U(A_1,RT_0,RC)$ and $U(A_1,RT_1,RC)$.

The utility function thus obtained is as follows:

$$U(A,RT,RC) = \{U(A,RT_0,RC_0) + [U(A,RT_0,RC_1) - U(A,RT_0,RC_0)] \cdot [\frac{U(A_1,RT_0,RC)-U(A_1,RT_0,RC_0)}{U(A_1,RT_0,RC_1)-U(A_1,RT_0,RC_0)}]\} \cdot \{1 - \frac{U(A_0,RT,RC_0)}{U(A_0,RT_1,RC_0)}\}$$

$$+ \{U(A,RT_1,RC_1)+[U(A,RT_1,RC_0)-U(A,RT_1,RC_1)] \cdot [\frac{U(A_1,RT_1,RC) - 1}{U(A_1,RT_1,RC_0) - 1}]\} \cdot \{\frac{U(A_0,RT,RC_0)}{U(A_0,RT_1,RC_0)}\} \qquad (8)$$

The complete proof of this theorem is provided as an Appendix to this paper. The points and conditional utility functions indicated are shown graphically in figure 1.

The above equation is not a unique determinant of the utility function, as is shown below. The theory allows us to derive several equations, each of which is an equivalent statement of the utility function. For example, a slightly different derivation (see Appendix) yields a function that can be specified by the following point and conditional utilities (see figure 2):

Points: (A_0,RT_0,RC_1), (A_0,RT_1,RC_0) and (A_1,RT_1,RC_0) and scaling points;

Conditional utility functions: $U(A,RT_0,RC_0)$, $U(A,RT_0,RC_1)$, $U(A,RT_1,RC_0)$, $U(A,RT_1,RC_1)$, $U(A_0,RT,RC_0)$, $U(A_0,RT_0,RC)$ and $U(A_1,RT_1,RC)$, which yield a utility equation as follows:

$$U(A,RT,RC) = \{U(A,RT_0,RC_0)\ [1 - \frac{U(A_0,RT_0,RC)}{U(A_0,RT_0,RC_1)}]$$

$$+ U(A,RT_0,RC_1)\ [\frac{U(A_0,RT_0,RC)}{U(A_0,RT_0,RC_1)}]\}\cdot\{1 - \frac{U(A_0,RT,RC_0)}{U(A_0,RT_1,RC_0)}\}$$

$$+ \{U(A,RT_1,RC_1)\ [1 - \frac{U(A_1,RT_1,RC)-1}{U(A_1,RT_1,RC_0)-1}]$$

$$+ U(A,RT_1,RC_0)[\frac{U(A_1,RT_1,RC)-1}{U(A_1,RT_1,RC_0)-1}]\}\cdot\{\frac{U(A_0,RT,RC_0)}{U(A_0,RT_1,RC_0)}\} \quad (9)$$

We can now see that the determination of a three-dimensional utility function for this class of users has been reduced to the determination of several (conditional) one-dimensional utility functions and point utilities - a fairly simple assessment process both for experimenter and subject.

By using a similar procedure for each class of users (and remember, we predict that this will be a small number), we can effectively determine utility functions for the entire user population.

A UTILITY FUNCTION

Figures 3 through 9 show computer-generated plots of conditional utility functions derived from Eq. (8). A time-sharing user whose main job was programming was interviewed (as described above) for the purposes of assessing actual point and conditional utility functions and/or determining the appropriateness of the utility independence relationships specified. After appropriate scaling (to bring utilities in the range 0 to 1000), the utility space was generated. The utility value is shown on the z coordinate and the two variable characteristics on the x and y. The third coordinate is held fixed as indicated.

The first plot (figure 3) shows the user's utility for various values of response to trivial requests (RT) and response to compute-bound requests (RC) when availability is at its maximum value (100% probability of login). As can be seen, the relative preferences for RT (that is, a conditional utility curve for RT) is independent (has the same shape) for all values of RC. On the other hand, it can also be seen quite clearly that the shape of conditional utility curves for RC vary as RT goes from its maximum to its minimum value. By noting the relative magnitude in change of utility between one set of values for RT and RC and another set of values, the decision-maker can determine where the greatest marginal improvement will occur. He might use this information to help him select the alternative which best assured him of the appropriate changes in the values of those characteristics.

Figures 4 and 5 show plots similar to figure 3 except that the value of availability is 50% in figure 4 and 10% in figure 5. We note that the minimum and maximum values of utility have changed, but the approximate shapes of the conditional utility functions have not (indicating the fact that both RT and RC are utility-independent of A).

Figures 6, 7 and 8 show utility space for A and RC at different values of RT. Again, we can look at the stairsteps in any particular figure to see that RC is utility-independent of A but not vice versa. By comparing the curve that these stairsteps trace in the different diagrams, we can also see that neither A nor RC is utility-independent of RT (the curves traced by any particular conditional utility function vary in shape in the three figures, that is, for different values of RT).

Finally, figure 9 shows the utility space for A and RT with RC fixed at a low to medium level. We note here the relatively flat preference for good values of RT which trail off fairly rapidly. The utility independence relations between A and RT (A not utility-independent of RT but RT utility-independent of A) can again be seen by comparing the relative shape of the conditional utility curves at different values of the other variables.

The determination of utility spaces as was done above also serves the useful purpose of directing a decision-maker's search for alternatives. Simple mathematical analysis can show where the "steepest gradients" are relative to changes in the values of the characteristics. This can be combined with the manager's knowledge of the probable consequences of various actions to arrive at a set of possible policies which would be most advantageous to users of the computation facility. Further exploratory work needs to be done in this area as well as others to be described below.

OBSERVATIONS AND CONCLUSIONS

The experimental determination of a utility function of three parameters was successfully accomplished as outlined above. The usefulness of the procedure, however, lies in the fact that this can now be used as a part of a mangement decision-making process in the computer center. If we assume that the utility function depicted in the diagrams represents the only one that a manager has to deal with, then we can give some examples of how it might be used in a decision-making process. In particular, we feel that utility functions can be used in the process of "problem finding" [6] or "intelligence" [9] as well as in the final "choice" [9] between alternatives.

In the first case, we wish to see if there is any way that existing resources can be used to provide better services. For example, there is a general relationship between the number of users of a time-sharing system and the response time. If we currently set the maximum number of users at the value n, we can then measure the response time to trivial requests, r. Generally, by setting a lower value for the maximum number of users we will achieve a decreased response time (basically because a smaller number of users will be demanding a smaller amount of the total system's resources). The question is whether that is a desirable thing to do in terms of increasing user utility, and if so, at what point we should stop. If we look, for instance, at figure 9 and assume some current operating point of A = 50% and RT = 5 seconds, we can then determine that moving to a point (40%, 4 seconds) has approximately the same utility value while (40%, 3 seconds) has a significantly higher utility value. Similarly, moving to a point (60%, 6 seconds) has a lower utility value and (70%, 6 seconds) has a higher utility value (although not as high as (40%, 3 seconds)). Thus, if our present resources can be arranged to take us to one or the other of the better operating points (or some other operating point that has a higher utility value), then we have improved our overall service at no additional cost. Similar analyses can be done for other variables as well.

If we now take the situation where an additional amount of money is to be spent and the determination of which resource to spend it on is in question, then we can again use the utility function to help us in our decision. If we look at our operating point in the overall utility surface, it should be possible to determine mathematically the direction of "steepest gradient" (fastest increase in utility value). We can determine "feasible" directions according to the characteristics which can be modified by the application of further resources (we may be able to change availability to some degree, response time to trivial requests to another degree, and response time to compute-bound requests not at all - thereby ruling out certain directions of moving). Finally, we can then compute that point in the feasible region which provides the greatest utility increases (this procedure is very similar to some used in the solution of nonlinear programming problems). With the aid of the computer the mathematical manipulation may be fairly easily performed.

We have seen that the combination of the mathematical theory of utility determination along with the experimental assessment of a user's utility function can be used to aid in the management decision-making process. As it becomes easier to extrapolate the fairly simple problem discussed above to problems involving numerous characteristics and several groups of users (and, hence, several utility functions), we see general applicability of these techniques. Examples of further research in these areas are given below.

FURTHER RESEARCH

The major deterrent to the use of multi-dimensional utility calculations for management decision-making is in the complexity of the utility assessment procedures. Management problems are characterized by

1. a large number of relevant variables
2. a small number of utility independence relations
3. a large number of conditional utility curves with thresholds, "knees", or other irregularities

all of which serve to increase the number or complexity of the utility points or conditional functions which must be assessed in order to specify the complete utility function. Assessment procedures currently require laborious and painstaking work to ensure the validity of the assessed values - work that, in general, must be repeated for each service facility user being interviewed.

Research areas which would foster the acceptance of the utility approach to complex managerial decision-making currently include:

1. "Assessment by questionnaire". If adequate questions could be formulated to cut down significantly on interviewer time, utility determination could be integrated into existing management information-gathering procedures.

2. Utility function approximation and sensitivity analysis. The issue is establishing procedures for easily determining the dominant variables and relationships in any problem. A side effect of this could be a decrease in number and dimension of conditional utility functions that must be assessed.

3. User utility equivalence determination. Ad hoc procedures were used in this paper to determine user "classes" on the basis of similar characteristics. More work needs to be done to make this determination more accurate.

4. Time variance of utility assessment. How often must a manager reassess utilities in his user community to ensure that his data truly reflect its preferences? Research could be devoted to controlled experiments over time periods relevant to the service facility in question (six months or more in the case of computer systems - about the length of time necessary for major changes).

Research in these and other areas should be yielding further insights into the use of utility functions in practice in the near future.

SUMMARY

We are concerned with the basic problem of providing a mechanism for decision-making regarding the provision of some service to a population of users. By using the methodology of utility function assessment, it is possible to determine a multi-dimension utility function for each group of users.

As an example, utility functions were assessed for a group of users regarding several system characteristics of a general-purpose time-sharing system. The decision-maker (the computer center management) must then go about the task of evaluating probable change in the various attributes due to any action he might take and determining the relative importance (to him) of each user group's utility for the action. The results of these analyses can then be combined to evaluate a "most desirable" action.

As stated earlier, it is felt that the procedures illustrated in this paper hold considerable promise for the future. As management must deal with more and more complex problems with more variables, more "interested parties" and more possible actions, use of an integrated set of procedures for evaluation of alternatives will become more prevalent. Incorporation of individual preference data can only increase the depth and breadth of management understanding.

BIBLIOGRAPHY

1. Betaque, N.E., and G.A. Gorry, "Automating judgmental decision making for a serious medical problem", Management Science 17, no. 8 (April 1971).
2. Grochow, J.M., "Architectural implications of user needs in time-sharing service", IEEE Convention Digest, March 1971.
3. Keeney, R.L., "Evaluating multidimensional situations using a quasi-separable utility function", IEEE Trans. Man-Machine Systems MMS-9, 25-28 (1968).
4. Keeney, R.L., "Multidimensional utility functions: theory, assessment and application", Tech. Rept. 43, Operations Research Center, M.I.T., October 1969.
5. Miller, J.R., III, "The assessment of worth: a systematic procedure and its experimental validation", doctoral dissertaion, Sloan School of Management, M.I.T., June 1966.
6. Pounds, W.F., "The process of problem finding", Indust. Manag. Rev. 11, no. 1 (fall 1969).
7. Raiffa, H., "Preferences for multi-attributed alternatives", RM-5868-DOT/RC, RAND Corp., April 1969.
8. Simon, Herbert A., "Reflection on time-sharing from a user's point of view", Comp. Sci. Res. Rev. 1966, Carnegie Institute of Technology.
9. Simon, H.A., The new science of management decision, Harper & Bros., New York, 1960.
10. Stimler, Saul, "Some criteria for time-sharing system performance", Comm. ACM 12, no. 1 (1969).
11. Winkler, R.L., "The quantification of judgment: some methodological suggestions", J. Amer. Stat. Assoc. 62, 1105-1120 (1967).
12. Yntema, D.B., and L. Klem, "Telling a computer how to evaluate multi-dimensional situations", IEEE Trans. Human Factors in Electronics HFE-6, 3-13 (September 1965).

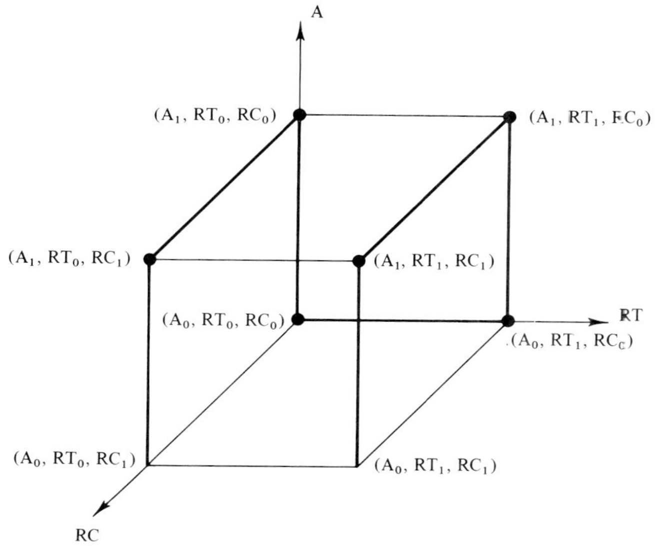

Fig. 1. Points and one-dimensional utility functions necessary to evaluate utility values within a cube under the following utility independence relations: RT ui A, RT ui RC, RC ui A.

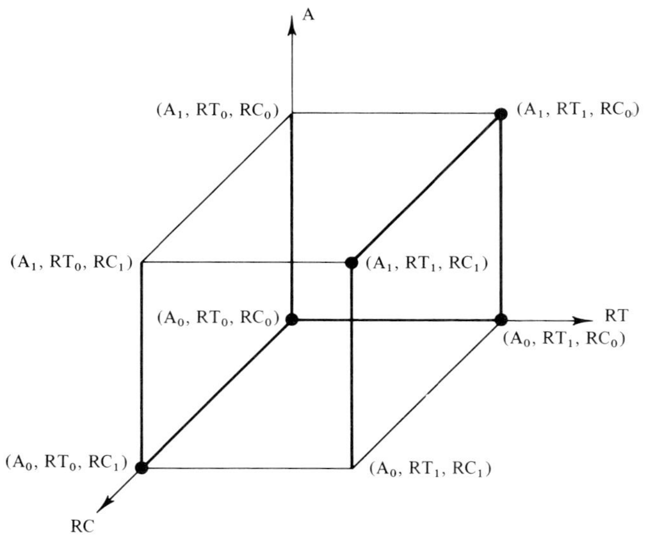

Fig. 2. Possible alternative requirements for calculation of utility space having independence relations: RT ui A, RT ui RC, RC ui A.

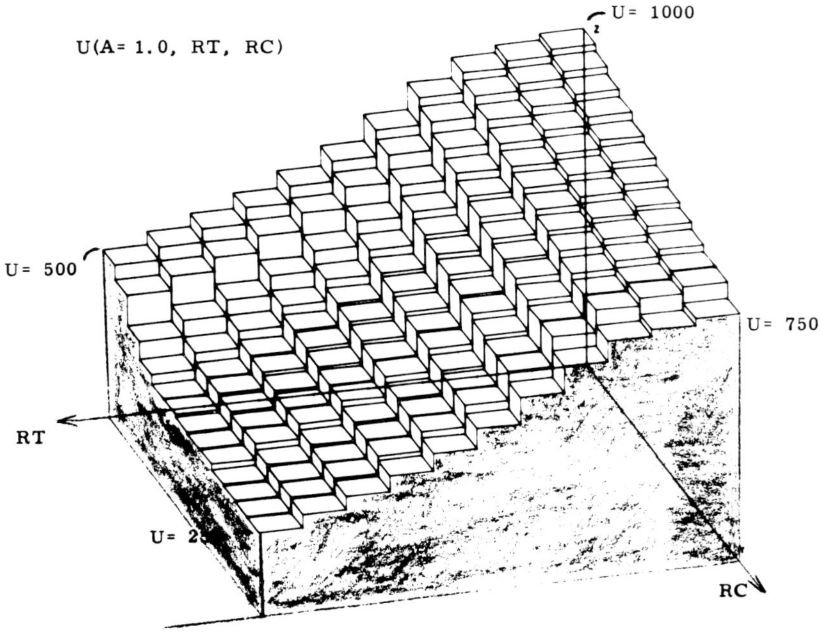

Fig. 3. Utility space calculated using data from point and conditional utility functions in Eq. (8) (three-dimensional view shown is for highest value of availability).

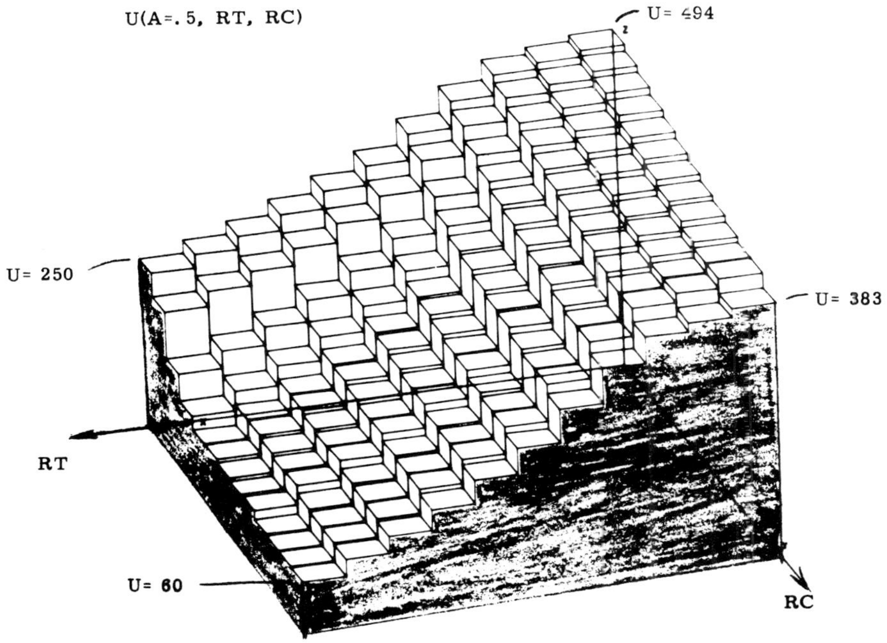

Fig. 4. Conditional utility space with availability 50%.

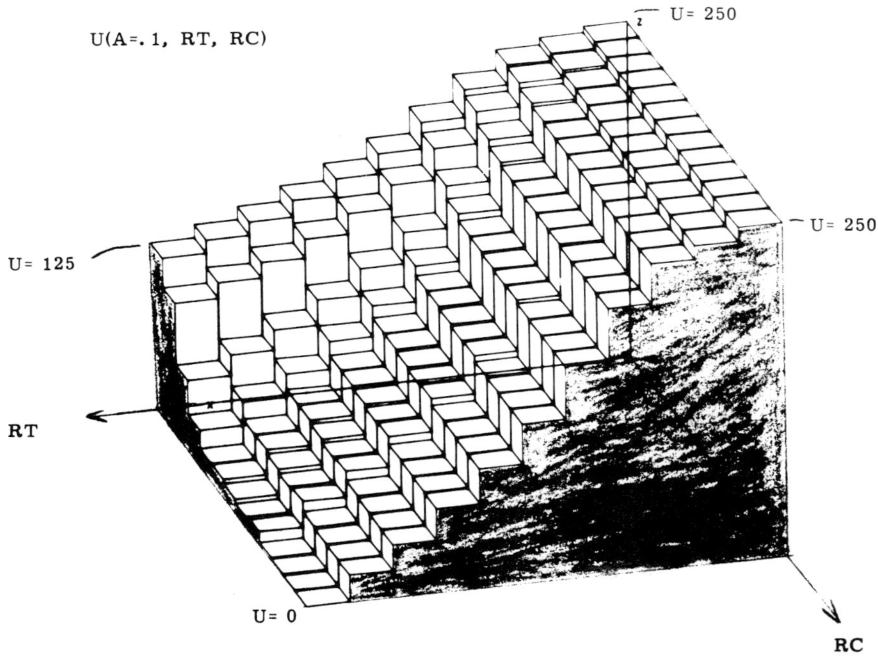

Fig. 5. Conditional utility space at minimum value of availability.

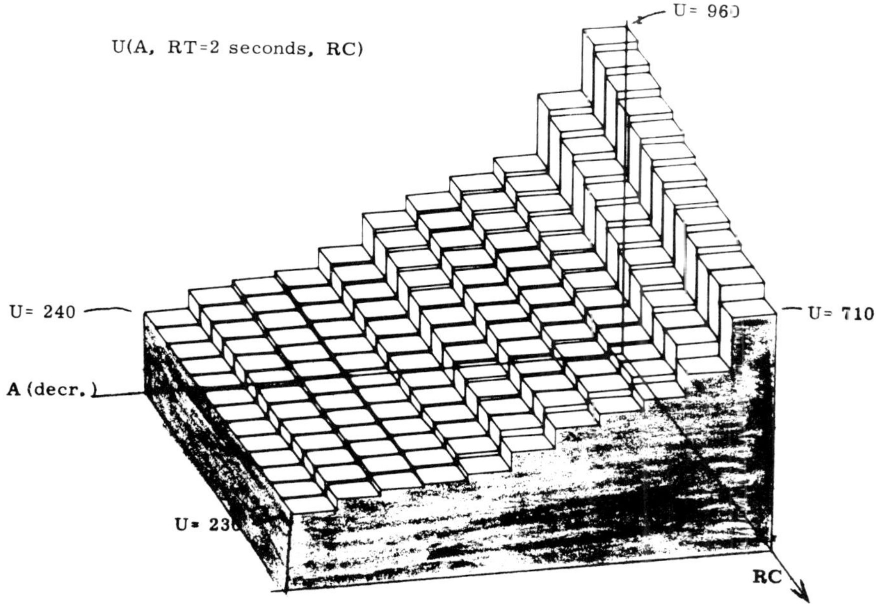

Fig. 6. Conditional utility space with response time to trivial requests at most favorable value.

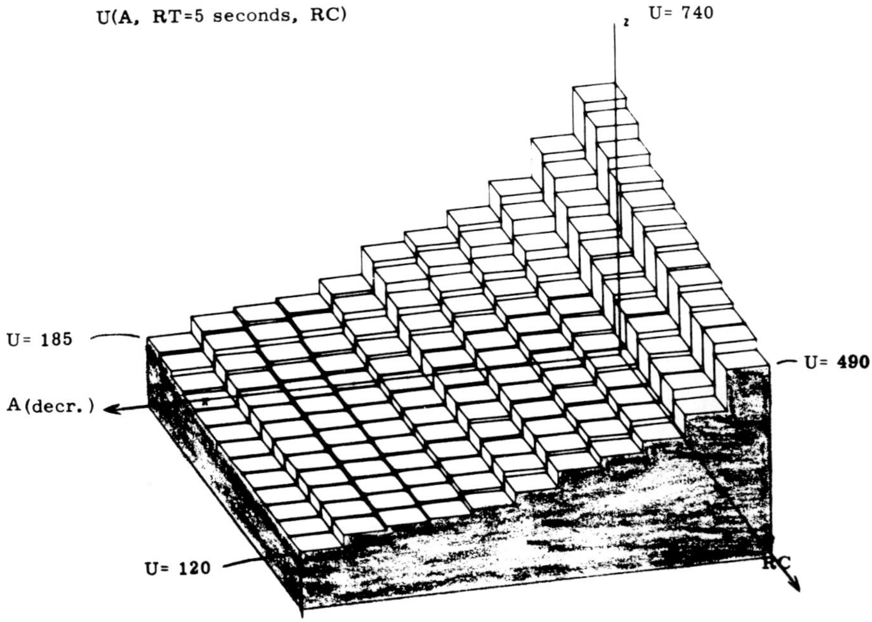

Fig. 7. Conditional utility space with response time to trivial requests at intermediate value.

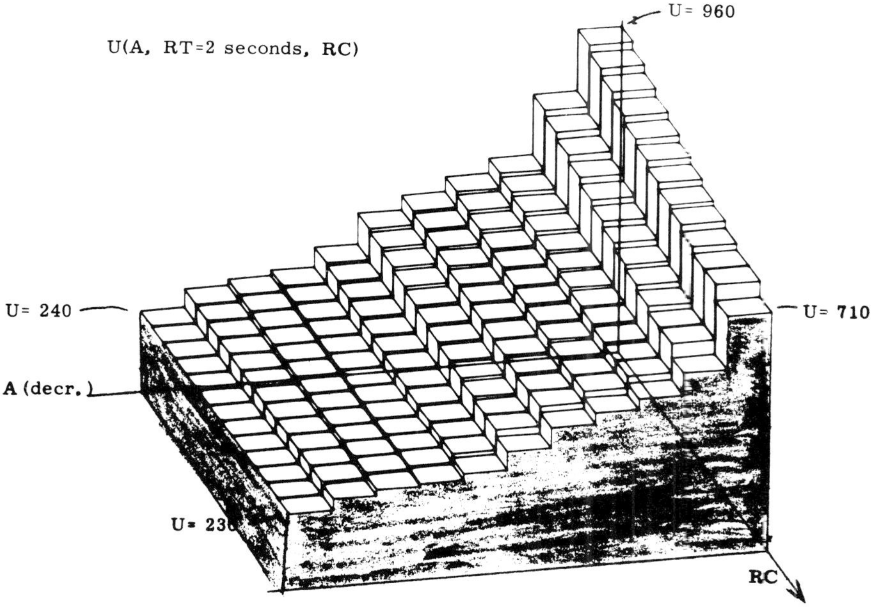

Fig. 6. Conditional utility space with response time to trivial requests at most favorable value.

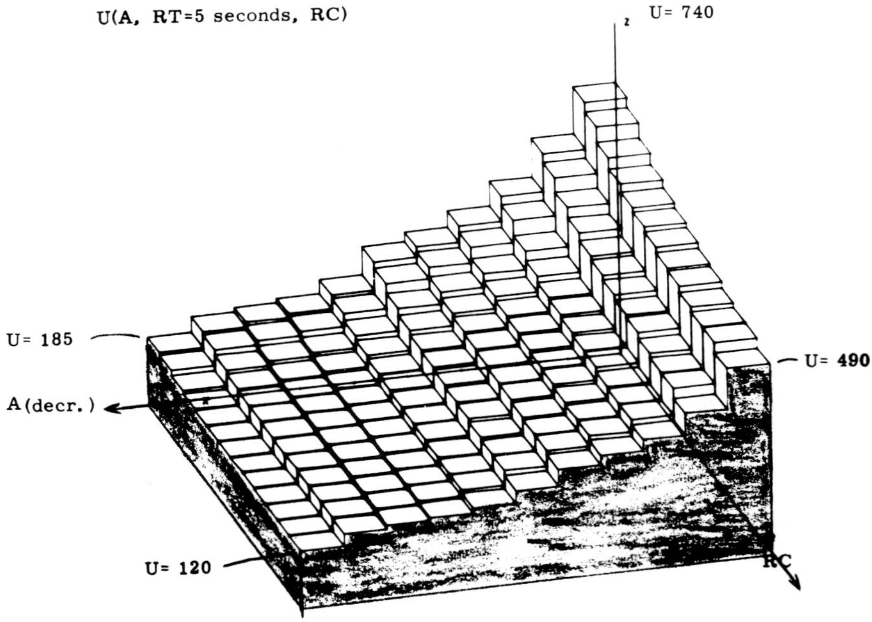

Fig. 7. Conditional utility space with response time to trivial requests at intermediate value.

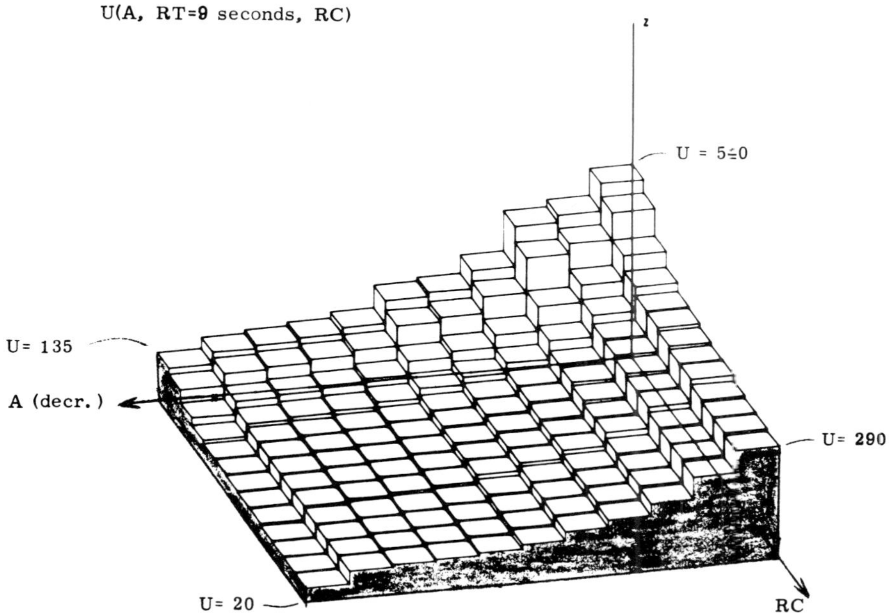

Fig. 8. Conditional utility space with response time to trivial requests at least favorable value.

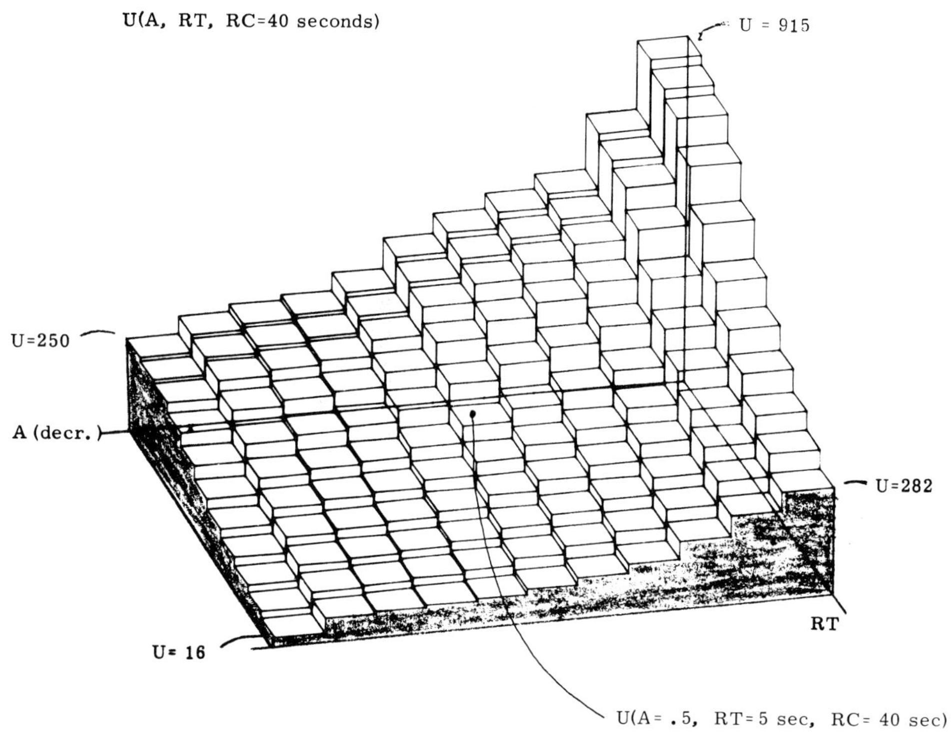

Fig. 9. Conditional utility space with response time to compute-bound requests at intermediate value.

APPENDIX

The following is the proof of the result stated in Eq. (8) in the text. We will show that, given the utility independence relations shown below, the utility function U(A,RT,RC) can be evaluated by the expression in Eq. (21) using only the indicated conditional and point utilities.

Utility independence relations:

$$\text{RT ui A} \tag{1}$$

$$\text{RT ui RC} \tag{2}$$

$$\text{RC ui A} \tag{3}$$

Given the definitions of utility independence and mutual utility independence given by Keeney [4] we can write the following equations.

Using (1) and (2) above:

$$U(A,RT,RC) = C_1(A,RC) + C_2(A,RC)U(A_0,RT,RC_0) \tag{4}$$

for some value of $RC = RC_0$ and $A = A_0$. Using (3) above, and setting $RT = RT_0$:

$$U(A,RT_0,RC) = D_1(A,RT_0) + D_2(A,RT_0)U(A_1,RT_0,RC) \tag{5}$$

for some value of $A = A_1$. Using (3) above, and setting $RT = RT_1$:

$$U(A,RT_1,RC) = E_1(A,RT_1) + E_2(A,RT_1)U(A_1,RT_1,RC) \tag{6}$$

for some value of $A = A_1$. We can now set the scale by assigning

$$\begin{aligned} U(A_0,RT_0,RC_0) &= 0 \\ U(A_1,RT_1,RC_1) &= 1 \end{aligned} \tag{7}$$

By evaluating Eq. (4) with $RT = RT_0$ we get

$$U(A,RT_0,RC) = C_1(A,RC) \tag{8}$$

By evaluating Eq. (5) with $RC = RC_0$ we get

$$U(A,RT_0,RC_0) = D_1(A,RT_0) + D_2(A,RT_0)U(A_1,RT_0,RC_0) \tag{9}$$

and by evaluating Eq. (6) with $RC = RC_1$ we get

$$U(A,RT_1,RC_1) = E_1(A,RT_1) + E_2(A,RT_1) \tag{10}$$

Substituting these results into our first equations, we arrive at:

$$U(A,RT,RC) = U(A,RT_0,RC) + C_2(A,RC)U(A_0,RT,RC_0) \quad (11)$$

$$U(A,RT_0,RC) = U(A,RT_0,RC_0) + D_2(A,RT_0)[(U(A_1,RT_0,RC)-U(A_1,RT_0,RC_0)] \quad (12)$$

$$U(A,RT_1,RC) = U(A,RT_1,RC_1) + E_2(A,RT_1)[U(A_1,RT_1,RC)-1] \quad (13)$$

By setting $RT = RT_1$ in Eq. (11), we get an expression for C_2:

$$C_2(A,RC) = \frac{U(A,RT_1,RC) - U(A,RT_0,RC)}{U(A_0,RT_1,RC_0)} \quad (14)$$

Similarly, we set $RC = RC_1$ in Eq. (12) to get an expression for D_2:

$$D_2(A,RT_0) = \frac{U(A,RT_0,RC_1) - U(A,RT_0,RC_0)}{U(A_1,RT_0,RC_1) - U(A_1,RT_0,RC_0)} \quad (15)$$

Similarly, we set $RC = RC_0$ in Eq. (13) to get an expression for E_2:

$$E_2(A,RT_1) = \frac{U(A,RT_1,RC_0) - U(A,RT_1,RC_1)}{U(A_1,RT_1,RC_0) - 1} \quad (16)$$

Substituting these results into our last set of equations, we get

$$U(A,RT,RC) = U(A,RT_0,RC) + [U(A,RT_1,RC) - U(A,RT_0,RC)] \frac{U(A_0,RT,RC_0)}{U(A_0,RT_1,RC_0)} \quad (17)$$

$$U(A,RT_0,RC) = U(A,RT_0,RC_0) + \frac{U(A,RT_0,RC_1)-U(A,RT_0,RC_0)}{U(A_1,RT_0,RC_1)-U(A_1,RT_0,RC_0)} \cdot [U(A_1,RT_0,RC) - U(A_1,RT_0,RC_0)] \quad (18)$$

$$U(A,RT_1,RC) = U(A,RT_1,RC_1) + \frac{U(A,RT_1,RC_0)-U(A,RT_1,RC_1)}{U(A_1,RT_1,RC_0) - 1} \cdot [U(A_1,RT_1,RC) - 1] \quad (19)$$

We rearrange Eq. (17) for convenience:

$$U(A,RT,RC) = U(A,RT_0,RC)[1 - \frac{U(A_0,RT,RC_0)}{U(A_0,RT_1,RC_0)}]$$

$$+ U(A,RT_1,RC) \cdot \frac{U(A_0,RT,RC_0)}{U(A_0,RT_1,RC_0)} \tag{20}$$

Now, by simply substituting Eq. (18) for $U(A,RT_0,RC)$ where it appears in Eq. (20), and Eq. (19) for $U(A,RT_1,RC)$ where it appears, we get our final result:

$$U(A,RT,RC) = \{U(A,RT_0,RC_0) + [U(A,RT_0,RC_1)-U(A,RT_0,RC_0)]$$

$$\cdot [\frac{U(A_1,RT_0,RC)-U(A_1,RT_0,RC_0)}{U(A_1,RT_0,RC_1)-U(A_1,RT_0,RC_0)}\} \cdot \{1 - \frac{U(A_0,RT,RC_0)}{U(A_0,RT_1,RC_0)}\}$$

$$+ \{U(A,RT_1,RC_1) + [U(A,RT_1,RC_0)-U(A,RT_1,RC_1)]$$

$$\cdot [\frac{U(A_1,RT_1,RC) - 1}{U(A_1,RT_1,RC_0) - 1}]\} \cdot \{\frac{U(A_0,RT,RC_0)}{U(A_0,RT_1,RC_0)}\} \tag{21}$$

It can now be seen that the following point and conditional utility relationships are sufficient to compute the utility function $U(A,RT,RC)$ given the independence relations in Eqs. (1), (2) and (3):

Point utilities: $U(A_0,RT_1,RC_0)$
$U(A_1,RT_0,RC_0)$
$U(A_1,RT_0,RC_1)$
$U(A_1,RT_1,RC_0)$

Conditional utility functions:

$U(A,RT_0,RC_0)$
$U(A,RT_0,RC_1)$
$U(A,RT_1,RC_0)$
$U(A,RT_1,RC_1)$
$U(A_0,RT,RC_0)$
$U(A_1,RT_0,RC)$
$U(A_1,RT_1,RC)$

This set of utilities is shown pictorially in figure 1.

In order to arrive at text Eq. (9), we simply allow A = A_0 instead of A_1 in Appendix Eq. (5) and continue the analysis from there. The point and conditional utilities necessary to calculate the utility space from this equation are depicted in figure 2.

DISCUSSION OF SESSION I

F.J. Anscombe

Drs. Grenander and Tsao have given us a stimulating survey of the subject of computer system evaluation, and have in fact defined the objectives of this conference. They have expressed themselves clearly and forcefully. I have nothing to add to their presentation nor any qualification to offer. I'd like to spend a few minutes expressing in different language something they have said, because it is important and may bear saying twice.

Drs. Grenander and Tsao have remarked that statistical methods of experimental design have proved useful in agriculture. I think there is a good parallel between agriculture and computer systems, utterly different though they seem at first glance. It is particularly appropriate to notice the similarity because so much of present-day statistical methodology was developed to meet the needs of agricultural research, and may in turn serve well in the study of computers and undergo further development in consequence.

The agricultural scientist deals with such matters as fields of wheat. The simple phrase, "A field of wheat", stands for a fantastically complicated set of phenomena. The soil of the field is a medium with fascinating physical and chemical properties, in which flourish earthworms, insects, bacteria, fungi. The medium is exposed to great diurnal and seasonal changes in sunlight, rain, wind. The wheat that is grown on the field is a population of organisms cohabiting with other plants (weeds), insect predators, diseases.

For a good deal more than a century, agricultural scientists have tried to devise more effective ways for farmers to grow crops. They have had two main modes of attack that have complemented and supported each other. Neither would have been much good without the other. One of them was basic research, conducted usually in the laboratory - in the test tube, through the microscope. In basic research some aspect of the whole natural process is isolated for study - nutrition, pathology, genetics. Understanding of basic relations is sought, and sometimes ideas for better farming are generated. The other main line of attack has been the field trial, the statistical study and testing out of methods under farming conditions. Careful observation in the field provides a stimulus for basic research, which in turn generates practical ideas that can be tested in the field.

Without basic research, the field trials, the statistical-type study under ordinary operating conditions, would be likely to be just blind groping, yielding little of interest. Without field trials, the relevance of basic research to practice would be altogether unclear.

I think that all of us who have ever tried to reconcile theory, or basic research, with practice - in whatever area of study - have noticed that in some instances they agree beautifully and in some instances they don't. When there is good agreement, we generally feel that the simplification made in the theoretical or basic study was the right one, presenting the essence of the real situation. When there is disagreement, we generally feel that the wrong simplification was made. Theory may say that when X is small enough Y will be proportional to X. Suppose the true relation of Y to X, unknown to us, is indeed a function,

$$Y = 1 - e^{-X} \quad (\text{for } X > 0) .$$

If the range of X of practical interest is from 0.1 to 0.2, practice and theory will closely agree. If the range is from 10 to 20, they will utterly disagree - the theory for small X was correct but irrelevant.

The general thrust of the Grenander-Tsao paper is that up till now much more effort has gone into fundamental or theoretical study of computer systems than into statistical study and evaluation of them. The imbalance needs to be redressed. For fundamental theoretical studies to achieve full effectiveness they must have the statistical support of "field trials".

The relation of research into computer systems (whether theoretical or statistical) to the everyday work of a computer manufacturer or of the director of a computer center is similar to the relation of agricultural research to the job of farming. Most farmers do not engage in agricultural research. The urgent need to do somehow takes precedence over discovering how to do better. Yet agricultural research does farming; improved practices, strains, materials are readily adopted by the farming community, for good economic reason. And so it is surely with computers.

Finally a brief comment on Mr. Grochow's paper. I am afraid my heart sinks when I hear that someone is applying utility theory, or decision theory, to some practical decision task. It is like someone claiming to use elementary mathematical logic in a deductive task. Mathematical logic presumably does somehow describe or characterize the way mathematicians think, and yet most mathematicians don't consciously use formal logic in their work. As far as I know, the proper user of formal logic is the computer, which indeed has to be self-conscious about its deductive reasoning. (Pardon the anthropomorphism - you know what I mean!) Similarly, I think it will only be computers that will make decisions by decision theory.

In his paper Mr. Grochow has focused attention on just one aspect of the determination of utilities. He wants to estimate utility functions of several variables, and seems to take for granted the estimation of utility functions of one variable. But it was my understanding when I was last hearing of these things that to determine numerical utilities for just a small finite set of possibilities is horrendously difficult - the subject usually shows that his preferences do not conform with the consistency axioms of decision theory and so no utilities exist. If Mr. Grochow has been determining complete utility functions, is there not a danger that he is being hoodwinked?

I suggest that attempting to determine numerical utilities in this problem is both too ambitious for success and also unnecessary - how could the computer center director decide even if he knew the complete utility functions for all his clients? I suggest it would be more practical and no less useful to attempt merely a preference ordering - if you like, utility indifference curves, not utility functions.

Certainly it is admirable that a computer center director should systematically attempt to inform himself of the preferences of his clients and not be influenced exclusively by engineering considerations. I suspect, however, that some clients will have difficulty in assessing in advance what their reaction will be to any substantial change in operating conditions or facilities.

SESSION II

EVALUATION OF CURRENT SYSTEMS

CHAIRMAN: N. RASMUSSEN

MODELLING THE PERFORMANCE OF THE OS/360 TIME-SHARING OPTION (TSO)

Edwin R. Lassettre and Allan L. Scherr

ABSTRACT. This paper will discuss the use of a relatively simple performance model during the development of the OS/360 Time-Sharing Option (TSO). This model was used in conjunction with measurements for early prediction of the ultimate performance of the system and for identification of performance problems during TSO implementation. Furthermore, it is useful in understanding the performance of the final system. The model is described and a useful graphical representation provided. The techniques of using the model for the above purposes are described. The predictions of the model are compared to actual system measurements. The experiences described in this paper represent an interesting case study in the melding of theory and practice.

INTRODUCTION

This paper describes some of the techniques used to predict and control the performance of the Time-Sharing Option (TSO) of Operating System/360. These techniques center around the use of a rather simple analytic model for time-sharing systems and measurements of various versions of the system under load.

TSO is built upon the MVT configuration of OS/360 and is based upon the technique of swapping the programs conversing with users at remote terminals. Swapping is simply a technique for sharing an area of the system's main storage among the programs associated with various users. In TSO, each user is associated with a program, and only one program may be in a predefined area of main storage or "region" at a time. Thus, when a user's program is given a time slice, it is first brought into main storage from a disk or drum storage device, allowed to execute, and then copied back out to the auxiliary storage device. Since this operation typically causes the replacement of one program in main storage by another, the technique is called "swapping". Up to fourteen regions of main storage may be used simultaneously for swapping. In this case, the total number of users would be divided into fourteen groups, each group permanently associated with a region. Regions may be of various sizes to accommodate the requirements of the system's users. Users entering the system, or "logging on", are assigned to whichever region is large enough to satisfy their size requirements and, where there is a choice, the region with the lightest usage is selected. In TSO, users may execute virtually any program that could run under OS/360 including compilers, assemblers, and their own programs. In addition, there is a broad range of conversational support. For further details, see [1] and [2].

In all of the discussion that follows, a single time-sharing region is assumed. Thus, all users' programs contend for the same main storage area. Unused main storage and other system resources can be used for additional work; however, this will not be taken into account here.

TSO was developed over the period starting with the original concept of the system in 1967 and ending with the availability of the program in June, 1971. During the initial design period in 1967 and 1968, performance projections were made for the final system based on the use of a combination of measurements of the base upon which TSO was to be built (OS/360 - MVT) and the analytic time-sharing system model. During 1969, an operational prototype of TSO was built on a Release 16 version of MVT, and the performance of the system was measured and compared to the predictions made in the previous year. The construction of the final system occurred during late 1970 and early 1971, and during this period calibrations and final performance verifications were made. Throughout this period, comparisons between the actual performance of the system and the predictions of the model were continually made. Discrepancies were typically corrected by modifying the system rather than the model. As will be seen in later parts of this paper, this technique proved to be extremely important.

The following sections describe the model used, the way it was employed to produce performance projections, the way in which system performance was measured, and finally, the correspondence of the predictions with the performance of the final system.

THE MODEL

Before describing the model, definitions for the parameters of time-sharing performance will be given. Figure 1 shows the typical sequence of operations at a time-sharing terminal and the corresponding processing performed by the system. First the user enters a request for processing in the form of input entered from the terminal. This phase is ended by the completion of the input, usually signalled by a "carrier return" or "end-of-block" character. The system places this input in the queue of work it must perform for all of the users in the system. When this particular user's input is selected, the program associated with him processes it. The execution of this program may result in output being sent to the user's terminal. Furthermore, this execution may be interleaved with the processing of other users' input using a time-slicing technique. Processing of an input ends with a request by the program for more input. Based on this operation, the following parameters are defined:

U - The user time. This is the elapsed time from the request for input made by the processing program to the completion of the requested input by the user. During this period, the system has essentially no work to do for the user. At the user's terminal, the user time is made up of the completion of output printing, idle time (or 'think' time), and input time.

R - Response time. This interval begins with the completion of input and ends when the program processing this input finishes and requests more input. The response time begins at the end of a user time and ends with the beginning of the next user time. Taken together, the user time and the response time constitute a complete cycle, called an 'interaction'.

S - Service time. This is the time required by the system to execute the program that processes the input entered by the user. The service time cannot be greater than the response time and is typically smaller due to the fact that the system interleaves the execution of many users' programs during the response time period. Any overhead resulting from this interleaving (e.g., swapping) is included in the service time.

The time-sharing performance model used is a simple adaptation of the classical queueing theory solution to the machine repair problem. A complete description of this model, as applied to a time-sharing system, can be found in [3].

The parameters of the model are:

n - the average number of time-sharing users actively interacting with the system
U - the average user time
R - the average response time
S - the average service time per instruction.

For the purposes of the model, the service time must be the amount of time that a user requires exclusive use of a single resource. In many systems, no such resource can be identified. However, in TSO, occupancy of main storage fulfills this requirement. Thus the service time in TSO is simply the amount of time that a program must tie up main storage for its execution and swapping.

For the time being, the assumption will be made that n is constant and that both U and S are exponentially distributed. The effect of deviating from these assumptions will be discussed later in this section.

Using this model, the following results are obtained:

1. The total number of interactions processed by the system per second is n/(U+R).

2. The probability that there are no requests for service in the system (i.e., that all users are in the user time portion of the interaction) is P_0:

$$P_0 = \frac{1}{\sum_{i=0}^{n} \frac{n!}{(n-i)!}\left(\frac{S}{U}\right)^i}$$

As n gets large, P_0 tends to zero.

3. The average response time is

$$R = \frac{nS}{1-P_0} - U$$

4. The average number of users waiting for a response (i.e., in the response time portion of the interaction) is

$$Q = \frac{nR}{r+U}$$

Given a fixed U and S, the relationship of response time to the number of interacting users is shown in figure 2. As previously stated, when the number of users is high, P_0 tends to zero, and the relationship between n and R becomes linear: R = nS - U. The point at which the straight-line portion of this curve intersects the horizontal axis is sometimes called the "saturation point". At this point n = U/S. This saturation point is easily calculated and is a fair approximation of the capacity of a time-sharing region.

This model has been carefully studied with respect to its sensitivity to the probability distribution of U and S. W.C. D'Avanzo of Purdue University [5] has obtained the following results:

1. The formula for the average response time above is generally true regardless of the distributions of U and S. However, P_0 will vary with the shape of these distributions.

2. With respect to the distribution of U, the key requirement is that the overall distribution of arrival from <u>all</u> of the users in the user time portion of the interaction be exponential. This requirement will be satisfied with <u>any</u> distribution of U provided that there are enough users in the system. Using observed distributions for U [3], n greater than about 20 satisfies this criterion.

3. For the purposes of the model, the primary measure of the distribution of S is its standard deviation. Generally, if the standard deviation is equal to the mean, the shape of the curve in figure 2 remains unchanged. If the standard deviation differs from the mean, the curve diverges from that shown. In any case, when n is one or large enough n for P_0 to be approximately zero, all distributions yield identical results. In the intermediate ranges of n, the response times are higher than shown if the standard deviation is greater than the mean and lower if the standard deviation is less than the mean. Figure 2 depicts this relationship graphically.

Most of the performance specification work relating to TSO was to determine the number of interacting users that the system could support with a certain processing load. The average user time and the required average response time were typically given. Figure 3 shows a logarithmic plot of n versus S for average response times varying from 1 to 10 seconds. The user time is assumed to be 35 seconds. Based on a number of measurements of this parameter in various environments, including at M.I.T.'s Project MAC [3] and IBM Research [4], this value is deemed to be reasonable and relatively conservative for users of typewriter-like terminals similar to the teletypes or IBM 2741. Figure 3 shows the number of terminals versus average service time for a fixed average user time (35 sec) and various average response times. This particular representation of the model was found to be extremely useful and eliminated the need for many calculations.

USE OF THE MODEL TO PROJECT PERFORMANCE

All of the TSO performance specifications were defined in terms of standard workloads on the system. Each terminal was assumed to be performing the same work and to have the same user time distribution. U was specified as being exponentially distributed with a mean of 35 sec. The actual workload was described in terms of a "script" which represented the actual input that would be entered from a terminal. A script consists of inputs corresponding to commands that request various services of the system and to replies made by the user to the interactive programs. The workload for a typical script would start with a "LOGON" wherein the user is signed onto the system, authorization and accounting checking is done, and an OS/360 environment in which he can run his programs is created. Next, the "EDIT" command might be invoked allowing the entry of a source program in FORTRAN. This program would then be compiled and executed interactively. Additionally, commands listing the content of certain data sets or files at the terminal, deleting data sets, etc., might be included. Finally, the script would end with a LOGOFF command terminating the session, performing accounting, etc.

A script performing the type of work described above was, in fact, one of the TSO benchmarks. This script consisted of 66 lines of input and represented a session of approximately 45 minutes' duration. Another aspect of the benchmark was that terminals were to execute these scripts in a staggered manner. That is, at any instant in time, all terminals are in different parts of the script. When a terminal completes the script, it starts it again from the beginning.

USE OF THE MODEL FOR PERFORMANCE PREDICTION

The key to using the model for predicting performance

is the determination of the value of S, the average service time per interaction. As previously stated, S is made up of two components: the average program execution time per interaction and the average swapping time per interaction. Both of these activities tie up the critical resource, the main storage region. In the single-region case, the execution time is simply the time that the program would take to execute in the absence of a time-sharing or multiprogramming environment. Thus, for each interaction in the script a measurement or estimate must be made of the execution time.

In the case of TSO, each interaction was analyzed to determine which functions would be used, which system services, etc. In may cases these functions or services already existed in the current version of OS/360-MVT. When possible, measurements were made to determine execution times for these functions. In all other cases, estimates were made assuming the computing power and input/output device characteristics corresponding to those of the measurement configuration. Summing all of these execution times and dividing by the number of interactions in the script yields the average.

Swapping time is computed by determining the size of the various functions called by the execution of the script and estimating the number of time slices (and therefore the number of swaps) required to execute each interaction. Every input in the script will cause at least two swapping actions, one at the beginning of the interaction and one at the end. Additional swapping would occur if the processing of the interacting program were interrupted by the expiration of a time slice. Under the assumption of a fixed slice, it is easy to make this estimate. Since programs vary in size during their execution, it was necessary to estimate initial, intermediate and final sizes for each interaction. As an example, take a program with an initial size of 18K bytes, an intermediate size of 80K bytes, a final size of 24K bytes and an execution time of 2.5 sec. Using a time slice size of 1 sec would imply that for this interaction the program would be swapped in and out three times: 18K in, 80K out and in twice, and 24K out. Using the performance characteristics of the storage device used for swapping, the total swap time can then be computed. In the case of the IBM 2301 drum storage unit, the revolution time is 17 millisec and 18K bytes can be transmitted per revolution. Thus, a swap of B thousand bytes can occur in $((1/2)+(B/18))\times 17$

millisec. The additional half-revolution is added to account for the fact that the drum must reach a starting point before recording or reading can occur. For a disk device with a moving recording head mechanism, the calculation is more complex and requires the addition of the time it takes to move this mechanism, the "seek" time. In addition, a large swap may require multiple seeks in order to record all of the information.

The above process is repeated for each of the interactions of the script to arrive at the average swap time per interaction. One further item must be considered: the overhead associated with the terminal support and the time slice end/start processing that must be performed to prepare a program to be swapped out and then restored to an executable state when it is subsequently swapped in. This was handled in the early performance estimates for TSO by increasing the mean service time by a fixed percentage. Numbers in the range of 5% to 25% were used, the latter to cover unforeseen contingencies.

SYSTEM MEASUREMENT

Large-scale system measurements were made to verify the performance estimates and to track implementation changes. The measured results were compared with the predictions and system changes were made as needed to correct any discrepancies. Most of the differences between TSO and its model were due to programming errors; a surprisingly small number of design changes were needed.

The performance of the actual system was measured using the techniques previously described and simulating the interacting users with a second computer system. A program was written which would send the script inputs from this second system to the TSO system, wait for the reply, check that it was correct, select an exponentially distributed random number for the user time, delay for the appropriate time, and send the next input in the script. This program performed this function for each of the interacting terminals. The hardware used for the TSO system was precisely the same used for live users with the only exception being that instead of lines going to typewriter-like terminals, these lines were cabled to another system. Figure 4 depicts a typical configuration.

Response times, as defined in the model, were measured

with a program that is a standard component of TSO. The user time calculated by the second system is based on its perception of the response time. Since it can only see the output generated by the interactive program and not its actual completion, there is a slight discrepancy. There was typically about a 0.5 sec increase in the user time seen by the second system compared to the time seen in the TSO system. This difference was ignored.

Workload simulation was also relatively straightforward, although considerable setup time was required. Benchmark terminal sessions were scissored into subsessions and subsession dependencies (primarily data generated prior to the subsession) were determined. These data and setup procedures needed to allow execution of the subsessions in arbitrary order were prepared and appended to the TSO system. Finally, weighting factors were assigned to the subsessions to achieve the desired workload and these factors, together with macro-like descriptions of the subsession interactions, were appended to the simulator. The execution of a performance run was performed in two phases. First, each of the simulated users was logged onto the system. Then a subsession was selected for each user and started after a delay equivalent to one user time. Whenever a subsession completed, another was chosen at random using the weighting factors described above. At least one subsession in the mix included a log-off/log-on sequence which, in effect, terminated one session and began another. Figure 5 shows this process graphically. Performance data were collected from both systems during measurement runs. On the TSO side, the trace facility built into the system was used for data collection. On the simulator side, a time-stamped log of inputs and outputs was collected.

Data reduction was done in two passes. In the first phase the TSO and simulator data were edited and merged to produce a single stream of annotated system events with a common time base. Annotation took the form of inserting additional events representing terminal events into the event stream collected from the TSO system. This stream of events was then reduced to obtain response times, users' times, service times, processor times, etc.

MEASUREMENT VERSUS PREDICTIONS

The performance of TSO is determined by the combination of the average service time and the ability of the system to share its resources among all of the interacting programs. The analytic model was used to discriminate between problems in the two areas. That is, the model represented an objective for the performance of the system given that the interacting programs produced a particular average service time. It should be noted that the analytic model is of such a simple nature that no real model calibration can be done. Rather, it was the system which was calibrated. Given that a particular service time average was achieved, it was the objective for the system to perform at least as well as the model would predict. This objective allowed us to distinguish performance problems in the system from problems with the interactive programs themselves.

Figure 6 shows a schematic diagram of the process that was used to compare the results of the model with system measurements. Once a measurement run was made, the observed average service time, average user time, and number of terminals were used with the model to predict a mean response time. This was then compared to the measured response time. If the observed response time was worse than that predicted by the model, attention was directed to the control program, i.e. the portion of the system that performs time slicing, swapping, multiprogramming, etc. The assumption is made that there is a problem in the area of how TSO multiplexes the use of the system resources among the interactive programs. Detailed measurements were used to analyze the control program, and in every case some modification was made. The process was then repeated. If the control program's performance matched the predictions of the model, and the overall performance of the system met or exceeded the specifications for the particular workload being run, the process was complete. If overall performance was inadequate, then the service times of the individual interactive programs were examined. Once again, modifications were made to improve performance. In this area, program structure (and swap size), algorithms, input/output devices and/or configuration, etc., were candidates for modification. Typically modifications to TSO involved the correction of "bugs" or the modification of the operating parameters or hardware configuration of the system.

During the phase where measurements were made and compared with model predictions, the following conclusions and observations were made:

1. The average execution time of a particular script was essentially independent of the number of interacting terminal users. However, the amount of swapping did increase as more terminals were added, bringing the average service time up.

2. In order to measure the performance of the interacting programs providing the processing for the various scripts, it was necessary only to run TSO with a single terminal. In practice a second terminal was also connected, but inactive so that swapping would occur.

3. The model assumes that only one critical resource exists in the system, namely main storage. In fact, many interlocked resources exist in TSO. For example, the use of certain control tables in the system is serialized. As a result of the fact that TSO performance was consistent with predictions of the model, the conclusion is that these other interlocks are relatively unimportant.

4. The most effective way to improve TSO performance is to improve the efficiency of the MVT base. During the period when the hardware configurations and operating parameters were being optimized for maximum TSO performance, it became apparent that the batch-processing performance of the MVT base was also substantially improved. Moreover, many of the design changes made to MVT to allow for better TSO performance also enhanced MVT.

Figure 7 shows the error between measured response times on TSO compared with predictions from the model. As can be noted from the diagram, the correspondence is extremely good, with errors in the range of +24.4% to -13.5%. The mean error was +2.5%. These measurements represent a total of 37 performance runs. The range of number of terminals was 17 to 95; service time from 360 millisec to 1.9 sec. The following table shows several examples of measurements compared with model predictions. The hardware systems ranged from S/360 model 50 with 2314 disk to a S/370 model 165 with 2305, 3330 drum access facilities. In all cases, the terminal workload is the same and is based on the script previously used as an example.

SUMMARY

In this paper we have described some of the techniques used to predict and manage the performance of TSO from its conception through to its completion. The use of the model

number of terminals	observed R	predicted R
15	3.72	4.40
22	4.47	5.12
35	3.62	3.45
35	2.53	2.66
15	3.47	3.69
30	4.38	3.70
22	4.81	5.81
47	4.67	4.99
70	4.18	4.12
24	4.72	6.03
60	5.29	5.02
95	2.88	2.66

played a key role in establishing the feasibility of the system as well as in helping to identify problem areas in the system as it progressed through development. It should be noted that the model relies on extension of measurements of the actual system and, therefore, its applicability might be limited in certain other environments.

The model was only one of the tools used during the development of TSO. Many other techniques were used to augment it. Included were a prototype of the system, an imitative simulation model, other analytic models, and extensive measurement both at a system level and of detailed component performance.

The authors would like to acknowledge the substantial contributions made by the following people: Mr. A.B. Goldschmidt, Mr. J.J. Kearney, Mrs. P.G. Rainey, Mr. J.E. Reilly and Mr. A. Sapiro, all of IBM Poughkeepsie.

REFERENCES

1. A.L. Scherr and D.L. Larkin, "Time sharing for OS", *Proc. FJCC*, 1970.
2. *IBM System/360 Operating System Time Sharing Option: Planning for TSO*, IBM Corp. Form CG28-6698, 1971.
3. A.L. Scherr, *An analysis of time-shared computer systems*, M.I.T. Press, Cambridge, 1967.
4. W.J. Dougherty, "Scheduling TSS/360 for responsiveness", *Proc. FJCC*, 1970.
5. W.C. D'Avanzo, Ph.D. thesis, Purdue University (in preparation).

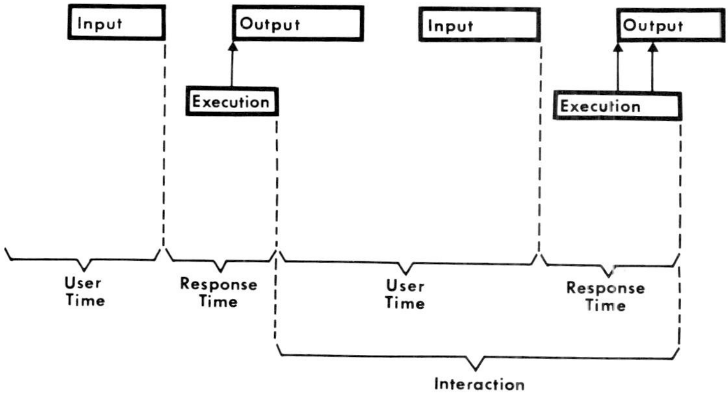
Input
Output
Execution
Input
Output
Execution
User Time
Response Time
User Time
Response Time
Interaction

Fig. 1.

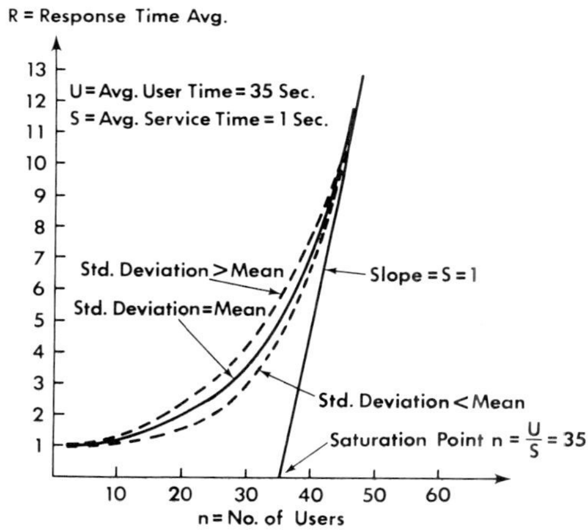
R = Response Time Avg.
U = Avg. User Time = 35 Sec.
S = Avg. Service Time = 1 Sec.
13
12
11
10
9
8
7
6
5
4
3
2
1
Std. Deviation > Mean
Std. Deviation = Mean
Slope = S = 1
Std. Deviation < Mean
Saturation Point n = U/S = 35
10
20
30
40
50
60
n = No. of Users

Fig. 2.

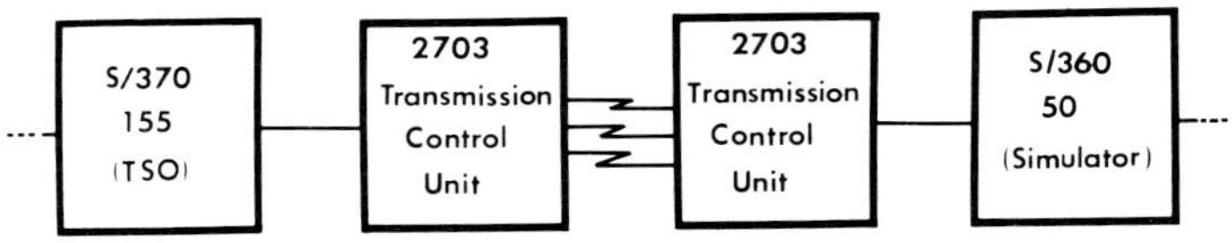
S/370
155
(TSO)
2703
Transmission
Control
Unit
2703
Transmission
Control
Unit
S/360
50
(Simulator)

Fig. 3.

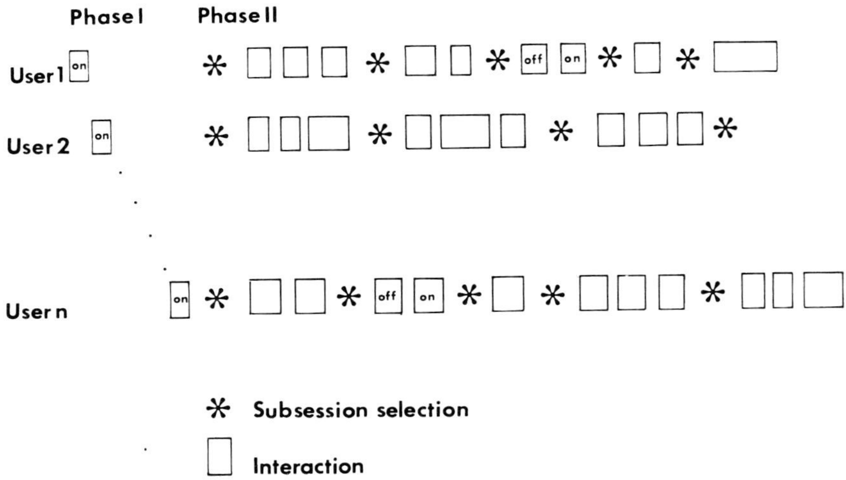
Phase I
Phase II
User 1
User 2
User n
on
off
Subsession selection
Interaction

Fig. 4.

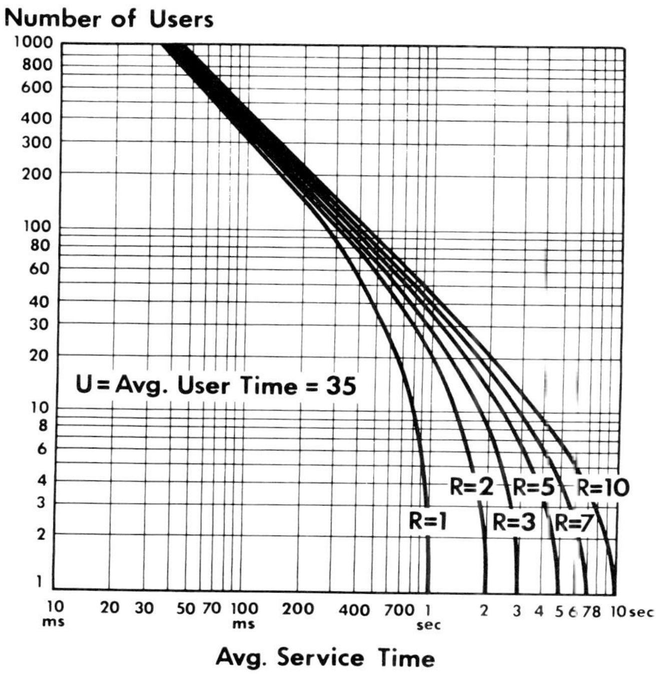
Number of Users
U = Avg. User Time = 35
R=1
R=2
R=3
R=5
R=7
R=10
Avg. Service Time

Fig. 5.

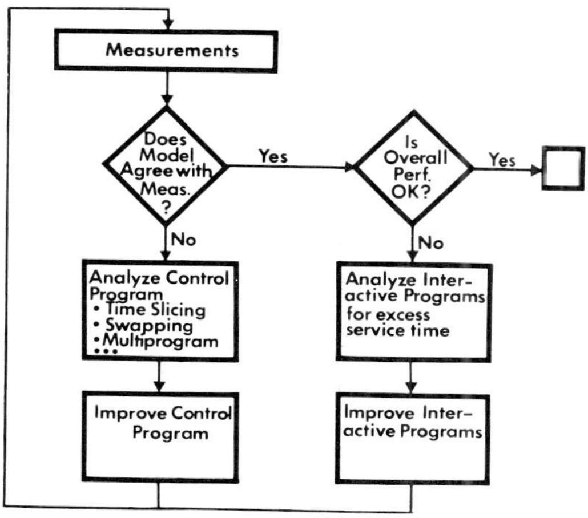
Measurements
Does Model Agree with Meas. ?
Yes
No
Is Overall Perf. OK?
Yes
No
Analyze Control Program
• Time Slicing
• Swapping
• Multiprogram
Analyze Interactive Programs for excess service time
Improve Control Program
Improve Interactive Programs

Fig. 6.

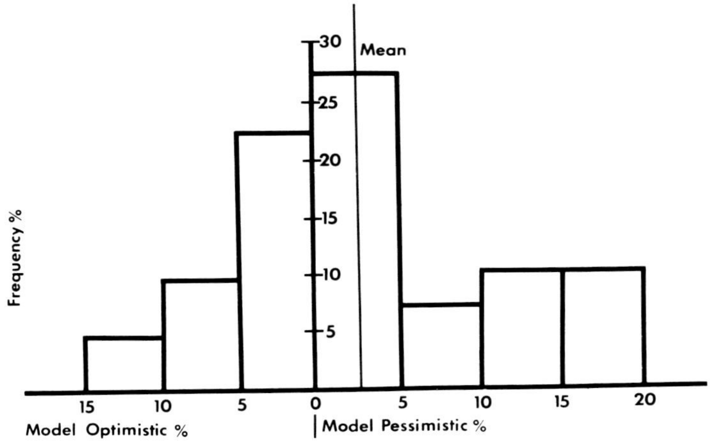
30
25
20
15
10
5
Mean
Frequency %
15
10
5
0
5
10
15
20
Model Optimistic %
Model Pessimistic %

Fig. 7.

A STATISTICAL EVALUATION OF THE SCHEDULER OF AN EXPERIMENTAL INTERACTIVE COMPUTING SYSTEM*

Harold A. Anderson, Jr. and Robert G. Sargent

ABSTRACT: This paper reports on a statistical evaluation of the scheduler of an experimental IBM APL\360 computer system using data collected under actual operating conditions. The objectives of the evaluation were twofold: (1) to determine the statistical properties of the input traffic, the users' computer service time demands, and the selected performance measures, and (2) to obtain linear least-square models relating the selected performance measures to the factors which measure the overall system activity and affect performance.

INTRODUCTION

Interactive computing systems were developed to increase the efficacy of the scientists' and engineers' use of the computer for problem solving [19]. Much of the performance evaluation effort to date for interactive computing systems has focused on investigating the interaction between programs and hardware rather than on the interaction of the users and the system [12]. This is especially true for the general programming systems (TSS/360, MULTICS) [6] where the problem has been our lack of knowledge about scheduling, resource allocation and program behavior. We do not have this problem with limited programming systems (JOSS, QUICKTRAN, APL\360) [6]. Limited programming systems can be evaluated to determine how well they have achieved the original goal of interactive computing. One fact which attests to the success of many of these systems is that over the years they have attracted a large population of users. This success has caused performance problems to develop since the peak periods of system utilization have been protracted out over a considerable part of the work day. Performance improvements through scheduler modifications are required to

*This research was partially supported by RADC contract AF 30602-70-C-0190.

obtain less severe performance degradation over these periods.

Queueing theory models and simulation models have been developed to investigate processor scheduling algorithms for interactive computing systems [1,11,12]. The insights into processor scheduling provided by these models have not been readily utilized by system designers since the assumptions upon which these models were based have not been validated [1,11]. In this work, the scheduler of an experimental limited programming system was evaluated in terms of characterizing the interactions of the users and the system. This was accomplished by including an event-stream-recording software monitor in the supervisor. The event stream recordings were used to provide data for statistically evaluating the scheduler and for validating many of the assumptions made in queueing theory and simulation models. The results of R.J. Van den Heever's theoretic investigation into scheduling for interactive computing were utilized in determining improvements to the scheduler of the monitored system [10]. The statistical evaluation of the relationships between the performance measures and the factors which measure system activity and affect performance provided the understanding needed to define scheduler improvements not obtainable from queueing theory models.

THE EXPERIMENTAL IBM APL\360 SYSTEM

The experimental IBM APL\360 system (hereinafter the system) whose scheduler was evaluated was a dedicated interactive computing system located at the IBM T.J. Watson Research Center [2]. It serviced a population of experienced users. This system was a limited programming system similar to the JOSS (RAND) system [3]. The hardware configuration consisted of an IBM 360/50 computer with 512K bytes of main memory, and one IBM 2314 disk unit. Users communicated with the system through the use of typewriter terminals, and vice versa.

The programs and data a user worked with at any time had to reside concurrently in an entity called a workspace. A workspace was limited in size to be no larger than 36,000 bytes. When a user submitted an input to the system it was stored in the user's workspace for servicing. His workspace was then flagged to be active and queued for service. A swap scheduling algorithm was used by the supervisor to load and unload the main memory with workspaces. The disk units were used for swapping workspaces and for storing the libraries of users' workspaces. There were four regions in the main memory used for storing active workspaces. A high degree of overlapped workspace swapping and servicing was accomplished by the system. Each region had a fixed size, which was large enough to store the maximum-size workspace.

The software subsystem was comprised of two main components, the supervisor and the interpreter [4] (figure 1). The evaluation of the interpreter-dispatching algorithm and the swap-scheduling algorithm implemented in the scheduler portion of the supervisor were of main concern. Earlier Hellerman and Ron conducted a limited investigation of the performance of a similar system [20]. The interpreter-dispatching algorithm was a modified foreground-background scheduling algorithm with two levels of queueing [1]. One cause of the modification to the foreground-background scheduling algorithm was a class of user inputs called special disk commands (commands) [5]. A command was invoked by a user to access a library of workspaces in order to change the status of one of the user's workspaces. It required the use of two main memory workspace regions in order to be serviced, and therefore received high-priority consideration in obtaining service. A second cause of the modification was the necessity of having a swap-scheduling algorithm to share the interpreter among all the active workspaces. The active workspaces residing in the main memory were serviced by a round-robin scheduling algorithm. A decision to swap a workspace from main memory to the swap disk resulted when one of the following events occurred: (1) an input arrived in the system from a user; (2) a library directory needed to be loaded into main memory; (3) ten time slices of service had been allocated without an input arriving or a directory being loaded into main memory; (4) a workspace in main memory required no further service and some active workspace resided on the swap disk. When an active workspace was swapped out of the main memory the one residing the longest was the one selected.

In this system the length of a time slice was a random variable which ranged from 0. to 0.27 sec. Table 1 summarizes the frequency counts of the observations on the length of a time slice for representative 10-minute sample intervals. The interpreter during the course of its execution at preselected points (for example, just before executing an operation on an array of data) would terminate the current time slice if the supervisor had requested it to do so. The supervisor would make such a request of the interpreter if one of the following events occurred: (1) the present time slice had taken over 0.1 sec; (2) the I/O channel used for all disk I/O had become idle; (3) an input had arrived from a user.

The software monitor that was implemented for the system was similar in design to the SIPE software monitor developed for the IBM TSS/360 system [7]. It was an event-stream recording system which recorded the occurrence of the following events: (1) the interpreter was dispatched to service a user's workspace; (2) a supervisor service was requested by the interpreter; (3) a disk I/O request was made by the supervisor; (4) a special disk command was recognized; (5) an input from a user entered the system; (6) the supervisor entered the wait state. Each recorded event was time-stamped, the user who caused the event was identified, and the main memory workspace region containing the user's workspace was identified. These events permitted the users' activity, the input traffic, the workload, the internal congestion and the swapping activity to be ascertained for any recorded time interval. System overhead which could be attributed to the software monitor was held to a minimum (<1%) since all information recorded was readily determinable. Over the course of a month, fourteen different time periods of system operation were recorded. In total, over seventy-eight hours of data were recorded.

CHARACTERIZATION OF THE USERS' COMPUTER SERVICE TIME DEMANDS

The computer service time demands of users (service times) of the system were widely variable (ranging from 1/60th of a second to 5 minutes). This was observed from the cumulative relative frequency distribution tabulated from the recorded demands of all user inputs. The long-tailed distribution has a large estimated coefficient of variation and positive skewness. To eliminate this large variability, the users' inputs were classified into three classes: compute transactions, commands, and output-bound transactions. A compute transaction required only servicing by the interpreter. A command required interpreter servicing and accesses to a workspace library. An output-bound transaction was a compute transaction which generated many lines of output while being serviced by the interpreter. After it had been allocated the full quota of output buffers the supervisor permitted, the scheduler terminated servicing it until all the output buffers were emptied. Table 2 summarizes the statistics for the service times for all the inputs and for each input class. These statistics indicate that the service times within each input class are still widely variable.

The different cumulative relative frequency distributions of service times for each monitored time period were tabulated. Visual comparison of these tables revealed no indication that the service time distributions varied significantly from monitored period to monitored period (table 3). The recorded input traffic was also broken down into 10-minute time intervals and the composition of the input traffic in terms of the percentage of each type of input present was calculated. These percentages were found not to vary significantly across all the 472 recorded 10-minute time intervals (figure 2). Thus, both of these observations imply that the user opopulation of this system can be characterized in terms of their demands for computer services even though they are widely varying. This type of characterization facilitates tailoring the scheduling algorithms to servicing a particular user population.

The service times for the inputs are characterized by calculating the following five functions: the cumulative relative frequency distribution, the histogram, the estimated completion rate, the estimated mean residual service time, and the estimated concentration curve. The completion rate (failure rate), $\phi_h(X)$, is the conditional probability of an input's service being completed in the time interval (X,X+h] having received X seconds of service:

$$\phi_h(X) = \frac{F_T(X+h)-F_T(X)}{1-F_T(X)}$$

where $F_T(X)$ is the d.f. of service time and $h > 0$. The mean residual service time (mean residual lifetime) is the conditional expectation of the service time remaining for an input which has received X seconds of service:

$$E[T|S=X] = \frac{\int_X^\infty (1-F_T(t))dt}{1-F_T(X)}$$

where T is the remaining amount of service required and S is the amount of service received. The concentration curve illustrates the percentage contribution to the mean value from the percentiles of the population of service times. It is a plot of

$$Q(t) = \frac{\int_0^t X dF_T(X)}{ET}$$

versus $F_T(t)$, where ET is the mean of the service times. These five characterizations of service times for all inputs are represented in figures 3 - 7. The cumulative relative frequency distribution and histogram show that a majority of the inputs require small amounts of computer service, i.e. 77.8% of the service times were $\leq$ 0.2 sec. This illustrates the advantage of a time-slicing strategy for this system. This observation is further amplified by the estimated concentration curve which indicates that the upper 4% of the service time samples contributed 90% of the estimated value of the mean. A similar observation has been made for the service times of other limited-purpose interactive computing systems. Figures 5 and 6 indicate that as an input receives increasing amounts of service its estimated completion rate decreases while the expected residual service time increases. These observations have a significant effect on scheduling for this system. At any time newly-arrived inputs have the shortest expected residual service time. Therefore, if your performance optimization criterion is to minimize the mean overall delay in the system, or equivalently to minimize the mean number of inputs in the system, then newly-arrived inputs should be serviced ahead of those queued inputs which have received some service [8,9]. This decision rule is included in the foreground-background scheduling algorithm. The optimality of this scheduling algorithm for interactive computer systems with increasing mean residual service times (lifetimes) was investigated by R.J. Van den Heever [10]. The same service time characterizations apply to each of the three input classes as applies to the distribution of the service times for all inputs (figures 8 - 10).

In queueing theory models of limited-purpose interactive computing systems, the assumption is made that the service times of all customers (inputs) are independent. The validity of this assumption for the system was partially investigated by estimating the serial correlation coefficients of the service times of inputs arriving over 32 randomly selected 10-minute time intervals. The estimated serial correlation coefficients were consistently close to zero in value for all 32 realizations.

No commonly employed distribution function has been successfully fitted to any of the empirical service time distributions. The number of observations for any one of them was quite large, thus requiring a very close fit between an empirical distribution and the selected distribution in order to pass any goodness-of-fit test. A composite of Weibull distributions

$$f_i(t)dt = \left(\frac{t}{\sigma_i}\right)^{n_i-1} \exp\left[-\left(\frac{t}{\sigma_i}\right)^{n_i}\right]\frac{dt}{\sigma_i} \quad a_i<t\leq a_{i+1},\ i=1,\ldots,m$$

$$a_1=0 \quad \text{and} \quad a_{m+1} = \infty$$

fit the empirical distributions satisfactorily (table 4). This family of distributions was selected based on the properties of the estimated completion rate and the estimated mean residual service time.

DESIGN OF EXPERIMENT CONSIDERATIONS

After examining the summaries of the recorded event streams, it was concluded that the system was being heavily utilized for long periods of time; i.e., over 75% of the time the central processor's utilization was greater than 90%. It was also concluded that the scheduling algorithms degrated performance to an unnecessary degree during these periods. In order methodically to evaluate the performance degradation due to increasing user activity, a set of experiments needed to be designed. Since controlled experimentation was not possible in this work, the experiments to be analyzed had to be derived from the set of event streams already recorded.

The performance measures

The three performance measures: the reaction time for all inputs, the response times for compute transactions, and the response times for commands, were selected as the variables to be used in the evaluation. Reaction time is the time that elapses from when an input arrives in the system until it receives its first time slice of service. It measures how effective a scheduler is in dispatching service to a newly-arrived input. Response time is the time that elapses from when an input (requiring less than 2 seconds of service) arrives in the system until the system is finished servicing it. Since compute transactions and commands are scheduled using different decision rules, the response times for each were analyzed separately. The two-second limitation on service times arises from the observations of psychologists that a response time of 15 seconds (which is not unreasonable for a service time demand greater than two seconds) causes a user doing computations at a terminal to no longer perceive he is working interactively with the computer [13].

For every recorded 10-minute time interval, the mean value, the standard deviation, and the maximum value of reaction time were estimated. In addition, the values of each of the following activity factors were determined: the number of users serviced, the number of inputs generated by the users, the total service time demand of the inputs, the central processor utilization, and the number of disk accesses that occurred. The value of the average reaction time was then used to determine the values of the activity factors when performance degradation was evident. These results served to define the region of the activity factor space for evaluating the system's performance. It was concluded that a 3^2 factorial design (f.d.) would be an appropriate design for the experiments. The two factors were the number of users serviced and the average input rate per user. These factors are desirable to incorporate in any model of the system and the results of their manipulation are easily interpretable. The factor, the number of users serviced, is different from the number of users signed-on the system. The former measures the number of users who had input something to the computer over the sample interval. The latter is not a good measure of user activity, since a user's input insertion time exceeding several minutes was observed 2.5% of the time. A sample time interval of 10 minutes was chosen because it was close to the median of the length of a user's terminal session and for time intervals of this length, the rate of occurrence of an input to the system generally appeared to be trend-free.

The input traffic

For a sample time interval which satisfied one of the

three levels for both factors to be selected as a candidate for a replication of an experiment, it had to satisfy the following additional criterion. The rate of occurrence of an input had to be trend-free over the sample time interval. This criterion was established to eliminate the effect of a changing input rate which might result from a significant change in the overall congestion within the system. In 75% of the 133 10-minute sample intervals examined, the null hypothesis of a trend-free rate of occurrence of an input was not rejected at the 5% significance level [14,15].

For each sample interval which passed the above criteria, the time series of the interarrival times of the input traffic was tested to see if the Poisson process was an appropriate model for it. The statistical methodology used was developed by P.A.W. Lewis in his work on the statistical analysis of series of events [14,15]. The following test statistics were computed for each time series analyzed: (1) the one-sided and two-sided Kolmogorov-Smirnov statistics; (2) the Anderson-Darling statistics; (3) Durbin's modification of (1) and (2); (4) the Moran test statistic; (5) the first three product moments for exponential scores; (6) the first four moments of the interarrival times, the standard deviation, skew, and kurtosis; (7) the serial correlation coefficients. The null hypothesis that a Poisson process models the generation of the trend-free input traffic was not rejected at the 5% significance level (table 5).

The sampling plan

From the large number of recorded sample intervals, eight replications of each experiment of the 3^2 f.d. were randomly selected. For each replication of an experiment, the following statistics were estimated: (1) for the reaction time based on 30 observations -

$$\overline{X},\ \overline{\log_{10} X},\ S_X^2 \text{ and } S_{\log_{10} X}^2;$$

(2) for the response time of compute transactions based on 30 observations -

$$\overline{X},\ \overline{\log X},\ S_X^2 \text{ and } S_{\log_{10} X}^2;$$

(3) for the response time of commands based on 20 observations -

$\overline{X}$, $\overline{\log_{10}X}$, S_X^2 and $S_{\log_{10}X}^2$.

The sampling scheme for selecting observations was determined by first choosing representative 10-minute intervals which satisfied the experimental conditions for the two factors, and second examining the estimated serial correlation coefficients of the time series of observations for each performance measure. This was done to assure that the selected observations were as little correlated as was practical to achieve. The estimated standard deviation of each performance measure was found to be proportional to the estimated mean value. For this reason the transformation $Y = \log_{10}(X + .5)$ was applied to each observation in order to reduce the effect of the heteroscadascity of the variances on performing statistical tests in the analysis of variance [16]. The logaritumic transformation has often been applied to observations on elapsed times by experimenters in other fields [17].

RESULTS OF THE 3^2 FACTORIAL DESIGN OF EXPERIMENTS

The two factors, the number of users serviced and the average input rate per user, were selected to investigate the performance degradation of the scheduling algorithm in terms representative of the interaction between the user population and the system. The levels for the number of users serviced were U_1 = 26 or 27 users, U_2 = 30 or 31 users, and U_3 = 34 or 35 users. The levels for the average input rate per user were I_1 = 1.36 inputs/minute/user, I_2 = 1.62 inputs/minute/user, and I_3 = 1.88 inputs/minute/user. Performance degradation had been previously observed over this range of the factors. Three levels were used to estimate the averages of the linear effects, the quadratic effects and the interactions between both factors. Table 6 contains the least-square estimates [16] $\hat{N}_{ij} = \overline{Y}_{ij}$ of the parameters of the linear model for the 3^2 f.d. $Y_{ijk} = N_{ij} + \varepsilon_{ijk}$ where in the present context Y_{ijk} is the average value of the performance measure calculated for the kth replication of the experiment when the number of users serviced (factor 1) was at its ith level and the average input rate per user (factor 2) was at its jth level. $\overline{Y}_{ij} = (1/8)\Sigma_k Y_{ijk}$, $\varepsilon_{ijk} \sim N(0,\sigma^2)$, $N_{ij} = \mu + \alpha_i + \beta_j + \gamma_{ij}$, and $i=1,2,3$; $j=1,2,3$ and $k=1,\ldots,8$. The parameters μ, α_i, β_j and γ_{ij} are the general mean of the performance measure, the main effect of the ith level of factor 1, the main effect of the jth level of factor 2 and the interaction of the ith level of factor 1 and

and the jth level of factor 2, respectively. The main effects and interactions are subject to the constraints $\Sigma_i \alpha_i = \Sigma_j \beta_j = \Sigma_{ij} \gamma_{ij} = 0$.

Examination of table 6 reveals that as the number of users serviced increased the average value of the performance measures increased. However, no similar consistent pattern emerges for increases in the average input rate per user. The conclusion derived from the analysis of variance based on the orthogonal contrasts for the 3^2 f.d. was that the average linear effect of the number of users serviced was statistically significant at the 5% significance level for all three performance measures, while all the other effects and interactions were not statistically significant (table 8). The lack of statistical significance for the effect of the average input rate per user was unexpected since this factor is directly related to the magnitude of the input traffic.

From the analysis of variance for each performance measure it was determined that the sum of the squares due to error was large relative to the sum of the squares explained by the effects and interactions of the two factors of the f.d. The main limitation of these factors in accounting for the observed behavior of the performance measures is that they represent only implicitly the workload being serviced by the computer. One can infer from them that as the activity and the number of users increases, the workload on the system is expected to increase. However, the correlation is not 1. The service time demands of the input traffice are samples from service time distributions with large variances. The service time demands (workload) generated by a given user population working at some average input rate per user will vary widely. This is made evident in table 7 where the total service time demand of the input traffic is shown to vary markedly from one 10-minute sample interval to another. Therefore, the average value of the performance measures will have a large unexplained variation unless the workload is accounted for by including additional pertinent factors in the model.

REGRESSION MODELS FOR THE PERFORMANCE MEASURES

Regression analysis was used to develop models for the performance measures which took into account the workload on the system as well as other factors which had a significant effect on performance. The behavior of the average values of the performance measures depends on the state of the system and on the decision rules of the scheduler. In viewing the system as a queueing network (figure 11), the state of the system is represented by the number of active workspaces in each queue. The expected value of the length of a queue, EL, contains much information about the queue and the service demands of the customers. For example, in the M|G|1 queue

$$EL = \frac{\lambda^2 ES^2}{2(1-\rho)}$$

where λ is the average input rate to the queue, ES^2 is the second moment of the service time distribution, and $\rho = \lambda ES$ is the utilization of the server. The equation for EL becomes significantly more complicated when a foreground-background scheduling rule, a time-slicing strategy, and a swapping algorithm are included in the queueing discipline [10]. Therefore, the regression models investigated contained the average length of the queues over the sample intervals. The average queue lengths include information not only on the workspaces which joined the queues during a sample interval but also on the backlog of active workspaces in the queues at the start of a sample interval.

The general form of the regression models was influenced by Little's formula for the G|G|1 queue,

$$EW = EL/\lambda$$

where EW is the expected waiting time in the queue. Other factors were incorporated into the models to account for the interactions among the queues and the disk I/O activity. The parameters of the models were estimated from the performance data contained in 64 sample intervals. The sample intervals were derived from those used in the factorial design. However, they had to satisfy the additional condition that the computer processor's utilization exceed 67%. A stepwise regression analysis technique was employed (with a partial F-test criterion at the 5% significance level) [18] to determine the set of performance factors to be incorporated in each model. The analysis of variance calculations are summarized in table 9. The mean sum of squares due to error was reduced considerably by each regression model from its previous value in the analysis of variance for the 3^2 f.d. A separate test of fit for each model was not possible due to the lack of replicated experiments [18].

The estimated regression models are

$$E[\log_{10}(R+.5)] = -0.240 + 0.09473W_{IN} + 0.003969\overline{L}_{IP} + 0.00007778D$$

$$E[\log_{10}(C+.5)] = 0.1563 + 0.09402W_{IN} + 0.01549W_{C} + 0.01048\overline{L}_{IP} + 0.00006098D$$

$$E[\log_{10}(W+.5)] = -.08968 + 0.07914W_{IN} + 0.005179\overline{L}_{C} + 0.00172W_{IP} + 0.0001073D$$

for the reaction time, the response time of commands, and the response time of compute transactions, respectively.

$$W_{IN} = \overline{L}_{IN}/\overline{\lambda}_{IN}, \quad W_{IP} = \overline{L}_{IP}/\overline{\lambda}_{IP}, \quad W_{C} = \overline{L}_{C}/\overline{\lambda}_{C},$$

where $\overline{L}_{IN}$, $\overline{L}_{IP}$, $\overline{L}_{C}$ and $\overline{\lambda}_{IN}$, $\overline{\lambda}_{IP}$, $\overline{\lambda}_{C}$ are the average queue lengths and average input rates for the respective queues (figure 11). D is the number of disk accesses which occurred over a sample interval. The queue statistics which appear in the regression model for a particular performance measure result from the scheduling decision rules which cause the queues to interact. Assuming the error terms in the regression analysis have a $N(0,\sigma^2)$ distribution, then confidence intervals were computed for both the estimated parameters of a regression model and for the observations for a given set of regressor values.

The actual values estimated for the models parameters were not the most important result of the regression analysis. Determining the statistical significance of different performance factors in explaining the observed behavior of the performance measures and investigating the interrrelationships among the different performance factors by regression analysis proved to be more valuable. A diagnostic approach to determining the effect of one performance factor on another as well as on a particular performance measure resulted from developing the regression models. As an example: the length of a time slice is a random variable. Figure 12 illustrates how $\overline{T}$, the average length of a time slice, decreased as D, the number of disk accesses in a sample interval, increased. Regressing $\overline{T}$ on D we obtain the equation

$$\overline{T} = 10.864 - 0.0008706D \quad \text{(in 60ths of a second)}$$

The decrease in $\overline{T}$ due to increases in D results in fewer inputs being processed in one time slice of service. Therefore system throughput is degraded. The throughput degradation due to this interrelationship was as much as 10% to 12% during periods of high system utilization.

In all the regression models investigated, the number of disk accesses was a significant factor. This pointed to deficiencies in both the interpreter-dispatching algorithm and the swapping algorithm. We summarized these deficiencies as follows: (1) the scheduling algorithms did not differentiate enough among the active workspaces based on the amount of service received, i.e. a foreground-background scheduler of more than two levels was required; (2) the length of a time slice was too closely controlled by the disk I/O activity instead of by the amount of service already received by the active workspace; (3) the workspace swapping activity became too high when there were more than 3 active workspaces in the inprocess queue. This occurred during prolonged periods of high input traffic because newly-arrived inputs typically required less than one complete time slice of service; therefore workspace regions in main memory became rapidly available to the active workspaces in the inprocess queue which were stored on disk. These deficiencies are evidenced by the interrelationships which exist among the performance factors and between the performance factors and the performance measures. The interrelationships have been quantified by regression analysis.

The statistical evaluation of the scheduling algorithms of the system primarily focused on investigating the behavior of the first moments of the performance measures. The standard deviations of the performance measures were estimated and found to be proportional to the average value of the performance measures. They were also regressed on the performance factors and the same sets of performance factors were judged to be significant as were statistically significant for the corresponding average values. Performance degradation became apparent to the users when response times became lengthened and highly variable. We believe a major cause of the variability of the performance measures was due to the cylical scanning of the sequentially ordered system control tables by the scheduler. This is to be investigated in future work.

CONCLUSIONS

The service time demands of the users of the system have a wide variability. This variability persists even after the users' inputs are classified into three classes of inputs: compute transactions, commands, and output-bound transactions. Due to the large sample sizes for the empirical service time distributions, a very accurate fit by a distribution function was required to pass a goodness-of-fit test. No commonly employed distribution was satisfactorily fit. Composite Weibull distributions were fit to the empirical service time distributions. Interactive computing is appropriate since 77.8% of the inputs serviced required $\leq$ 0.2 seconds of service. The service time demands of successively arriving inputs are uncorrelated. Each empirical service time distribution has an increasing mean residual service time (lifetime). Therefore, the foreground-background scheduling algorithm modified for swapping is an appropriate scheduling algorithm for the system. The input traffic to the system with a trend-free input rate can be modelled by a Poisson process for time intervals of at least as long as 10 minutes. The number of users serviced by the system is a statistically significant factor in predicting the performance of the system. However, it is not a very reliable factor due to the wide variability of the service time demands of the users over different 10-minute intervals. The average lengths and average input rates of the internal queues of the system and the disk I/O activity over a 10-minute sample interval are the performance factors for which an adequate linear regression model can be constructed for predicting the expected value of the logarithm of the performance measures (reaction time and response time). Regression analysis has provided a diagnostic technique for quantifying the interrelationship among the performance factors and between the performance factors and performance measures which become significant during periods of high system utilization.

Future work

As pointed out previously, the variances of the performance measures increase with increasing system utilization. This increased variability in performance has an immediate deleterious effect on the concurrent users of the system. An improved scheduling algorithm for the system will be synthesized from the results of a simulation model. The various scheduling algorithms investigated will be evaluated in terms of the effects of their decision rules on the mean and variances of the performance measures.

Acknowledgements. We sincerely appreciate the cooperation of Richard Lathwell and Alex Morrow of the IBM APL 360 project at the IBM Philadelphia Scientific Center; without their technical support this work could not have been accomplished. In addition, we appreciate the benefits we have derived from discussing systems evaluation methodology with Hisashi Kobayashi and Po Yue of the Computer Science Department of the IBM T.J. Watson Research Center.

BIBLIOGRAPHY

1. McKinney, J., "A survey of analytical time sharing models", Comp. Surveys 1, 105-116 (1969).
2. Breed, L., and Lathwell, R., "The implementation of APL 360", Proc. ACM Symposium on Experimental Systems for Interactive Applied Mathematics, 1967.
3. Bryan, G.E., "JOSS 20,000 hours at a console, a statistical summary", AFIPS Conf. Proc. FJCC 1967, vol. 31, pp. 769-777.
4. IBM APL\360 Program Listings (Program Number 5734-XM6), IBM Corporation.
5. APL 360-OS and APL 360-DOS System Manual (LY20-0678), first edition, 1971, IBM Corporation.
6. Wegner, P., Programming languages, information structures and machine organization, McGraw-Hill, 1968.
7. Denniston, W.R., "SIPE: A TSS/360 software measurement technique", Proc. 24th Nat. Conf. ACM, 1969, pp. 229-245.
8. Schrage, L., "Some queueing models for time-shared facilities", Ph.D. dissertation, Cornell University, 1966.
9. Conway, R., Maxwell, L, and Miller, L., Theory of scheduling, Addison-Wesley, 1967.
10. Van den Heever, R.J., "Computer time-sharing priority systems", Ph.D. dissertation, University of California, Berkeley, 1969.
11. Estrin, G., and Kleinrock, L., "Measures, models and measurements for time-shared computer utilities", Proc. ACM 1967 Nat. Conf., pp. 85-96.
12. Anderson, H., and Sargent, R., "Bibliography of modeling, evaluation and performance measurements of time-sharing computer systems", unpublished.
13. Miller, R., "Response time in man-computer conversa-

tional transactions", AFIPS Conf. Proc. FJCC 1968, vol. 33, pp. 267-277.

14. Cox, D.R., and Lewis, P.A.W., The statistical analysis of series of events, Methuen, 1966.
15. Lewis, P.A.W., Katcher, A., and Weis, A., "SASE IV: an improved program for the statistical analysis of series of events", IBM Research Report RC-2365, IBM, Yorktown Heights, 1969.
16. Scheffé, H., The analysis of variance, John Wiley, 1959.
17. Zellen, M., "Factorial experiments in life testing", Technometrics 1 (1959) 269-288.
18. Draper, N., and Smith, H., Applied regression analysis, John Wiley, 1966.
19. Watson, R., Timesharing system design concepts, McGraw-Hill, 1970.
20. H. Hellerman and Y. Ron, "A time-sharing system simulation and its validation", IBM Tech. Report 320-2984, IBM New York Scientific Center, Data Processing Division, April 1970.

Time Slice Length (sec.)	I. Frequency	II. Frequency	Time Slice Length (sec.)	I. Frequency	II. Frequency
.0	34	93	.15	91	315
.017	54	194	.167	97	257
.033	57	189	.183	103	216
.050	57	191	.2	86	169
.067	53	190	.217	110	146
.083	52	191	.233	1432	650
.1	62	280	.25	7	13
.117	294	1080	.267	2	1
.113	32	115	.283	1	0

Table 1. The Frequency Counts of the Length of a Time Slice for Two 10 Minute Sample Intervals.

	All Inputs	Compute Transactions	Commands	Output-Bound Transactions
Mean (sec.)	2.064	0.826	0.741	20.910
Variance	387.8	96.092	23.679	4542.9
Std. Dev.	19.69	9.803	4.866	67.40
Skew	21.07	33.69	25.61	6.22
Num. of Samples	120765	99868	13395	7502

Table 2. The Statistics of the Service Times for All Inputs and for the Three Classes of Inputs.

Service Time (sec.)	8/20/70	8/27/70	9/11/70
0.2	.841	.810	.800
0.4	.888	.868	.863
0.6	.901	.896	.890
0.8	.916	.914	.905
1.0	.928	.925	.912
2.0	.953	.953	.939
3.0	.965	.965	.954
4.0	.972	.972	.964
5.0	.976	.977	.971
10.0	.983	.985	.990

Table 3. The Cumulative Relative Frequency Distributions of the Service Times for All Inputs for 3 Separate Recording Periods.

		Weibull Parameters for Composite D.F.	
Input Type	Range	σ	n
All Inputs	$0 \le X \le .4$	0.05927	0.3361
	$.4 < X < \infty$	0.02278	0.2261
Compute Transactions	$0 \le X \le .4$	0.04405	0.3705
	$.4 < X < \infty$	0.00902	0.2182
Commands	$0 \le X \le .1$	0.02646	0.1103
	$.1 < X \le .4$	0.03817	0.2017
	$.4 < X \le 1$	0.05393	0.2392
	$1 < X < \infty$	0.1214	0.3314
Output-Bound Transactions	$0 \le X \le 10$	2.366	0.2739
	$10 < X < \infty$	2.448	0.2792

Table 4. The Set of Composite Weibull Distributions Which Fit the Empirical Service Time Distribution Functions.

	Sample Interval Input Statistics			
# of Inputs	358	478	573	688
Trend Statistic-U	0.15	-.66	1.22	0.58
Mean	1.72	1.25	1.04	0.87
Variance	2.72	1.36	1.11	0.70
Std. Dev.	1.65	1.17	1.05	0.84
Coef. of Var.	0.96	0.93	1.01	0.96
Skew	1.64	1.62	1.99	1.71
Kurtosis	6.48	6.06	8.24	6.54
ES1	348	429	544	702
ES2	317	481	574	682
ES3	333	515	559	692
KS+	0.20	0.74	0.31	0.76
KS-	1.04	0.20	0.95	0.90
KS	1.04	0.74	0.95	0.90
WN2	1.61	0.53	1.04	1.42
MORAN	382	506	624	750
DN+	0.55	0.48	0.61	0.37
DN-	0.70	1.19	0.89	0.98
DN	0.70	1.19	0.89	0.98
DWN2	0.57	1.49	0.54	0.80

ES1,ES2,ES3: First three product moment statistics for exponential scores.
KS,KS-,KS+: Kolmogorov-Smirnov statistics.
WN2: Anderson-Darling statistic.
DN,DN-,DN+: Durbin's modification to the Kolmogorov-Smirnov statistics.
DWN2: Durbin's modification to the Anderson-Darling statistic.

Table 5. Examples of the Statistics for the Input Traffic of Four 10 Minute Sample Intervals.

Average Reaction Times

	Transformed Observations			Untransformed Observations		
	I_1	I_2	I_3	I_1	I_2	I_3
U_1	0.01554	-0.07806	0.002495	0.657	0.371	0.612
U_2	0.04081	0.07326	0.03294	0.695	0.827	0.678
U_3	0.04282	0.07177	0.07743	0.691	0.834	0.878

Average Response Times of Compute Transaction

	Transformed Observations			Untransformed Observations		
	I_1	I_2	I_3	I_1	I_2	I_3
U_1	0.08312	0.004276	0.0613	1.019	0.760	0.860
U_2	0.1300	0.1879	0.1439	1.138	1.449	1.202
U_3	0.1618	0.1719	0.1968	1.380	1.474	1.971

Average Response Times of Commands

	Transformed Observations			Untransformed Observations		
	I_1	I_2	I_3	I_1	I_2	I_3
U_1	0.2736	0.2318	0.2933	1.824	1.851	2.021
U_2	0.3896	0.3404	0.3376	3.338	2.271	2.885
U_3	0.3446	0.3943	0.4068	2.466	4.219	4.177

Table 6. The Overall Averages of the Performance Measures Based on 8 Replications per Experiment of the 3^2 F.D.

	U_1I_1		U_3I_3	
Replication	# of Inputs	Total Service Time Demand (sec.)	# of Inputs	Total Service Time Demand (sec.)
1	364	644	667	645
2	387	547	664	622
3	356	885	650	466
4	386	365	632	469
5	378	443	689	496
6	384	326	698	551
7	369	383	655	443
8	384	240	710	832

U_1I_1 - 26 or 27 users inputting at an average input rate of 1.4 inputs/min/user.

U_3I_3 - 34 or 35 users inputting at an average input rate of 1.9 inputs/min/user.

Table 7. The Number of Inputs in the Input Traffic and the Total Service Time Demand for Two of the 3^2 F.D. Experiments.

		Reaction Time			Compute Transaction Response Times			Command Response Times		
Source	D.F.	S.S.	Mean S.S.	F-Ratio	S.S.	Mean S.S.	F-Ratio	S.S.	Mean S.S.	F-Ratio
Linear Effect of U	1	.0847	.0847	7.901*	.1944	.1944	11.89*	.1606	.1606	7.944*
Quadratic Effect of U	1	.0012	.0012	.112	.0265	.0265	1.621	.0162	.0162	.807
Linear Effect of I	1	.0003	.0003	.028	.0010	.0010	.061	.0012	.0012	.060
Lin. U × Lin. I	1	.0045	.0045	.420	.0065	.0065	.398	.0030	.0030	.179
Quad. U × Lin. I	1	.0009	.0009	.084	.0001	.0001	.006	.0230	.0230	1.145
Quadratic Effect of I	1	.0027	.0027	.252	.0011	.0011	.067	.0056	.0056	.279
Lin. U × Quad. I	1	.0260	.0260	2.425	.0098	.0098	.599	.0131	.0131	.652
Quad. U × Quad. I	1	.0195	.0195	1.819	.0280	.0280	1.713	.0002	.0002	.010
Error	63	.6751	.0107		1.0300	.01635		1.2656	.02009	
Total (Correlated)	71	.8254			1.2973			1.4890		

*Significant at the 95% Significance Level

Table 8. The 3^2 F.D. Analysis of Variance Tables for the Transformed Performance Measures.

	Reaction Time				Response Time of Compute Transactions				Response Time of Commands			
Source	D.F.	S.S.	Mean S.S.	F-Ratio	D.F.	S.S.	Mean S.S.	F-Ratio	D.F.	S.S.	Mean S.S.	F-Ratio
S.S. due to mean	1	.1521			1	.0313			1	8.009		
S.S. due to regression	3	.5565	.1855	128.4*	4	.8447	.2112	86.8*	4	.9652	.2413	51.00*
S.S. due to error	60	.0867	.0014		59	.1435	.0024		59	.2791	.0047	
S.S. due to total	64	.7953			64	1.020			64	9.254		

*Significant at the 95% significance level.

Table 9. The Regression Models' Analysis of Variance Tables for the Transformed Performance Measures.

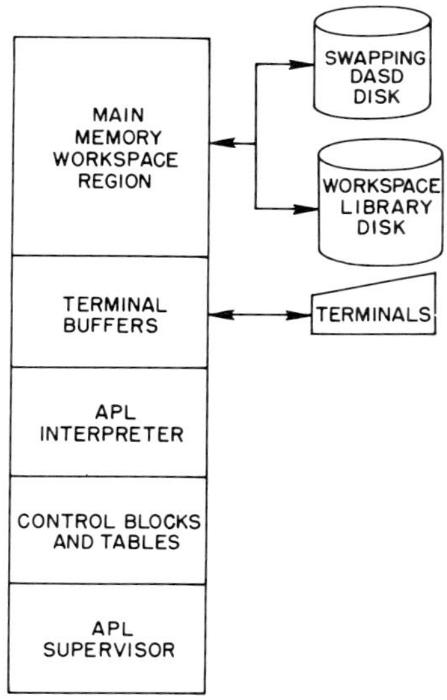

Fig. 1. The experimental IBM APL\360 system

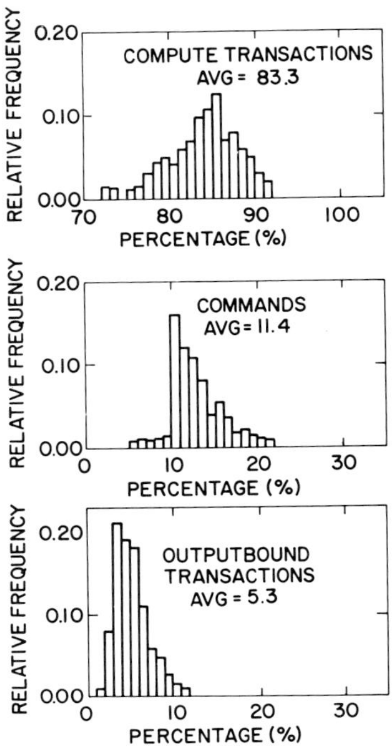

Fig. 2. The histograms of the percentage of inputs in the input traffic samples which were compute transactions, commands, and output-bound transactions.

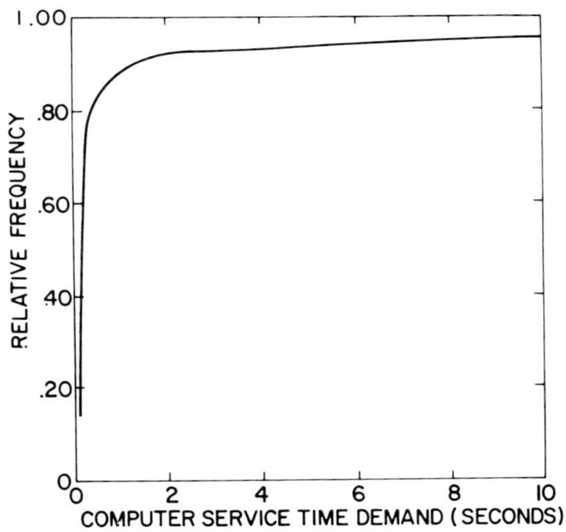

Fig. 3. The cumulative relative frequency distribution of the computer system time demand for all inputs.

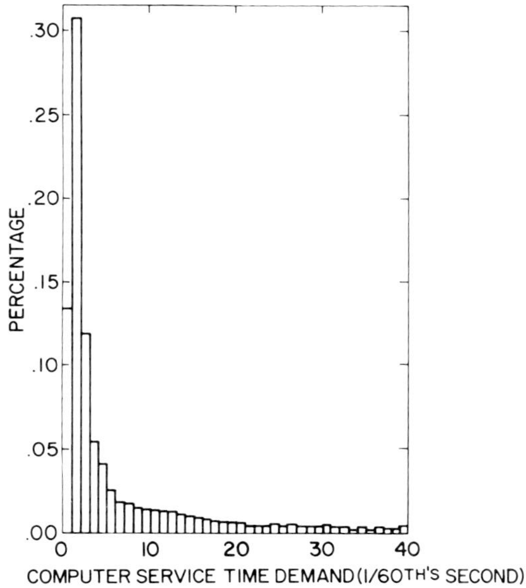

Fig. 4. The histogram of the computer service time demand for all inputs.

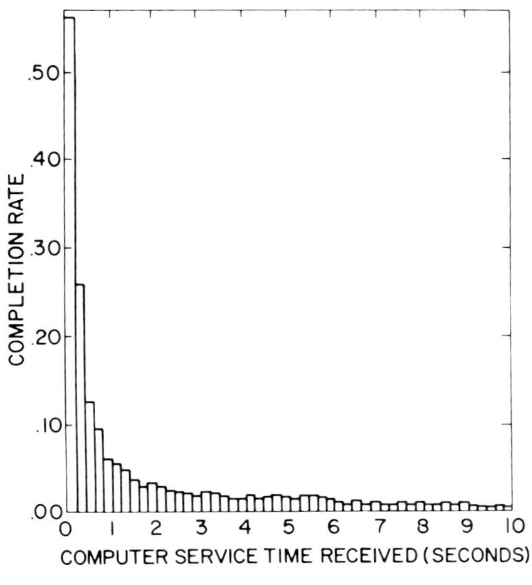

Fig. 5. The estimated completion rate for all inputs.

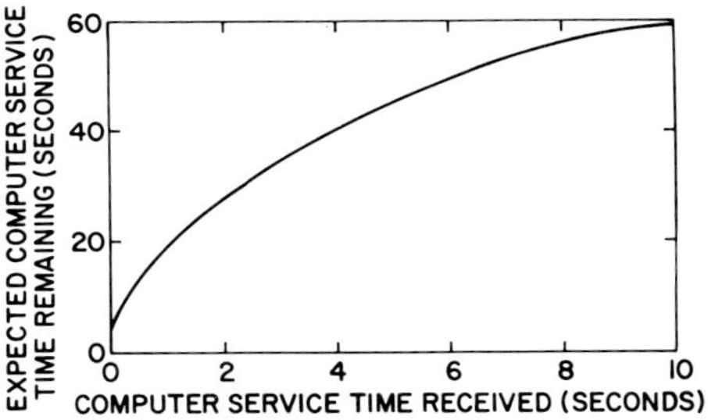

Fig. 6. The estimated mean residual service time for all inputs.

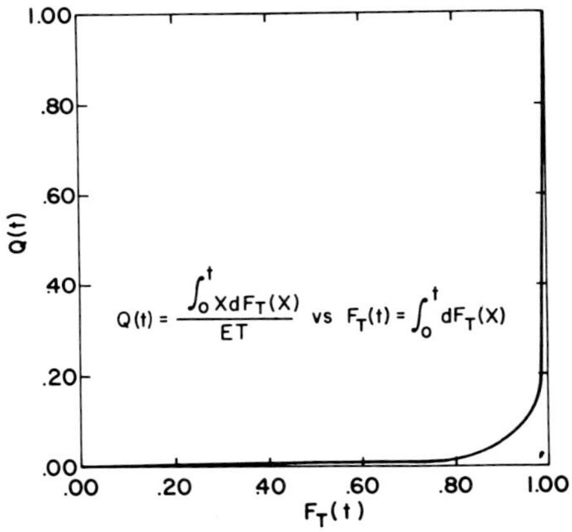

Fig. 7. The estimated concentration curve for all inputs.

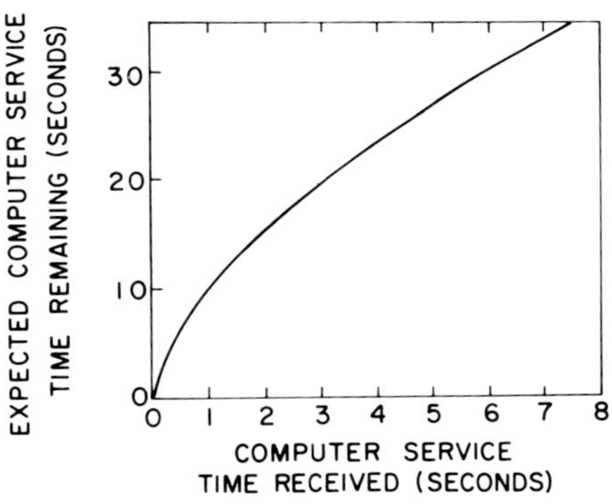

Fig. 8. The estimated mean residual service time for compute transactions.

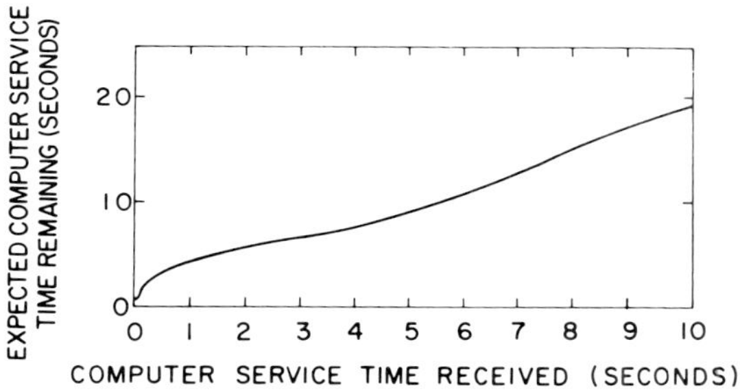

Fig. 9. The estimated mean residual service time for commands.

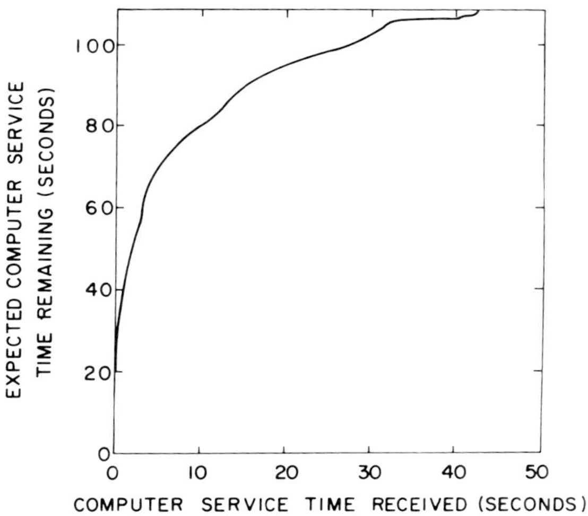

Fig. 10. The estimated mean residual service time for output-bound transactions.

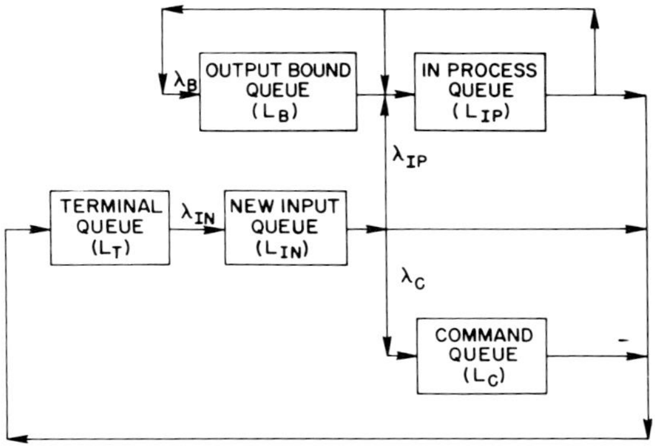

Fig. 11. The queueing network for the experimental IBM APL 360 system.

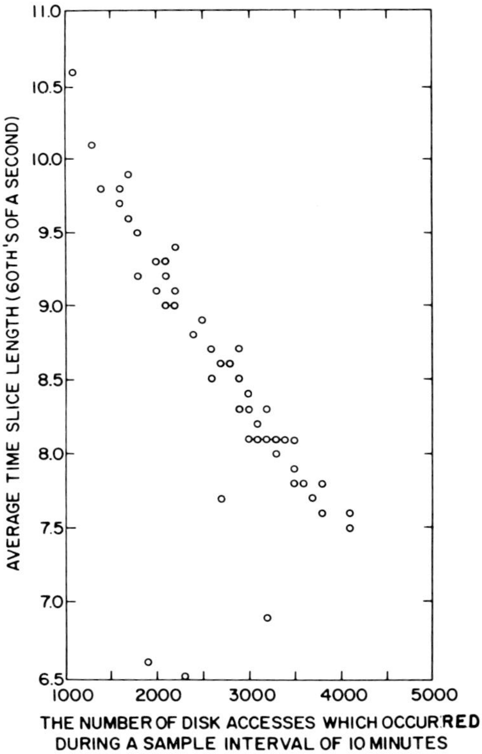

Fig. 12. The average length of a time slice versus the disk I/O activity.

DISCUSSION OF SESSION II

by H. Hellerman

Both papers in this session were worthwhile contributions since they not only discussed models of time-sharing systems but also compared results from the models with measured system behavior.

Lassettre and Scherr's paper is based on a mathematical model derived from classical queueing theory. It assumes a single server, Poisson arrival and service time mechanism, no time slicing, no overlap of swapping and compute. The sole measure of performance is mean response time. Needless to say, almost no actual time-sharing system (including TSO) operates under these simplified conditions. Service times, especially, are not exponentially distributed. So the question is, can useful results, at least for the authors' purpose of system performance debugging, still make good use of this model? The authors say yes. They cite as one item of evidence certain theoretical studies (done elsewhere) showing that the mean response time predicted by the model is relatively insensitive to details of the service time distribution (for given mean and variance). Even accepting this, it does not address a more fundamental question: how valuable is mean response time as a measure of time-sharing system performance? In my view, this measure is a poor one (for non-trivial interactions). A better performance indicator is a statistic on the response relative to the service time required for the request, e.g. the "stretch factor" defined as the ratio of response to service time (or its reciprocal). Unfortunately, statistics on such a performance measure do not emerge from analytic performance models. Furthermore, a measure like mean stretch factor is much more likely to be sensitive to the service distribution (and, more important, to user satisfaction).

For the above reasons, the paper leaves something to be desired as a computer-science contribution. It does, however, have one highly significant and important aspect. The model was actually used to help run down performance bugs under real system development conditions; i.e., we have one of the very few reported cases of a system being forced to meet analytically-derived performance objectives. For moving us a step in this important direction, the authors are to be commended.

The Anderson/Sargent paper is a careful study of an APL time-sharing system. It verifies results reported earlier (by Hellerman and Ron) that the inter-arrival distribution is nearly exponential but the service times depart very significantly from the exponential. Since the system is a heavily used and highly successful one, the distribution functions cited in the paper, including those for service times, and the fact that service times can be approximated by a two-parameter Weibull distribution, should be of interest to people doing modelling and simulation of time-sharing systems.

The performance measures used in the Anderson/Sargent paper are, I believe, in general realistic and appropriate. As done by earlier workers, they consider system responsiveness in three classes: 1) compute (i.e. fairly long CPU demand), 2) command (little CPU demand but fast response required) and 3) output-bound. Although these categories are worthwhile, as I understand it, category (2), which should include all the significant man-machine functions, in fact did not, since it only measured one important class of these. Yet enough was included of highly man-machine-interactive functions for the results to be significant. Although the three-way breakdown of response type is a useful one (with mean and standard deviation given separately for each class), a measure that gives response conditional on service time (or the ratio) would enhance the study. A final section of the paper presents a linear regression model on measured performance to expose some relationships between performance factors and performance measures.

SESSION III

DESIGN OF EXPERIMENTS

CHAIRMAN: F. J. ANSCOMBE

A MULTI-FACTOR PAGING EXPERIMENT:
I. THE EXPERIMENT AND THE CONCLUSIONS

R.F. Tsao, L.W. Comeau and B.H. Margolin

ABSTRACT. This paper reports on a statistically designed and analyzed experiment performed to study the effects of four factors, memory size, problem program, load sequence of system program subroutines and replacement algorithm, upon the paging process. The usefulness of the number of page swaps as a measure for comparing replacement algorithms is evaluated and alternative measures are proposed. The experiment and the conclusions drawn are presented in this paper, while the statistical methodology used to derive these conclusions is presented in a companion paper.

1. INTRODUCTION

Time-sharing and multiprogramming, which aim at simultaneously increasing the efficiency of the system and the response to the user, have created a need for improved system management of storage. Paging machines [7] have become a popular means of effecting this improvement. In a paging environment, both main memory and secondary storage are divided into equal-size units called pages. Program execution halts any time some needed information is not in main memory; this generates a page exception, signalling the need to bring in from secondary storage the page containing the necessary information. If needed, a replacement algorithm then determines which page resident in core is to be "swapped" with the page from secondary storage needed in main memory.

The effect of various replacement algorithms has been the subject of many papers (e.g., [1,4,9]); numerous attempts have been made to construct a good, if not optimal, replacement algorithm. Most papers, however, contain only simulated small experiments, with variation of only one or two factors influencing paging; empirical evidence is limited or entirely lacking. The authors know of no well-planned and well-executed experiment that investigates paging. This might be due to the large amount of machine time required for such an investigation. With the exception of the important study of Belady et al. [2], analytic studies of paging algorithms have been almost non-existent.

Many factors other than the replacement algorithm are known to influence paging, e.g. the available main memory and the load sequence of the system program subroutines. Two different approaches to the experimental exploration of a multi-factor system are the classical "one-at-a-time" experimentation, sometimes mislabelled the "scientific method", and the statistical design of experiments. In the former, only one factor of interest is varied at a time, all other factors being held constant. This supposedly "isolates" the factor's effect from all other considerations. The latter approach varies the factors simultaneously, thereby permitting estimation of not only the effect of each factor individually but also the synergistic effects or interactions of the factors. The statistically designed experiment will typically be more efficient, be more comprehensive, and its conclusions will have greater validity and applicability. These points will be discussed in a future paper by two of the authors (Margolin and Tsao) on the general subject of the design of experiments for computer system evaluation.

The present paper reports on a statistically designed and analyzed experiment performed to study the effects of four factors on the paging process; these four factors were the main memory size, the program being executed, the load sequence of system program subroutines and the replacement algorithm. The present paper will discuss only the experiment performed and the conclusions drawn; a companion paper [11] on the statistical methodology used to derive the conclusions presented herein also appears in the present volume.

The present paper, in its outline, follows a standard form for reporting upon designed experiments. It consists of:

(a) an explicit statement of the problems under investigation;
(b) a description of the experimental environment, specifying those aspects of the study which were constant throughout the experimental period;
(c) a discussion of the factors varied in the course

of the experiment;

(d) a description of the design of the experiment;

(e) a description of the responses monitored during the experiment, plus the basic data needed to derive these responses;

(f) a presentation of the conclusions drawn, written for the statistical layman;

(g) tables of the basic data and the resulting responses;

(h) a presentation of the statistical analyses performed upon the responses, which led to the conclusions presented.

Aspect (g) is presented partly in appendices and (h), as previously mentioned, appears in a companion paper [11].

2. PROBLEMS UNDER INVESTIGATION

It has generally been agreed that the comparison of replacement algorithms in a paging system should be based on a figure of merit called the total number of pages swapped or PS count; i.e., the total number of times a page was brought into main memory from secondary storage during a given job execution. The smaller the PS count, the better the algorithm. A major difficulty in using the PS count to evaluate replacement algorithms in an operational setting is that this measure also reflects factors other than the replacement algorithm. For example, it is very difficult to infer whether a large PS count is attributable to a heavy workload or to a poor replacement algorithm.

In the first phase of this study, the following problems were formulated for investigation:

(a) Is the PS count useful at all as a measure for the comparison of various replacement algorithms? If not, are other measures which the authors have formulated any improvement?

(b) What are the important factors that affect the paging, and what are their effects, both qualitative and quantitative?

3. EXPERIMENTAL ENVIRONMENT

The experiment was conducted in the IBM Cambridge Scientific Center with an IBM 360/40 of the following configuration:

(a) 256K bytes of high-speed main memory;

(b) two selector channels, each having 2311 disk drives attached;
(c) a multiplexor channel with the standard card equipment;
(d) as many as sixteen 2741-1050 terminals attached;
(e) a 64 word × 16 bit associative memory.

The software package for this system was called CP/40. This is the predecessor of the current CP/67, which supports virtual machines. Because of the structure of the hardware and software, it was relatively easy to determine the paging action of a program by recording and examining the contents of the array in associative memory [10].

4. DESCRIPTION OF THE EXPERIMENT

The planning of the paging experiment dealt with three main concerns: the selection of the experimental factors, the design of the experiment, and the determination of the set of responses to be measured.

4.1 Experimental factors

Experimental factors are controllable elements believed to be influential in the operation of the process under investigation. The number of possible factors can sometimes be too large for all to be incorporated in the initial experiment, which is limited by time and money considerations. The experimentor, then, must select for study those factors he believes to be potentially most important. In the paging experiment, four factors were selected for investigation:

(A) the replacement algorithm,
(B) the load sequence of system subroutines,
(C) the set of problem programs,
(D) the main memory size.

The various settings of any one factor are known as the levels of that factor. The selection for experimentation of both the factors and their levels is a complex task. A more detailed discussion of the reasoning behind factor and level selection will be included in the expository paper on the design of experiments by Margolin and Tsao.

For each of the four factors above, three levels of interest were chosen.

4.1.1 Factor A: the replacement algorithm

The three algorithms chosen, i.e. the three levels of this factor, were

Level 1: LRUV - this is essentially the "least recently used" procedure, where the least recently referenced page has highest priority for replacement [9];

Level 2: FIFO - this is the standard "first-in-first-out" procedure which replaces the page that was brought into core earliest;

Level 3: RAND - this procedure selects the page for replacement via a random number generator.

These algorithms were chosen because they had indicated high, medium and low performances respectively in a previous experiment with a single program and a single memory size.

4.1.2 Factor B: the load sequence of system program subroutines

Comeau [5] has shown (with a particular memory size, algorithm and program) that paging actions are significantly influenced by the load sequence of system subroutines. For this experiment, three different load sequences or deck arrangements of the monitor system, CMS (Cambridge Monitor System) were chosen:

Level 1: GROUP - this was the operating version of CMS at the time of the experiment; this sequence grouped together the subroutines which called each other frequently [8];

Level 2: FREQY - this sequence resulted from a past program execution, whose paging demands were evaluated; those subroutines causing the most page swaps were grouped together;

Level 3: ALPHA - this sequence was the alphabetical arrangement of the subroutines by their names.

4.1.3 Factor C: problem programs

In this experiment, the problem programs were compiled, but not executed. Three programs were chosen from a set of 28 FORTRAN jobs frequently selected as a standard mix for performance measurements.

Level 1: SMALL - this is the smallest of the 28 FORTRAN jobs in terms of the number of FORTRAN statements and the compiling time; it contains 55 statements and takes 3.88 seconds to compile with no paging;

Level 2: MEDIUM - this program contains 215 statements and takes 18.05 seconds to compile with no paging;
Level 3: LARGE - this is the largest of the 28 FORTRAN jobs; it contains 595 statements and takes 149.55 seconds to compile with no paging.

4.1.4 Factor D: main memory size

With the page size fixed at 4096 bytes, main memory size can be measured by the number of available pages. The choice of memory size is critical, since if too few or too many pages are available in core to the program, the significance of paging disappears. There was no way of knowing beforehand exactly which number of pages would fall into these two extremes; previous experimentation, however, had indicated that 20±4 pages would probably be a suitable range for exploration. Thus, the following three levels of main memory size were chosen:

Level 1: 24 available pages,
Level 2: 20 available pages,
Level 3: 16 available pages.

It should be noted that the levels within each factor above were ordered in such a fashion that the higher the level, the more paging expected.

4.2 Design of the experiment

For the four experimental factors above, each with three levels of interest, there are 81 (3^4) possible combinations of one replacement algorithm, one load sequence, one problem program and one main memory size. The experimental design consisting of all 81 combinations or runs is known as a complete 3^4 factorial experiment [6]; this was the experiment run. The investigation was conducted in the dedicated system mode, and was completed in under 60 CPU hours. The number of factors and levels of interest, and the specific design employed in the experiment should not be considered as standard; the 3^4 factorial experiment is merely one design among a great many in the statistics literature [3,6] that are useful in the experimental exploration of a multifactor system.

4.3 Responses and alternative measures

Responses or observations are the experimental results obtained in a study; they depend upon the particular combinations of the factor levels chosen for investigation. Each response is either a basic measurement or a function of a set of basic measurements taken directly on the system.

The PS count was the first response selected for study; the 81 observed values of PS are in Table 1.

Intuitively, a good algorithm should have the following two properties:

(a) It should leave in main memory those pages which will be referenced in the near future, where "near future" is taken as the time interval between two successive page swaps;
(b) It should replace those pages in main memory which would then stay in secondary storage as long as possible from that point on; the length of a page's stay in secondary storage may be measured in terms of the number of page exceptions occurring before that page is brought back to main memory.

Prior to the experiment, intuitively appealing measures labelled the activity count (AC) and the inactivity count (IC) were formulated by the authors as potential improvements over the PS count in terms of properties (a) and (b) above, respectively. The activity count measures the average number of pages in main memory referenced at least once between successive page swaps, relative to the total number of pages in main memory. It may be viewed as the average percentage of core active, i.e. referenced at least once, between successive page swaps. Thus it measures the effective core utilization. AC required recording, for each of the 81 experimental combinations, the number of pages active in core between the first and the second page swap, between the second and the third page swap, etc. Note that these individual numbers cannot exceed the number of pages in main memory.

One summary of this activity count data is a frequency distribution, f(x), which specifies for each of the 81 combinations the frequency with which x pages were active in core between successive page swaps. These 81 frequency distributions appear in Appendix I.

If the number of pages available in main memory is M, then the PS count is related to the corresponding frequency distribution as follows:

$$PS = \sum_{x=0}^{M} f(x) .$$

If x pages are referenced between successive swaps, then (100x/M) is the percentage of core active between the swaps. Thus the average percentage of core activity between swaps is determined as:

$$AC = \frac{\sum_{x=0}^{M} (\frac{100x}{M}) f(x)}{\sum_{x=0}^{M} f(x)} = \frac{100}{M} \cdot \frac{\sum_{x=0}^{M} xf(x)}{PS} \qquad (4.3.1)$$

The 81 observed values of AC for this experiment are presented in Table 2. As an example of the computation needed for AC, consider f(x) in Appendix I for the combination (24 pages, GROUP, SMALL, LRUV): $\Sigma xf(x) = 309$, $M = 24$, $\Sigma f(x) = 32 = PS$, and hence $AC = (100/24) \cdot (309/32) = 40\%$.

The third and last response observed was the inactivity count (IC), which measures the average time (in terms of the number of page swaps) of a page in secondary storage after it has been swapped out of main memory, relative to the core size M. This response required recording for each of the 81 experimental combinations, and for each page swapped out of core, the number of page swaps occurring before that page reentered core. Appendix II contains the 81 frequency distributions, g(y), for these data, where y represents the number of page swaps between a page's exit from and reentry into main memory. Theoretically, IC should be computed as:

$$\frac{100}{M} \cdot \frac{\sum_{y=0}^{\infty} yg(y)}{\sum_{y=0}^{\infty} g(y)} = (\frac{100}{M}) \frac{\sum_{y=0}^{\infty} yg(y)}{PS} , \text{ since } PS = \sum_{y=0}^{\infty} g(y) .$$

For practical purposes, in this experiment all values of y greater than M were recorded as M. Thus

$$g(M) = PS - \sum_{y=0}^{M-1} g(y)$$

and hence IC was computed as

$$IC = \frac{100}{M} \cdot \frac{\sum_{y=0}^{M} yg(y)}{PS} \quad . \qquad (4.3.2)$$

As an example of the computation needed for IC, consider g(y) in Appendix II for the combination (24 pages, GROUP, SMALL, LRUV):

$$M = 24$$

$$\sum_{y=0}^{23} g(y) = 21, \quad \sum_{y=0}^{24} g(y) = PS = 32$$

$$g(24) = 32 - 21 = 11$$

$$\sum_{y=0}^{24} yg(y) = 343$$

Thus

$$IC = (100/24) \cdot (343/32)$$

$$= 44.7.$$

The 81 values observed for IC are in Table 3.

5. CONCLUSIONS

The three responses observed in this experiment were statistically analyzed. Empirical models were constructed for each response and the effects of the four factors, both individual and synergistic, were estimated; then the models were calibrated for their goodness of fit to the data. The conclusions drawn from the analyses are presented below with no attempt to substantiate them. The statistical analyses and their justification of the inferences appear in the companion paper [11]. The conclusions apply only to the factors and levels studied. If new factors or new levels of the original four factors are introduced, there will be a need for further experimentation to validate the findings of this study.

Conclusion 1: AC is a better measure than PS or IC for comparison of replacement algorithms.

The PS count was influenced more by memory size, the problem program or the deck arrangement than by the replacement algorithm. This clearly limits its usefulness for the purposes of comparing replacement algorithms to cases where the three other factors can be controlled.

AC is a decided improvement over PS for purposes of comparing replacement algorithms. This is true because AC is:

(a) most sensitive to the replacement algorithm among the four factors studied;
(b) not significantly influenced by the problem program, thus allowing comparison of two replacement algorithms under different workloads;
(c) a percentage, and thus bounded between 0 and 100; hence, it can be used to judge an algorithm's performance relative to an ideal of 100% efficiency;
(d) a measure of core utilization in relation to overhead (i.e. paging).

As far as IC is concerned, AC is preferable to it as a measure for comparing replacement algorithms because AC is less influenced by the problem program.

Conclusion 2: Factors affect paging multiplicatively.

The four factors of interest each had a multiplicative or proportionate effect on the amount of paging; i.e., changes in factor levels increased or decreased the number of page swaps by (approximately) some fixed proportion. This suggests that the logarithm of the PS count (LPS) is a more appropriate response for statistical analysis because it affords a far simpler description and summary of the 81 experimental results; i.e., in the logarithmic scale, the four factors studied could be modelled linearly [11].

Conclusion 3: The significance of the effects in decreasing order was main memory, problem program, deck arrangement and replacement algorithm.

All four factors significantly affected paging. Although the effect of the replacement algorithm was the smallest of the four factors studied, it is appreciable nonetheless.

Conclusion 4: Effects of main memory size and deck arrangement are interdependent.

The magnitude of the effect of memory size was approximately unaffected by the problem program or the replacement algorithm, but was affected appreciably by the deck arrangement. Paging always increased in going to a smaller main memory, but the percentage of the increase depended upon which deck arrangement was under consideration.

In a similar fashion, the magnitude of deck arrangement effect did not depend on the problem program or the replacement algorithm, but did depend upon the main memory size. An increase in paging accompanied the change from GROUP to FREQY to ALPHA for all three main memory sizes, but the proportion of the increase depended upon which memory size was under consideration.

In the language of statistics, an interaction was found between main memory size and the deck arrangement (this same interaction was found in the analysis of AC and IC). This offers a plausible explanation for the anomalous finding of Belady et al. [2]. From the system point of view, if this interaction is sufficiently large, optimizing deck arrangement may be counterproductive for a variable memory size, i.e. a multiprogramming environment.

Conclusion 5: LRUV was the preferred replacement algorithm.

The magnitude of replacement algorithm effect was not dependent upon any other factor. LRUV substantially decreased paging when compared with either FIFO or RAND. FIFO only slightly decreased paging when compared with RAND.

Conclusion 1 above answers the questions (a) under investigation in section 2. The other four conclusions deal with question (b). All these conclusions are qualitative in nature. The quantification of these conclusions appears in the companion paper [11].

Acknowledgements. The authors wish to thank N. Rasmussen, Manager of the IBM Scientific Centers, for his support in conducting this experiment in the IBM Cambridge Scientific Center, and P. Morkepetrics for his assistance in programming. We are also grateful to L. Belady and L.Y. Yiu for their valuable comments on the manuscript.

REFERENCES

1. Belady, L.A. (1966). A study of replacement algorithms for virtual storage computers. IBM Syst. J. 5, 2.
2. Belady, L.A., Nelson, R.A. and Shedler, G.S. (1969). An anomaly in space-time characteristics of certain programs running in a paging machine. Comm. ACM 12, 6.
3. Cochran, W.G. and Cox, G.M. (1957). Experimental designs. Wiley, New York.
4. Coffman, E.G. and Varian, L.C. (1968). Further experimental data on the behavior of programs in a paging environment. Comm. ACM 10, 2.
5. Comeau, L.W. (1967). A study of the effect of user program optimization in a paging system. ACM symposium on operating system principles, Gatlinburg, Tenn.
6. Davies, O.L. (1956). The design and analysis of industrial experiments. Hafner, New York.
7. Denning, P.J. (1970). Virtual memory. Computing surveys 2, no. 3.
8. Hatfield, D.J. and Gerald, J.A. (1970). Automatic page fitting of program modules: a means of improving the performance of paging systems. IBM Cambridge Scientific Center Tech. Rep. G320-2056.
9. Mattson, R.L., Gecsei, J., Slatz, D.R. and Traiger, I.W. (1970). Evaluation techniques for storage hierarchies. IBM Syst. J. 9, 2.
10. Seawright, L.L. and Kelch, J.A. (1969). An introduction to CP67/CMS, IBM Cambridge Scientific Center Tech. Rep. 320-2032.
11. Tsao, R.F. and Margolin, B.H. (1971). A multi-factor paging experiment. II. Statistical methodology. These proceedings.

Table 1. PS (total number of page swaps).

A \ C \ D \ B		Group			Freqy			Alpha		
		24 P.	20 P.	16 P.	24 P.	20 P.	16 P.	24 P.	20 P.	16 P.
LRUV	S	32	48	538	52	244	998	59	536	1,348
	M	53	81	1,901	112	776	3,621	121	1,879	4,639
	L	142	197	5,689	262	2,625	10,012	980	5,698	12,880
FIFO	S	49	67	789	79	390	1,373	85	814	1,593
	M	100	134	3,152	164	1,255	4,912	206	3,394	5,838
	L	233	350	9,100	458	3,688	13,531	1,633	10,022	17,117
RAND	S	62	100	1,103	111	480	1,782	111	839	2,190
	M	96	245	3,807	237	1,502	6,007	286	3,092	7,654
	L	265	2,012	12,429	517	4,870	18,602	1,728	8,834	23,134

Table 2. AC (activity counts). (Computed from eq. (4.3.1).)

A \ C \ D \ B		Group			Freqy			Alpha		
		24 P.	20 P.	16 P.	24 P.	20 P.	16 P.	24 P.	20 P.	16 P.
LRUV	S	40.0	40.5	42.5	42.1	39.5	38.1	40.4	35.0	33.1
	M	41.7	46.0	42.5	40.4	38.5	36.9	42.9	36.5	32.5
	L	44.2	48.5	41.2	41.7	36.5	36.2	42.5	34.0	32.5
FIFO	S	31.7	38.5	38.1	35.8	38.5	33.1	35.0	32.0	31.2
	M	37.5	40.0	36.2	35.4	34.5	32.5	35.8	24.5	30.6
	L	36.7	41.5	35.6	33.7	33.5	31.9	35.8	28.5	29.4
RAND	S	34.6	36.5	35.6	34.2	41.0	33.7	37.5	33.0	29.4
	M	36.2	37.5	35.6	35.8	42.5	33.1	39.6	32.5	28.7
	L	39.2	42.5	35.6	40.4	40.0	33.7	41.2	32.5	29.4

Table 3. IC (inactivity counts). (Computed from eq. (4.3.2).)

A \ C \ D \ B		Group			Freqy			Alpha		
		24 P.	20 P.	16 P.	24 P.	20 P.	16 P.	24 P.	20 P.	16 P.
LRUV	S	44.7	53.6	13.1	45.9	17.2	22.7	43.4	19.0	30.7
	M	34.3	45.2	9.6	36.2	13.9	20.6	37.9	16.4	30.7
	L	26.9	42.5	12.7	36.6	15.2	23.9	13.7	18.3	32.3
FIFO	S	32.0	42.0	11.1	34.9	14.8	20.3	35.7	14.2	27.8
	M	19.8	35.2	8.3	30.3	9.1	19.1	27.9	10.8	28.6
	L	20.7	31.9	11.1	26.7	13.4	21.1	10.0	12.5	27.7
RAND	S	29.2	31.8	19.4	29.3	13.2	22.7	26.4	16.5	29.1
	M	25.9	24.2	18.8	21.5	9.6	23.2	18.9	14.5	29.7
	L	26.0	8.4	19.1	20.2	11.2	22.7	9.2	16.8	28.3

APPENDIX I: FREQUENCY DISTRIBUTION OF AC, f(x).

(A) Replacement Algorithm: LRUV

(D) Memory Size: 24 P.

(C) \ (B)	Group			Freqy			Alpha		
x	S	M	L	S	M	L	S	M	L
0	1	3	7	4	9	28	4	9	109
1	0	0	0	0	1	1	1	0	1
2	0	0	1	2	4	5	2	2	9
3	0	2	5	0	0	7	2	6	75
4	1	1	7	0	3	7	3	6	42
5	1	3	7	3	14	18	2	5	33
6	7	7	9	3	4	12	0	5	32
7	3	3	5	2	2	11	5	6	55
8	1	2	4	3	3	6	2	7	64
9	3	4	15	5	13	24	7	8	43
10	1	1	5	6	7	15	7	12	38
11	3	4	12	5	16	24	5	8	61
12	2	3	7	5	5	7	3	5	35
13	1	5	12	1	3	14	2	5	50
14	5	4	7	1	2	19	1	5	36
15	0	2	12	4	7	17	4	7	33
16	1	3	4	1	3	7	1	3	52
17	0	3	9	4	6	13	2	4	[illegible]0
18	2	3	8	0	3	10	2	8	42
19	0	0	4	2	4	3	2	1	39
20	0	0	2	0	1	2	0	4	26
21	0	0	0	0	1	5	1	2	22
22	0	0	0	1	1	3	1	2	21
23	0	0	0	0	0	4	0	1	12
24	0	0	0	0	0	0	0	0	0
PS	32	53	142	52	112	262	59	121	980

(A) Replacement Algorithm: LRUV

(D) Memory Size: 20 P.

(C) \ (B)	Group			Freqy			Alpha		
x	S	M	L	S	M	L	S	M	L
0	3	6	17	29	107	342	43	133	396
1	1	1	1	3	15	63	12	44	150
2	0	0	3	7	10	86	13	58	176
3	1	2	6	2	7	68	67	214	773
4	2	3	6	10	40	155	47	194	626
5	4	6	9	31	99	373	35	146	473
6	9	10	16	20	56	184	36	124	339
7	5	3	6	12	48	143	52	144	447
8	3	4	10	15	42	108	38	125	420
9	3	7	13	23	66	254	40	121	365
10	2	2	8	26	59	125	31	101	293
11	5	8	16	15	58	188	35	111	326
12	2	4	16	14	49	129	16	49	121
13	3	11	22	4	16	68	20	86	252
14	2	1	13	5	11	61	12	58	139
15	0	5	10	12	36	96	11	32	105
16	1	3	9	5	20	71	14	69	136
17	0	3	6	2	13	41	7	32	69
18	2	2	6	1	5	23	6	31	73
19	0	0	4	8	19	47	1	7	19
20	0	0	0	0	0	0	0	0	0
PS	48	81	197	244	776	2625	536	1879	5698

(A) Replacement Algorithm: LRUV

(D) Memory Size: 16 P.

(B) / (C) x	Group			Freqy			Alpha		
	S	M	L	S	M	L	S	M	L
0	53	194	546	101	346	977	99	337	959
1	8	31	148	42	181	442	62	223	602
2	6	10	90	64	241	708	100	307	965
3	98	343	1093	17	65	264	192	649	1899
4	23	80	321	108	413	1237	202	762	2052
5	56	201	529	193	777	2141	148	549	1452
6	27	92	214	56	185	507	73	198	623
7	18	38	147	65	192	603	106	410	1109
8	43	167	464	53	210	515	86	303	848
9	32	118	414	81	266	767	106	345	802
10	31	114	284	56	237	469	94	283	643
11	23	109	342	81	249	561	23	71	320
12	55	186	417	28	91	276	10	39	132
13	45	154	484	6	16	121	32	125	333
14	16	59	135	12	51	130	6	13	54
15	4	5	61	35	101	294	9	25	82
16	0	0	0	0	0	0	0	0	0
PS	538	1901	5689	998	3621	10012	1348	4639	12880

(A) Replacement Algorithm: FIFO

(D) Memory Size: 24 P.

(C) \ (B) x	Group			Freqy			Alpha		
	S	M	L	S	M	L	S	M	L
0	0	0	0	2	5	16	2	2	14
1	4	3	16	5	9	35	8	12	55
2	2	4	16	7	14	38	4	13	117
3	2	7	12	2	7	46	3	22	138
4	4	8	17	4	11	32	7	21	189
5	3	5	12	4	10	30	3	7	125
6	9	10	18	4	8	13	4	4	66
7	4	10	12	2	6	18	6	18	145
8	2	5	9	4	8	19	6	10	91
9	4	3	15	8	16	25	5	11	64
10	3	2	6	10	12	24	7	15	67
11	1	8	21	7	10	29	7	9	59
12	4	6	14	3	10	18	6	11	75
13	2	8	18	2	5	15	3	8	82
14	2	5	9	5	6	19	3	7	77
15	0	5	8	2	7	22	3	8	43
16	1	5	9	3	7	13	3	5	19
17	1	2	6	1	5	18	2	3	34
18	1	4	11	0	3	5	0	5	36
19	0	0	4	3	2	8	1	6	34
20	0	0	0	0	1	5	1	3	35
21	0	0	0	0	1	5	0	2	39
22	0	0	0	1	1	2	0	2	20
23	0	0	0	0	0	3	1	2	9
24	0	0	0	0	0	0	0	0	0
PS	49	100	233	79	164	458	85	206	1633

(A) Replacement Algorithm: FIFO

(D) Memory Size: 20 P.

(B) / (C)	Group			Freqy			Alpha		
x	S	M	L	S	M	L	S	M	L
0	0	0	0	14	51	133	3	3	3
1	4	9	14	17	175	315	38	221	949
2	4	5	22	28	109	374	81	308	920
3	1	6	27	22	39	261	97	489	1491
4	5	8	27	26	83	276	98	490	1146
5	4	12	33	53	131	498	100	433	1336
6	12	19	25	8	52	191	55	106	282
7	7	13	23	18	54	196	77	320	1088
8	3	7	19	34	75	215	48	290	786
9	3	4	12	35	114	261	46	148	627
10	6	6	20	17	63	226	42	137	379
11	2	12	30	37	100	162	26	190	283
12	5	11	17	20	49	118	16	47	143
13	8	4	23	11	20	70	27	56	253
14	2	6	21	10	27	50	17	34	51
15	0	2	7	15	26	96	8	54	70
16	0	3	9	10	22	81	14	13	55
17	1	6	12	5	19	43	8	15	70
18	0	1	7	5	20	37	5	27	35
19	0	0	2	5	26	85	8	13	55
20	0	0	0	0	0	0	0	0	0
PS	67	134	350	390	1255	3688	814	3394	10022

(A) Replacement Algorithm: FIFO

(D) Memory Size: 16 P.

(C) \ (B) x	Group			Freqy			Alpha		
	S	M	L	S	M	L	S	M	L
0	0	0	2	15	50	144	3	3	3
1	98	470	1232	160	568	1664	182	615	1876
2	9	57	899	195	699	1973	179	692	2099
3	107	435	1533	76	280	890	248	872	2655
4	80	303	396	141	536	1456	246	832	2657
5	84	341	1054	269	1044	2795	222	764	2256
6	52	215	307	48	174	463	91	278	735
7	72	214	775	110	362	917	166	619	1842
8	72	336	363	79	258	729	141	431	1149
9	55	171	560	104	349	954	91	318	703
10	60	268	544	85	265	597	46	135	375
11	57	155	488	21	78	233	21	60	124
12	11	88	352	16	41	113	29	131	434
13	16	68	501	10	28	112	11	47	60
14	14	23	57	27	121	347	5	17	47
15	2	8	37	17	59	144	12	24	102
16	0	0	0	0	0	0	0	0	0
PS	789	3152	9100	1373	4912	13531	1693	5838	17117

(A) Replacement Algorithm: RAND

(D) Memory Size: 24 P.

(C) \ (B)	Group			Freqy			Alpha		
x	S	M	L	S	M	L	S	M	L
0	0	0	0	0	1	1	1	1	5
1	2	3	10	4	16	17	6	15	88
2	3	7	17	5	8	25	8	17	71
3	3	9	19	11	9	23	5	11	89
4	1	4	11	9	10	23	4	19	114
5	3	1	15	10	19	30	3	19	76
6	10	9	18	4	20	28	7	13	79
7	5	3	6	3	11	30	7	20	127
8	7	10	16	9	28	34	10	16	123
9	5	7	12	13	21	41	11	16	89
10	5	6	27	8	12	42	10	20	95
11	4	7	19	14	22	35	5	19	108
12	4	8	15	4	10	29	9	15	102
13	4	7	16	3	8	27	5	18	70
14	1	4	14	1	11	38	4	11	76
15	3	4	13	5	7	30	2	11	65
16	1	3	14	2	14	22	6	7	70
17	0	3	14	4	3	9	2	9	63
18	1	1	6	0	2	5	2	12	56
19	0	0	3	1	2	16	0	6	55
20	0	0	0	0	1	3	1	2	35
21	0	0	0	1	2	4	2	3	46
22	0	0	0	0	0	4	0	4	17
23	0	0	0	0	0	1	1	2	8
24	0	0	0	0	0	0	0	0	0
PS	62	96	265	111	237	517	111	236	1728

(A) Replacement Algorithm: RAND

(B) Memory Size: 20 P.

(C) x \ (B)	Group			Freqy			Alpha		
	S	M	L	S	M	L	S	M	L
0	1	2	9	3	7	24	2	9	37
1	6	15	125	33	85	262	65	245	647
2	5	18	97	25	93	305	63	238	730
3	10	19	128	19	63	285	65	251	683
4	6	14	126	32	87	335	80	374	998
5	11	19	147	30	101	391	65	287	753
6	5	15	125	30	84	344	73	230	732
7	9	21	120	28	84	291	77	244	777
8	4	22	119	43	143	386	81	280	817
9	9	16	130	39	132	422	69	235	645
10	7	19	126	45	92	357	59	197	613
11	10	18	116	39	101	345	52	185	435
12	9	12	147	29	114	262	36	116	363
13	4	12	128	20	60	240	27	72	256
14	2	10	104	21	79	203	7	63	167
15	0	7	105	18	59	171	12	39	102
16	1	3	91	16	50	135	3	13	48
17	1	3	69	9	47	92	1	10	23
18	0	0	0	1	20	13	1	4	6
19	0	0	0	0	1	7	1	0	2
20	0	0	0	0	0	0	0	0	0
PS	100	245	2012	480	1502	4870	839	3092	8834

(A) Replacement Algorithm: RAND

(D) Memory Size: 16 P.

(C) \ (B)	Group			Freqy			Alpha		
x	S	M	L	S	M	L	S	M	L
0	10	34	102	9	51	142	21	70	216
1	103	337	1186	153	570	1788	235	875	2691
2	83	255	918	178	598	1851	271	1050	2960
3	149	537	1627	167	549	1700	268	923	2847
4	127	422	1349	244	817	2408	368	1271	3661
5	90	364	1102	254	851	2582	222	811	2463
6	84	293	919	142	506	1524	181	617	1942
7	92	336	1149	145	513	1671	205	675	2081
8	100	324	1083	171	565	1628	169	583	1639
9	101	322	1015	138	420	1280	117	371	1128
10	67	297	859	93	301	924	79	253	811
11	59	159	667	49	152	576	34	92	391
12	29	99	374	29	87	337	13	41	180
13	9	28	79	10	31	173	6	13	100
14	0	0	0	0	2	16	0	2	23
15	0	0	0	0	0	2	1	1	1
16	0	0	0	0	0	0	0	0	0
PS	1103	3807	12429	1782	6007	18602	2190	7654	23134

APPENDIX II: FREQUENCY DISTRIBUTION OF IC, g(y).

(Note: g(y=T) denotes the number of pages which never reenter the main memory.)

(A) Replacement Algorithm: LRUV

(D) Memory Size: 24 P.

(C) \ (B)	Group			Freqy			Alpha		
y	S	M	L	S	M	L	S	M	L
0	3	12	35	6	24	73	9	39	675
1	5	6	12	5	12	19	8	9	57
2	5	7	12	7	6	20	5	6	38
3	2	3	14	1	2	10	1	2	23
4	1	5	13	4	8	7	2	2	20
5	1	3	12	0	4	8	4	6	14
6	2	2	1	1	2	5	2	3	10
7	0	0	4	0	6	5	1	2	9
8	0	0	7	2	3	4	0	3	7
9	0	0	4	2	3	5	1	0	4
10	0	0	0	1	1	6	1	2	4
11	0	0	0	1	3	5	0	2	3
12	0	0	2	0	4	11	0	2	2
13	0	0	2	2	4	3	2	1	5
14	0	0	0	0	3	3	0	1	1
15	1	0	1	0	1	5	4	1	5
16	0	0	1	2	1	6	0	1	4
17	0	0	1	0	2	3	2	5	5
18	0	0	0	2	2	2	0	3	1
19	0	0	0	0	0	3	0	2	1
20	0	0	0	0	0	9	0	4	0
21	0	0	0	1	0	4	0	0	2
22	1	0	0	0	0	3	0	1	2
23	0	0	0	1	2	3	0	0	2
≥ 24	2	5	8	2	6	24	5	11	70
T	9	10	13	12	13	16	12	13	16
PS	32	53	142	52	112	262	59	121	980

(A) Replacement Algorithm: LRUV

(D) Memory Size: 20 P.

(C) \ (B)	Group			Freqy			Alpha		
y	S	M	L	S	M	L	S	M	L
0	7	10	34	115	371	979	275	883	2073
1	3	4	17	49	175	502	38	206	910
2	2	5	8	14	63	382	37	151	538
3	0	2	8	8	28	200	40	155	430
4	4	7	10	2	31	94	15	99	335
5	2	4	8	3	7	82	19	62	250
6	2	8	4	1	7	38	11	45	192
7	1	4	12	1	3	41	9	25	122
8	1	0	7	4	5	27	6	24	86
9	2	4	13	4	4	25	3	14	68
10	2	3	5	0	5	23	4	16	58
11	1	5	6	1	3	10	7	14	44
12	0	1	8	1	1	12	4	9	46
13	0	2	3	4	6	13	2	3	44
14	0	0	6	0	3	14	2	10	34
15	2	0	3	4	3	11	1	9	30
16	0	2	3	0	1	7	6	10	16
17	0	1	2	0	0	10	5	8	16
18	2	0	2	1	0	1	0	6	13
19	0	0	1	2	3	3	4	7	17
≥ 20	4	5	20	14	40	131	32	106	356
T	13	14	17	16	17	20	16	17	20
PS	48	81	197	244	776	2625	536	1879	5698

(A) Replacement Algorithm: LRUV

(D) Memory Size: 16 P.

(C) y \ (B)	Group			Freqy			Alpha		
	S	M	L	S	M	L	S	M	L
0	223	824	1988	211	721	1900	248	736	2290
1	150	575	1447	165	644	1599	166	539	1432
2	62	218	911	208	785	1782	163	555	1398
3	30	97	466	112	471	1302	117	488	1136
4	15	59	275	70	235	843	123	426	1087
5	8	29	158	45	155	560	82	309	882
6	4	16	84	24	104	389	82	266	760
7	4	7	61	23	71	218	60	232	607
8	4	5	47	29	60	169	43	207	507
9	0	5	26	5	49	115	26	130	371
10	1	4	12	5	26	93	33	107	282
11	1	1	14	3	24	77	13	79	217
12	4	2	6	4	11	50	21	69	209
13	2	3	11	0	10	56	15	58	181
14	1	0	6	0	12	36	11	38	159
15	0	1	4	2	12	42	11	37	110
≥ 16	15	37	155	72	210	757	114	342	1228
T	17	18	21	20	21	24	20	21	24
PS	538	1901	5689	998	3621	10012	1348	4639	12880

(A) Replacement Algorithm: FIFO

(D) Memory Size: 24 P.

(C) \ (B) y	Group			Freqy			Alpha		
	S	M	L	S	M	L	S	M	L
0	19	51	74	30	63	264	38	109	1082
1	1	7	22	3	19	19	3	7	229
2	4	8	37	2	5	13	1	12	106
3	3	7	21	1	2	7	2	2	34
4	2	0	18	6	6	9	3	4	16
5	0	6	8	3	5	5	3	4	8
6	1	3	7	3	3	6	2	6	7
7	1	1	5	1	5	3	0	3	3
8	5	1	2	2	1	5	1	2	5
9	0	1	3	1	3	1	2	1	6
10	1	0	0	0	5	6	1	0	2
11	0	0	0	0	4	2	0	2	1
12	0	0	1	0	2	2	1	2	3
13	0	0	1	0	2	1	0	0	2
14	0	0	1	3	2	4	1	0	2
15	0	0	2	0	2	4	1	3	2
16	0	0	1	3	0	6	1	1	3
17	0	0	2	2	0	2	0	0	2
18	0	0	2	1	1	3	0	2	1
19	0	0	0	0	2	2	1	0	3
20	0	0	1	1	0	0	1	1	1
21	0	0	1	2	1	2	1	1	2
22	0	0	1	1	1	3	1	8	0
23	0	0	1	0	0	2	0	1	4
≥ 24	3	5	9	2	17	71	9	22	93
T	9	10	13	12	13	16	12	13	16
PS	49	100	233	79	164	458	85	206	1633

(A) Replacement Algorithm: FIFO

(D) Memory Size: 20 P.

(C) \ (B)	Group			Freqy			Alpha		
y	S	M	L	S	M	L	S	M	L
0	23	54	166	201	537	1390	506	2196	4435
1	5	3	23	109	503	825	51	66	2019
2	2	7	10	16	71	795	24	88	628
3	0	5	6	3	15	177	82	650	758
4	2	8	6	4	18	51	31	73	1136
5	3	4	4	4	3	39	14	35	218
6	3	3	12	3	6	35	7	31	50
7	0	2	7	0	4	21	0	7	27
8	1	1	4	0	2	26	13	15	20
9	1	4	2	1	1	7	4	4	25
10	1	1	6	0	5	4	1	1	18
11	0	2	5	1	6	12	2	4	21
12	2	1	5	0	3	8	0	7	23
13	0	1	1	0	0	8	2	13	30
14	1	0	2	1	1	13	0	5	10
15	1	3	6	0	0	8	0	7	6
16	1	2	5	3	3	11	1	2	9
17	1	0	3	0	0	11	1	4	12
18	0	2	7	3	3	12	1	7	20
19	0	5	5	2	1	6	2	1	22
≥ 20	7	12	48	23	56	209	56	161	515
T	13	14	17	16	17	20	16	17	20
PS	67	134	350	390	1255	3688	814	3394	10022

(A) Replacement Algorithm: FIFO

(D) Memory Size: 16 P.

(B) / (C) y	Group			Freqy			Alpha		
	S	M	L	S	M	L	S	M	L
0	135	584	1492	276	890	2808	314	1134	3549
1	551	2348	4134	137	521	1776	193	571	1833
2	40	69	2582	611	2340	4071	147	481	1166
3	2	17	403	61	224	1900	118	396	1638
4	6	21	110	48	103	544	502	1631	3725
5	3	5	26	49	238	384	74	265	1778
6	3	1	17	14	61	265	35	113	396
7	3	9	10	18	39	106	24	124	176
8	1	2	13	12	54	123	22	137	252
9	0	7	8	7	62	132	27	99	313
10	0	1	8	11	12	108	10	91	105
11	5	5	5	6	6	71	9	100	74
12	2	1	7	4	3	37	11	53	90
13	1	4	6	1	23	24	5	76	103
14	0	5	14	3	7	57	12	43	111
15	1	10	7	5	14	74	8	39	73
≥ 16	19	45	237	90	294	1027	157	464	1696
T	17	18	21	20	21	24	20	21	24
PS	789	3152	9100	1373	4912	13531	1693	5838	17117

(A) Replacement Algorithm: RAND

(D) Memory Size: 24 P.

(C) \ (B)	Group			Freqy			Alpha		
y	S	M	L	S	M	L	S	M	L
0	33	51	134	59	136	314	58	180	1313
1	3	8	16	4	19	38	9	21	102
2	4	3	9	2	9	11	5	7	53
3	1	1	8	2	4	9	3	5	41
4	1	2	6	3	4	9	4	3	24
5	0	1	3	3	5	6	0	5	14
6	1	0	6	2	5	11	1	4	10
7	1	3	7	1	3	2	3	2	19
8	0	0	5	3	1	5	0	2	3
9	0	2	3	0	2	6	0	1	8
10	0	1	1	0	0	1	1	3	5
11	1	0	3	1	1	2	0	2	4
12	0	1	2	0	0	1	0	1	4
13	0	1	2	1	1	4	1	1	5
14	0	0	1	1	1	2	1	1	4
15	0	1	5	1	2	1	0	3	2
16	1	0	2	1	1	1	0	1	3
17	0	1	1	0	3	1	0	1	0
18	0	1	4	1	2	2	0	2	5
19	1	1	3	1	0	4	1	0	3
20	0	1	1	1	0	1	0	0	2
21	0	1	1	1	1	4	0	0	1
22	1	0	1	0	1	1	0	1	0
23	0	1	3	1	0	1	0	1	1
≥ 24	5	5	25	10	23	64	12	26	86
T	9	10	13	12	13	16	12	13	16
PS	62	96	265	111	237	517	111	286	1728

(A) Replacement Algorithm: RAND

(D) Memory Size: 20 P.

(C) \ (B)	Group			Freqy			Alpha		
y	S	M	L	S	M	L	S	M	L
0	54	160	1562	323	1098	3040	401	1408	3887
1	3	7	115	36	108	522	89	387	1020
2	3	9	48	29	64	313	63	266	764
3	3	1	33	12	43	195	49	212	578
4	0	2	34	15	27	143	41	153	442
5	2	3	24	3	21	96	32	127	295
6	0	0	19	2	12	62	20	80	256
7	2	1	7	1	7	44	23	89	215
8	0	1	7	4	5	47	18	58	160
9	2	2	7	2	4	25	9	39	162
10	0	0	5	1	3	22	8	43	130
11	0	0	9	3	3	19	9	23	112
12	3	2	7	0	2	13	3	31	101
13	1	3	2	1	1	13	5	17	76
14	0	3	2	1	1	11	3	10	63
15	2	2	5	1	4	10	8	16	48
16	0	2	1	1	6	12	2	7	40
17	0	0	1	2	3	10	2	5	30
18	1	1	5	0	2	10	3	4	35
19	1	0	3	2	3	8	0	5	29
≥ 20	10	32	99	25	68	235	35	95	371
T	13	14	17	16	17	20	16	17	20
PS	100	245	2012	480	1502	4870	839	3092	8834

(A) Replacement Algorithm: RAND

(D) Memory Size: 16 P.

(B) / (C) / y	Group			Freqy			Alpha		
	S	M	L	S	M	L	S	M	L
0	569	1882	6631	815	2634	8629	830	2876	9379
1	81	325	1036	173	519	1709	206	649	1944
2	70	268	762	130	439	1236	139	496	1486
3	57	222	576	80	328	1010	118	393	1221
4	41	151	461	81	312	771	103	383	1025
5	38	152	331	63	286	681	82	336	888
6	41	115	320	52	188	544	94	275	771
7	20	91	268	39	148	456	73	238	626
8	38	83	201	40	138	353	48	181	561
9	20	90	221	36	111	296	34	153	447
10	11	63	176	30	84	281	33	148	355
11	15	56	169	22	72	236	31	137	318
12	10	46	142	12	68	216	32	110	286
13	9	44	148	11	60	171	26	92	251
14	7	37	104	15	64	160	30	79	241
15	7	19	90	18	54	163	23	61	211
≥ 16	52	145	772	145	483	1666	268	1027	3100
T	17	18	21	20	21	24	20	20	24
PS	1103	3807	12429	1782	6007	18602	2190	7654	23134

A MULTI-FACTOR PAGING EXPERIMENT: II. STATISTICAL METHODOLOGY

R.F. Tsao and B.H. Margolin

ABSTRACT: This paper reports on the statistical methodology applied in the analysis of a multi-factor paging experiment by Tsao, Comeau and Margolin. First, the approach employed by the authors in that experiment in particular, and computer system evaluation in general, is described. Then aspects of experiment planning and preliminary data scanning are discussed. Finally, the process of empirical model construction and calibration is illustrated.

1. INTRODUCTION

This paper reports on the statistical methodology employed in a multi-factor paging experiment conducted by Tsao, Comeau and Margolin [1]. The experiment and the conclusions drawn are discussed in the above reference. The present paper, which may be viewed as a companion paper, presents the statistical analyses performed upon the data gathered in the paging experiment that led to the conclusions in [1].

Grenander and Tsao [2], in a critique of the state of the art, conclude that empirical (statistical) methods are among the most promising quantitative tools for computer system performance evaluation. They point out that these methods, which have been successfully employed in many other areas of research, have had limited application to computer science. In that paper, the authors advocate the statistical design and analysis of experiments as potentially one of the most important techniques for system evaluation. Using the multi-factor paging experiment as a case study, the present paper shows in detail how this technique can be applied to system evaluation in general.

2. METHODOLOGICAL CONSIDERATIONS

This paper will not repeat the detailed description of the paging experiment presented fully in [1]. It is assumed that the reader has already read the earlier companion paper and is interested in the methodology that led to the conclusions drawn from the paging experiment. The approach employed by the authors in the paging investigation in particular, and system evaluation in general, can be described as follows.

1. Acknowledgement of computer system complexity. As opposed to the simplifying assumptions adopted in an analytic approach to the study of computer systems, the authors choose to acknowledge the complexity of computer systems as processes for study. This complexity is twofold. First, there appear to be many variables of importance, and second, many of these variables appear to be interrelated. Part of the complexity may be due to observing variables in the wrong scale, or to dealing with the "wrong" set of variables. In these cases, statistical methodology may yield different (and, hopefully, smaller) sets of variables for study, thereby increasing understanding on the basis of observation.

2. Use of statistically designed experiments. A key empirical tool for investigating complex processes is the statistical design of experiments. The designed experiment provides a framework for the formulation of a study through the selection and specification of such matters as factors or variables for study, levels or conditions of the factors of interest, principal responses for monitoring, and the number of observations to be taken initially. The statistically designed experiment usually results in a cost reduction for equivalent information over a competing undesigned investigation. Moreover, the designed experiment affords greater validity to the conclusions since it allows consideration of the interrelation between or interactions of factors. Finally, a major advantage of the designed experiment approach is its active interference with the system, as opposed to passive observation of it. This permits exploration of the system beyond present or previously experienced operating conditions.

3. Interest in indication as opposed to formal statistical inference. The goal in this experimental work is to gather system knowledge. The approach is extremely non-dogmatic. As opposed to formal statistical inference, no distributional assumptions are made before the data are at hand. It is only through the data that evidence is built up for a thesis or shifted from one thesis to another. The

adoption of unjustified (albeit possibly correct) assumptions, such as is frequently the case, for example, in queueing theory work in this area, is the weakest link in the current applications of quantitative methodology to computer system research. Finally, clear delineation is made between what might be judged to be of statistical significance and what might be judged to be of scientific significance, i.e. of interest to a computer system analyst.

4. Preference for "empirical" modelling over "mechanistic" modelling. One of the most commonly used techniques in research is that of modelling. The approach adopted by the authors stresses "empirical" rather than "mechanistic" modelling. Although both types of models are in the form of mathematical equations, they differ in the way the model is constructed. In constructing a mechanistic model one starts with a body of preconceived theory and attempts to approximate the real-world problem by this theory. An empirical model, on the other hand, is more in the spirit of curve-fitting; one relies on data to suggest what form the model should take. Researchers in certain areas such as physics and chemistry are at times fortunate in having previously tested theory available. In these situations, mechanistic modelling can be a powerful tool. Computer science, however, is still in its quantitative infancy. The researcher in this area should resort, at least initially, to empirical models, which are purely data-driven. The class of such models is fairly wide, providing flexibility, but one must remember that they represent only empiricism. Nevertheless, these models will aid understanding and stimulate theoretical investigations.

5. Emphasis on model calibration and validation. Given any model, a reasonable question is "how well does this model fit?". Current thinking on this problem in the field of computer system evaluation seems highly questionable at best. A common practice is to compare some output of a model, e.g. CPU times, with a small set of corresponding "true" values, obtained by direct system observation. The average relative difference of these sets of values is then taken as a measure of goodness of fit of the model. No attention is paid to the fact that a "true" value may vary considerably the next time it is observed. This approach, which compares only averages and pays no attention to the variability of the averages, is too simplistic. In empirical modelling, the emphasis is on the following two phases: the calibration of how well the

model fits the data in hand and the validation of how well the model fits future observations made on the system. Points to note are that both calibration and validation are stochastic rather than deterministic considerations, and that statistical techniques will have to be used to reach satisfactory results. Both may be viewed as empirical credibility checks on the model.

3. PLANNING OF THE EXPERIMENT

The first step in the planning of an experiment is the selection of appropriate factors and levels for study. The choice of the factors is governed by conjecture and previous experience; the number of factors is invariably limited by cost in time and money. In an exploratory study, levels of the factors are frequently chosen to cover the range of feasible operating conditions. Then the remaining question is which design to use in the experiment. Many designs, satisfying different requirements, exist in the statistical literature. Considerations such as interest in studying interactions among the factors or the number of observations one can afford will help determine the choice of design. A general discussion of designed experiments for computer system research will be contained in a future paper by the authors.

In the study of the paging process, it was felt that investigating interrelationships among the factors studied, often neglected in computer system research, was important. This consideration of interaction tended to limit the number of factors and levels included in the experiment. Four factors, replacement algorithm (A), deck arrangement (D), problem program (P) and main memory size (M), were chosen for experimentation; three levels per factor were determined to cover high, medium and low performance of the factor, as indicated in previous simpler experiments.

All 81 possible combinations of the four factors were observed; this constitutes a full 3^4 factorial experiment [3]. The responses recorded were PS, AC and IC, as defined in [1]. These three sets of 81 responses are included in Appendix A (Tables A.1, A.2 and A.3) to this paper.

4. INDICATION: PRELIMINARY DATA SCANNING

The preliminary step in the analysis of data from an empirical investigation is scanning the data for clues or suggestions that will guide the later steps. Some of the techniques employed at this stage by seasoned data analysts include:

(1) rearrangement of the data, which then exhibit properties not previously noted,
(2) computation of simple summary statistics, such as means, medians, ranges or frequency distributions,
(3) construction of graphs and pictures in general, which allow the human eye to be employed as a pattern detector.

All these techniques are intended to be easy, quick and exploratory in nature. No hard and fast rule specifying a set of techniques is proposed; much depends on the researcher's conjectures and the characteristics of the data.

The data for PS in Table A.1 exhibit a striking monotonicity. In most cases, as one proceeds from Level 1 to Level 2 and on to Level 3 for any of the four factors (with the levels numbered as in [1]), holding all other factors fixed, the PS count increases. Had these data been presented in any other fashion, judicious rearrangement would have disclosed this fact. Not all multi-factor data can be so arranged; the fact these data can is reassuring, since prior knowledge suggested that this would be the case.

Table 4.1 Factor averages and their ranges for PS

Factor \ Level	Averages 1	2	3	Range of Averages
Replacement algorithm	2056	2986	3781	1725
Deck arrangement	1584	2913	4326	2742
Problem program	592	2047	6185	5593
Memory size	305	2006	6512	5207

To gain some indication of the relative importance of the factors on PS, the sets of averages and the sets' ranges shown in Table 4.1 were computed. This table suggests that for PS, the replacement algorithm has the smallest effect of the four factors, because its averages have the smallest range. It further suggests that the specific replacement algorithm LRUV, Level 1, is the best of the three. These indications will be the subject of scrutiny in later sections.

It is worthwhile noting that all 81 responses for PS are positive integers and that the overall range of the 81 observations, 23102, is nearly eight times as large as 2941, the overall average of the 81 observations. This is the first hint that the PS measurements may be in need of rescaling.

To get a feel for the effect on PS of a factor, say problem program, observe from Table A.1 that the effect of changing the problem program from SMALL to MEDIUM when the main memory size is 24 pages, the deck arrangement is GROUP and the paging algorithm is LRUV, is to increase PS from 32 to 53, an increase of 21 page exceptions. If this computation is repeated for the remaining eight pairings of main memory size and deck arrangement, where LRUV remains the paging algorithm, i.e. for all nine pairs in the top two rows of Table A.1, the following is found:

Table 4.2 Relationship between PS for SMALL (X) and the change in PS from SMALL to MEDIUM (Y) for LRUV.

X	32	48	538	52	244	998	59	536	1348
Y	21	33	1363	60	532	2623	62	1343	3291

A plot of Y against X in Figure 1 shows that Y, the magnitude of the increment in PS when changing from SMALL to MEDIUM, increases nearly linearly with X, the PS for SMALL. This strongly suggests that the paging process studied in this experiment behaves multiplicatively; that is, changes in a factor's levels change the PS count (approximately) proportionately rather than by some fixed additive amount.

Turning to the second response considered, AC, the 81 observed values are found in Table A.2. No rearrangement of this table will yield the striking monotonicity noted for PS, nor does rearrangement seem to yield anything else notable. Again, averages for each factor's levels and the corresponding ranges were computed and appear in Table 4.3. Table 4.3 suggests that the problem program has the smallest effect on AC, its range of averages being smallest. The overall average and range of the 81 AC values are 36.7 and 20.0 respectively. Finally, if the study of the effect of changing the problem program from SMALL to MEDIUM is repeated for the corresponding nine pairs in Table 4.2, the data shown in Table 4.4 are obtained. Plotting Y against X

Table 4.3 Factor averages and their ranges for AC

Factor \ Level	Averages 1	2	3	Range of Averages
Replacement algorithm	39.5	24.6	36.0	4.9
Deck arrangement	39.1	36.8	37.0	2.3
Problem program	36.3	36.7	37.0	0.7
Memory size	38.2	37.4	34.4	3.8

Table 4.4 Relationship between AC for SMALL (X) and the change in AC from SMALL to MEDIUM (Y) for LRUV.

X	40.0	40.5	42.5	42.1	39.5	38.1	40.4	35.0	33.1
Y	1.7	5.5	0	-1.7	-1.0	-1.2	2.5	1.5	-.6

(figure 2), now for AC, indicates no apparent relationship between the change in AC in going from SMALL to MEDIUM and the AC observation for the SMALL program. This is true for all other factors under study. Hence no change of scale is suggested for AC.

5. EMPIRICAL MODEL BUILDING

Empirical model building involves an iterative search for a mathematical expression relating a dependent variable Y to a set of independent variables $X_1, X_2, \ldots, X_p$, on the basis of observed data. Given a set of data, the data analyst is motivated by a desire to select the simplest possible mathematical function which enables him adequately to summarize the data at hand. This simplicity can be achieved by: (1) selecting a small but relevant subset of the p independent variables for inclusion in the expression, and/or (2) choosing a simple but plausible mathematical form to express the relationship. At times (1) and (2) can be facilitated by a rescaling of some of the variables. Generally speaking, a model that contains few parameters and has a simple mathematical form is easy to "interpret"; this, in turn, may reveal some fundamental relationship between the variables. The utility of empirical modelling ranges from a mere economical representation of the given data to a quantitative tool for scientific prediction. The statistical techniques for choosing an empirical model and estimating the parameters in the chosen model all come under a general heading of regression analysis. A discussion of regression analysis can be found, for example, in [4].

In the analysis of a designed experiment, analysis of variance (ANOVA), which can be viewed as a special case of regression, is often employed. ANOVA simply decomposes the variation of the dependent variable or response Y into components that can be attributed to various independent variables (factors) and their interactions plus noise or an unexplainable component. This decomposition permits an examination of the relative importance of the effects of the various factors and interactions in the modelling of the response. In this fashion, ANOVA may be considered a technique for empirical model building in experimental design studies. A detailed discussion of ANOVA can be found, for example, in [5].

The full set of 81 PS values can be fitted exactly by a model that includes all four factor main effects (A,D,P,M) and all their interactions; i.e., six first-order interactions (denoted by AD,AP, etc.), four second-order interactions (e.g., ADP) and the single third-order interaction (ADPM). Surely these components cannot all be important in modelling the response; moreover, the interpretation of the higher-order interactions is difficult. At the other extreme, an additive model, which contains only factor main effects, is extremely easy to interpret. Under the additive model, differences of PS between two levels of a factor, for example LRUV and FIFO, are the same regardless of the levels of the other factors. One of the purposes of the experiment was to determine if an additive model sufficed to fit the data, or if interactions were present in the paging process.

In the previous section there was an indication that the process under study, PS, behaved approximately multiplicatively. If this were exactly the case, then the logarithm of PS, LPS, should behave additively. The 81 values of LPS are in Table 5.1. The formal analysis of variance for LPS is presented in Table A.4 of Appendix A. The essence of this table, for our purposes, is summarized by Figure 3, which depicts the decomposition of the variation in LPS.

Table 5.1 Logarithm (base 10) of total number of page swaps (LPS)

A \ P	M \ D	GROUP 24P.	GROUP 20P.	GROUP 16P.	FREQY 24P.	FREQY 20P.	FREQY 16P.	ALPHA 24P.	ALPHA 20P.	ALPHA 16P.
LRUV	S	1.51	1.68	2.73	1.72	2.39	3.00	1.77	2.73	3.13
	M	1.72	1.91	3.28	2.05	2.89	3.56	2.08	3.27	3.67
	L	2.15	2.29	3.76	2.42	3.42	4.00	2.99	3.76	4.11
FIFO	S	1.69	1.83	2.90	1.90	2.59	3.14	1.93	2.91	3.23
	M	2.00	2.13	3.50	2.21	3.10	3.69	2.31	3.53	3.77
	L	2.37	2.54	3.96	2.66	3.57	4.13	3.21	4.00	4.23
RAND	S	1.79	2.00	3.04	2.05	2.68	3.25	2.05	2.92	3.34
	M	1.98	2.39	3.58	2.37	3.18	3.78	2.46	3.49	3.88
	L	2.42	2.30	4.09	2.71	3.69	4.27	3.24	3.95	4.36

It is clear from Figure 3 that the nearly additive model containing the four main effects, A, D, P, M, and the deck by memory interaction, DM, accounts for 99% of the variation in LPS. This model is given in Table A.5 with parameters estimated by standard least squares methods [6].

The analysis of variance decomposition is repeated for AC in Table A.6 and is summarized in Figure 4. From this figure, we note:

(1) As with LPS, the DM interaction is the only interaction of import;

(2) As opposed to LPS, problem program P is no longer an important main effect, accounting for less than 1% of the variation;

(3) The model containing three main effects A,D,M and the DM interaction accounts for 80% of the variation in AC. The model for AC is given in Table A.7.

These percentages of variation accounted for by the models above are but one measure of fit; calibration is needed to check the pattern of fit to the 81 individual observations.

6. CALIBRATION OF THE FITTED MODEL

Analysis of the experiment does not end with construction of an empirical model. Still to be answered is the skeptic's question, "How well does this model fit the data?". The statistical methodology for dealing with this matter is residual analysis.

A residual is defined as the deviation of the model's fitted value from the experimentally observed response. The set of residuals can be used to check the validity of any distributional assumptions made in the analysis, and, more importantly, it can provide an indication of departures of the model from the data. These departures can be classified into two categories: (a) a systematic pattern of departure over the data set, and (b) a few aberrant observations (commonly called outliers). Both (a) and (b) contribute to lack of fit of a model.

For the 3^4 experiment, the model of LPS in Table A.5 was used to compute fitted values and residuals for all 81 responses; the residuals are presented in Table 6.1. (The corresponding calibration of AC is omitted from this report.)

Table 6.1 Residuals computed from the model given in Table A.5.

A \ P	M \ D	GROUP			FREQY			ALPHA		
		24P.	20P.	16P.	24P.	20P.	16P.	24P.	20P.	16P.
LRUV	S	.18	.08	-.07	.11	-.04	-.02	-.05	-.04	.01
	M	-.05	-.13	.04	.01	.02	.10	-.18	.07	.11
	L	-.13	-.26	.01	-.14	.04	.03	.22	.04	.04
FIFO	S	.17	.04	-.09	.11	-.02	-.07	-.08	-.04	-.08
	M	.05	-.10	.07	-.02	.04	.05	-.13	.14	.02
	L	-.10	-.20	.02	-.00	.00	-.03	.25	.09	-.02
RAND	S	.16	.09	-.06	.14	-.05	-.07	-.08	-.08	-.08
	M	-.10	.04	.04	-.02	.00	.01	-.11	-.02	-.02
	L	-.17	.44	.04	-.15	.00	-.01	.16	-.08	-.01

One pattern exhibited by the residuals might be clustering of their signs, which may suggest an omitted but important interaction effect. A "quick and dirty" tabular method for discovering such clusters is a table solely of the signs of the residuals. For example, the residuals computed for LPS from a purely additive model, i.e. without the DM term,

Table 6.2 Sign table for residuals from purely additive model.

A \ P \ M \ D		GROUP			FREQY			ALPHA		
A	P	24P	20P	16P	24P	20P	16P	24P	20P	16P
LRUV	S	⊕	-	+	+	+	-	-	+	-
	M		⊖	⊕	-	+	+	⊖	⊕	-
	L	-	⊖	+	-	+	-	+	⊕	-
FIFO	S	⊕	⊖	+	+	+	-	-	+	⊖
	M	+	⊖	+	-	⊕	+	-	⊕	-
	L	+	⊖	+	-	+	-	+	⊕	-
RAND	S	⊕	-	+	+	+	-	-	+	⊖
	M	+	⊖	+	-	+	-	-	+	-
	L	-	-	+	-	+	-	+	+	-

would exhibit the pattern of signs in Table 6.2; the circled signs denote the residuals with large magnitudes. Note the striking pattern of clustering of the signs in the nine columns. These columns correspond to the nine combinations of deck and memory, thereby suggesting the need for the DM term. On the other hand, the signs of the residuals in Table 6.1 exhibit no such clustering.

Another property of the residuals one might check is whether their distribution can be approximated by a normal or Gaussian distribution [6]. This consideration is crucial to formal inference and is of value to us in that it might point to specific aberrant observations. A graphical procedure for assessing the normality approximation is the normal plot [7]; if the normality is tenable, the plot should appear to be a straight line on normal probability paper. The normal plot of the residuals in Table 6.1 is in Figure 5. This plot appears to be a nearly straight line with one glaring exception: the residual corresponding to the observation taken at the combination of factor levels

A = RAND, D = GROUP, P = LARGE, M = 20.

Statistical techniques for handling outliers can be found in [8]. A crude check of this residual (computed as +.442) shows it to be nearly four standard deviations away from zero, the average of the 81 residuals. This observation is suspect as an outlier, and signals the need to check for a recording error. This was done and no error was found. Two alternative explanations for the suspicious response are:

(1) the random number generator used in RAND for this particular observation may have produced an atypical sequence of page exceptions that caused the excessive paging;

(2) this particular combination of factor levels may be a <u>singular</u> point in the set of 81 combinations; i.e., something about this combination's true behavior differentiates it from the remaining 80 points as far as our model is concerned.

The feeling of the systems analysts was that (2) is an unlikely explanation.

Aberrant observations can have a substantial effect on the model constructed; they can increase the apparent variability, and can alter all parameter estimates and <u>all</u> fitted values. The effects are a function of the number and magnitude of these aberrations, and the total number of observations in the experiment. To guard against these possibilities, one should exclude observations judged to be outliers and recalibrate the model on the remainder of the data. This was done and the new model did not change substantially from that given in Table A.5.

Until now we have concentrated on calibrating the model with the data observed. Because these data were used in the construction of the model, the degree of calibration may be unduly optimistic. One way further to strengthen the model's credibility is to validate it against a <u>new</u> set of data. This is especially important when the model is intended for prediction. The techniques for validation are those for calibration, namely residual analysis. The residuals in this case, however, are computed by subtracting the <u>fitted values</u> of the <u>calibrated model</u> from the corresponding <u>new observations</u>. Calibration and future validation for any model can simultaneously improve to a point; past that point improvement in calibration is achieved at a sacrifice in future validation. Much statistical effort is expended in trying to locate this theoretical point. System knowledge and insight, where available, are extremely helpful in this effort. It is important to note that validation may fail for reasons other than overcalibration of the model. If the system undergoes an important change, it is not surprising that models based on data from a previous state of the system may fail to validate in the system's new state.

Validation of the results reported in this paper must await further experimentation.

7. QUANTIFICATION OF EXPERIMENTAL CONCLUSIONS

The purpose of this experiment, as stated in [1], was to attempt to answer the following questions:

(a) Is the PS count useful at all as a measure for the comparison of various replacement algorithms? If not, are other measures which the authors have formulated any improvement?

(b) What are the important factors that affect the paging, and what are their effects, both qualitative and quantitative?

Question (a) is answered by Table 7.1.

Table 7.1 Comparison of LPS, AC and IC

		Percent variation explained by: replacement algorithm	problem program
MEASURES	LPS	2.6	25.2
	AC	23.3	0.4
	IC	8.9	7.3

It is desirable that a measure for comparing replacement algorithms be sensitive to the algorithms and insensitive to other factors, such as problem program. One indication of sensitivity of a response to a factor is the percent variation in the response explained by the factor. It is clear from Table 7.1 that AC is a better measure than PS or IC for comparison of replacement algorithms; this was stated in Conclusion 1 of [1].

Question (b) is answered by Table 7.2. The multipliers in this table quantify the multiplicative effects of the individual factors upon paging, as indicated in Conclusion 2 of [1]. Note that the right-most multiplier in each factor's set of three is the largest and is decreasing in magnitude as one goes from memory to program to deck to algorithm. This is, in fact, the ordering of the factors in terms of decreasing importance, as stated in Conclusion 3

Table 7.2 Effect of individual factors on average* page swaps.

1.	Memory	24	20	16
	Average	162	776	4074
	Multiplier	1	4.8	24.7
2.	Program	SMALL	MEDIUM	LARGE
	Average	275	759	2455
	Multiplier	1	2.8	9.0
3.	Deck	GROUP	FREQY	ALPHA
	Average	347	955	1585
	Multiplier	1	2.8	4.6
4.	Algorithm	LRUV	FIFO	RAND
	Average	550	851	1122
	Multiplier	1	1.5	2.0

*Antilogarithm of arithmetic mean of LPS or, equivalently, the geometric mean of PS.

of [1]. Furthermore, Table 7.2 shows that FIFO and RAND generated 50% and 100% more page swaps than LRUV, respectively; clearly, LRUV is the preferred replacement algorithm, as stated in Conclusion 5 of [1].

One major discovery of this experiment is the interdependence of the effects of main memory and deck arrangement, as stated in Conclusion 4 of [1]. Its statistical importance is illustrated in Figures 3 and 4 for LPS and AC respectively via components of variability explained. For LPS, the DM component of variability, in addition to being almost five times larger than the total of the other five first-order interaction components, is larger than the component due to the algorithm. For AC, the DM component is not only larger than the total of the other five first-order interactions but is nearly equal to the sum of the components for problem program and memory size. The quantification of interaction is not easily understood; Appendix B is devoted to this topic.

It should be remembered that the answers provided above to questions (a) and (b) depend on the specific factors and levels chosen in this experiment. Further study is needed to test their wider applicability.

8. FINAL REMARKS

(1) Extension of this uni-programmed experiment to a multi-programming environment is a natural step to consider.

(2) Both the evaluation of system design and system tuning are fertile fields for the application of factorial experiments.

(3) The methodology employed in arriving at the conclusions is as important as the conclusions themselves; nevertheless, this methodology constitutes only a small portion of statistical methodology applicable to computer system research.

REFERENCES

1. Tsao, R.F., Comeau, L.W. and Margolin, B.H. (1971). A multi-factor paging experiment: I. the experiment and the conclusions. These Proceedings.
2. Grenander, U. and Tsao, R.F. (1971). Quantitative methods for evaluating computer system performance: a review and proposals. These Proceedings.
3. Davies, O.L. (1956). The design and analysis of industrial experiments. Hafner Publishing, New York.
4. Draper, N.R. and Smith, H. (1967). Applied regression analysis. John Wiley, New York.
5. Snedecor, G.W. and Cochran, W.G. (1967). Statistical methods. Iowa State University Press, Ames, Iowa (sixth edition).
6. Graybill, F.A. (1961). An introduction to linear statistical models, Vol. 1, McGraw-Hill, New York.
7. Daniel, C. and Wood, F.S. (1971). Fitting equations to data. Wiley-Interscience, New York.
8. Anscombe, F.J. and Tukey, T.W. (1963). The examination and analysis of residuals. Technometrics 5,2.

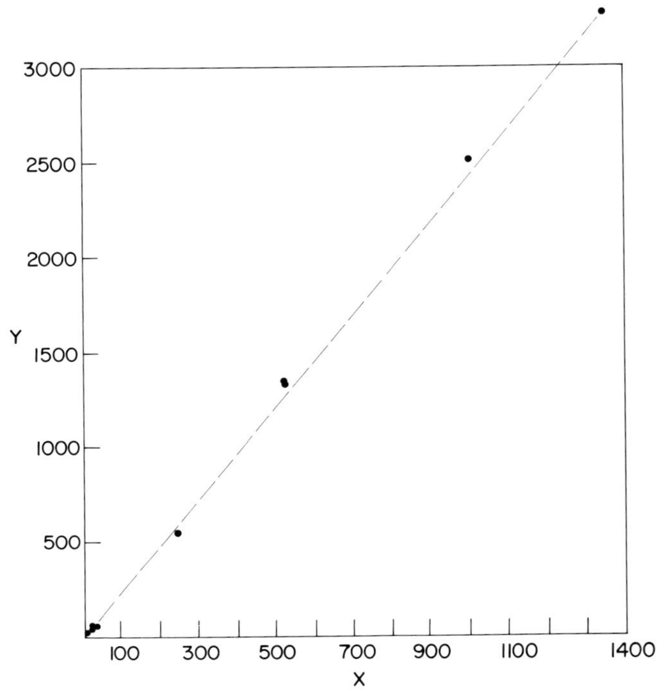

Fig. 1. Plot of Table 4.2.

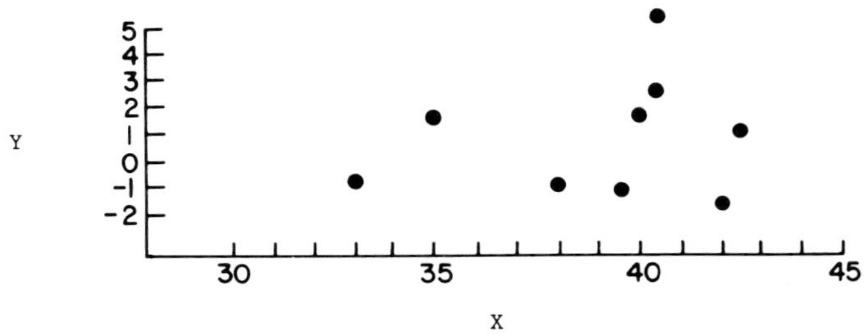

Fig. 2. Plot of Table 4.4.

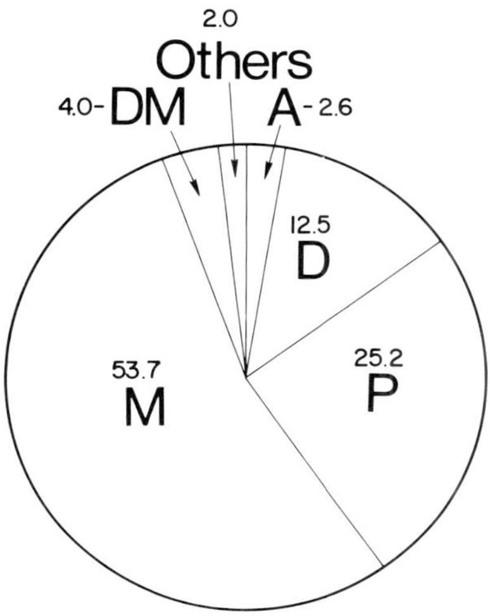

Fig. 3. Decomposition of variation in IPS.

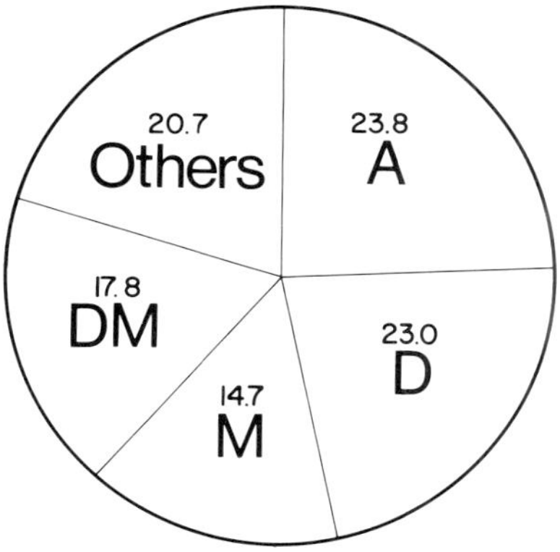

Fig. 4. Decomposition of variation in AC.

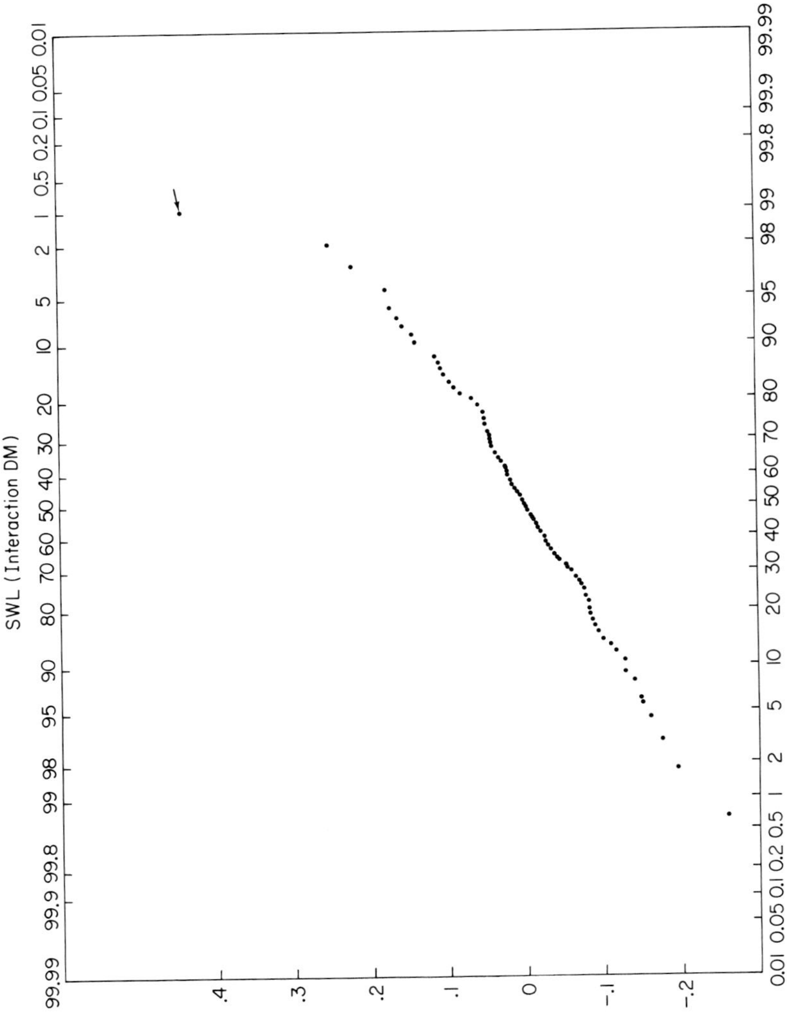

Fig. 5. Normal plot of the residuals in Table 6.1.

APPENDIX A.

Table A.1 Total number of page swaps

PA \ MD		GROUP			FREQY			ALPHA		
		24P.	20P.	16P.	24P.	20P.	16P.	24P.	20P.	16P.
LRUV	S	32	48	538	52	244	998	59	536	1348
	M	53	81	1901	112	776	3621	121	1879	4639
	L	142	197	5689	262	2625	10012	980	5398	12880
FIFO	S	49	67	789	79	390	1373	85	314	1693
	M	100	134	3152	164	1255	4912	206	3394	5838
	L	233	350	9100	458	3688	13531	1633	10022	17117
RAND	S	62	100	1103	111	480	1782	111	839	2190
	M	96	245	3807	237	1502	6007	286	3092	7654
	L	265	2012	12429	517	4870	18602	1728	8834	23134

Table A.2 Activity counts (AC)

PA \ MD		GROUP			FREQY			ALPHA		
		24P.	20P.	16P.	24P.	20P.	16P.	24P.	20P.	16P.
LRUV	S	40.0	40.5	42.5	42.1	39.5	38.1	40.4	35.0	33.1
	M	41.7	46.0	42.5	40.4	38.5	36.9	42.9	36.5	32.5
	L	44.2	48.5	41.2	41.7	36.5	36.2	42.5	34.0	32.5
FIFO	S	31.7	38.5	38.1	35.8	38.5	33.1	35.0	32.0	31.2
	M	37.5	40.0	36.2	35.4	34.5	32.5	35.8	29.5	30.6
	L	36.7	41.5	35.6	33.7	33.5	31.9	35.8	28.5	29.4
RAND	S	34.6	36.5	35.6	34.2	41.0	33.7	37.5	33.0	29.4
	M	36.2	37.5	35.6	35.8	42.5	33.1	39.6	32.5	28.7
	L	39.2	42.5	35.6	40.4	40.0	33.7	41.2	32.5	29.4

Table A.3 Inactivity counts (IC)

M \ D	P	GROUP 24P.	GROUP 20P.	GROUP 16P.	FREQY 24P.	FREQY 20P.	FREQY 16P.	ALPHA 24P.	ALPHA 20P.	ALPHA 16P.
LRUV	S	44.7	53.6	13.1	45.9	17.2	22.7	43.4	19.0	30.7
	M	34.3	45.2	9.6	36.2	13.9	20.6	37.9	16.4	30.7
	L	26.9	42.5	12.7	36.6	15.2	23.9	13.7	18.3	32.3
FIFO	S	32.0	42.0	11.1	34.9	14.8	20.3	35.7	14.2	27.8
	M	19.8	35.2	8.3	30.3	9.1	19.1	27.9	10.8	28.6
	L	20.7	31.9	11.1	26.7	13.4	21.1	10.0	12.5	27.7
RAND	S	29.2	31.8	19.4	29.3	13.2	22.7	26.4	16.5	29.1
	M	25.9	24.2	18.8	21.5	9.6	23.2	18.9	14.5	29.7
	L	26.0	8.4	19.1	20.2	11.2	22.7	9.2	16.8	28.3

Table A.4 Analysis of variance for LPS*

Source of variation	df	SS	MS
Total	80	48.8	
Main effects	8	45.8	5.7
A	2	1.3	
D	2	6.1	
P	2	12.3	
M	2	26.2	
First-order interaction	24	2.4	.1
AD	4	.07	
AP	4	.02	
AM	4	.03	
DP	4	.15	
DM	4	1.96	
PM	4	.14	
Second-order interaction	32	.48	.015
ADP	8	.05	
ADM	8	.13	
APM	8	.04	
DPM	8	.26	
Third-order interaction (ADPM)	16	.07	.004

*A reference for this standard ANOVA is [5].

Table A.5 Fitted model for LPS

1. Notation. Let

$Y_{ijk\ell}$ = LPS for the ith algorithm, jth deck arrangement, kth problem program and ℓth memory size, where $i,j,k,\ell = 1,2,3$. (e.g., $Y_{1231} = 2.42$).

μ = overall average of 81 observations

A_i = main effect of algorithm level i

D_j = main effect of deck level j

P_k = main effect of program level k

M_ℓ = main effect of memory level ℓ

$DM_{j\ell}$ = first-order interaction between deck level j and memory level

$\hat{Y}_{ijk\ell}$ = fitted value for corresponding LPS, $Y_{ijk\ell}$.

2. Fitted Model

$$\hat{Y}_{ijk\ell} = \mu + A_i + D_j + P_k + M_\ell + DM_{j\ell}.$$

3. Estimated Parameters

$\mu = 2.90$

$A_1 = -.16$ $A_2 = .02$ $A_3 = .14$
$D_1 = -.36$ $D_2 = .07$ $D_3 = .29$
$P_1 = -.47$ $P_2 = -.02$ $P_3 = .49$
$M_1 = -.69$ $M_2 = -.01$ $M_3 = .70$

$DM_{11} = .11$ $DM_{12} = -.30$ $DM_{13} = .19$
$DM_{21} = -.05$ $DM_{22} = .09$ $DM_{23} = -.04$
$DM_{31} = -.06$ $DM_{32} = .21$ $DM_{33} = -.15$

(e.g., $\hat{Y}_{1231} = 2.90+(-.16)+(.07)+(.49)+(-.69)+(-.05)$
$= 2.56$)

Table A.6 Analysis of variance for AC

Source of variation	df		SS		MS
Total	80		1472.0		
Main effects	8		910.5		112.6
A		2		350.2	
D		2		338.5	
P		2		6.0	
M		2		215.8	
First-order interaction	24		440.3		18.3
AD		4		46.7	
AP		4		21.4	
AM		4		38.2	
DP		4		39.5	
DM		4		262.0	
PM		4		32.5	
Second-order interaction	32		105.2		3.3
ADP		8		9.4	
ADM		8		45.2	
APM		8		12.4	
DPM		8		38.2	
Third-order interaction (ADPM)	16		16.0		1

Table A.7 Fitted model for AC

1. Notation. Let
   $Z_{ijk\ell}$ = AC for the ith algorithm, jth deck arrangement kth problem program and ℓth memory size, where $i,j,k,\ell = 1,2,3$. (e.g., $Z_{1231} = 41.7$)
   $\hat{Z}_{ijk}$ = fitted value for corresponding AC, $Z_{ijk\ell}$.
   The rest of the notation is the same as in Table A.5.

2. Fitted Model
   $$\hat{Z}_{ijk} = \mu + A_i + D_j + M_\ell + DM_{j\ell}.$$

3. Estimated Parameters

   $\mu = 36.7$

$A_1 = 2.8$	$A_2 = -2.1$	$A_3 = -.7$
$D_1 = 2.4$	$D_2 = .1$	$D_3 = -2.5$
$M_1 = 1.5$	$M_2 = .7$	$M_3 = -2.2$
$DM_{11} = -2.7$	$DM_{12} = 1.4$	$DM_{13} = 1.3$
$DM_{21} = -.6$	$DM_{22} = .8$	$DM_{23} = -.2$
$DM_{31} = 3.3$	$DM_{32} = -2.2$	$DM_{33} = -1.1$

   (e.g., $\hat{Z}_{1231} = 36.7+(2.8)+(.1)+(1.5)+(-.6)$
   $= 40.5$

APPENDIX B: DISCUSSION OF INTERACTION

When two factors are judged to interact in terms of their effects on a response variable, this is another way of saying that the effect of the different levels of either one of the factors depends on the specified level of the other. In such a situation it is somewhat misleading to report and discuss only the factor main effects. It is more revealing to present the averages for each different combination of the two factors and to explicitly demonstrate the dependence of the effect of one factor on the level of the other.

Table B.1 Averages for deck-memory combinations

		Memory	
DECK	24	20	16
GROUP	91	170	2669
FREQY	171	1136	4430
ALPHA	281	2487	5585

Table B.1 contains the geometric averages of PS (because of the multiplicative model) for the nine combinations of deck and memory. The multipliers for the deck main effects were reported in Table 7.2 as (1,2.7,4.6). If there were no deck-memory interaction, the estimate of the deck multipliers for a specific memory size should change little from (1,2.7,4.6); this is not the case. From Table B.1 we see that the relative superiority of GROUP over FREQY or ALPHA is greatest at 20 pages of main memory; the multipliers for FREQY and ALPHA relative to GROUP at 20 pages are nearly three times the corresponding multipliers reported in Table 7.2. This relative advantage of GROUP over FREQY or ALPHA shrinks considerably at either 24 or 16 pages.

There is a corresponding dependenc of the estimated memory multipliers on the specified deck arrangement. Table 7.2 reports the memory multipliers as (1,4.8,24.7). From Table B.1 we see that for the GROUP arrangement "squeezing" the memory size from 24 to 20 pages doubles the number of page swaps; for FREQY and ALPHA, the same memory squeeze increases page swaps nearly seven-fold and nine-fold, respectively.

DISCUSSION OF SESSION III

1: by Cuthbert Daniel

Tsao, Comeau and Margolin have given us the first comprehensive data showing the effects of varying four operating factors on the page-swapping activity of a large computer system. The two responses measured (number of pages swapped, PS, and activity count, AC) were planned beforehand and give interesting though differing results.

$$(1)\ \text{LPS} \hat{=} 2.90 + A\begin{cases}1.-.16\\2.+.02\\3.+.14\end{cases} + D\begin{cases}1.-.36\\2.+.07\\3.+.29\end{cases} + P\begin{cases}1.-.47\\2.-.02\\3.+.49\end{cases} + M\begin{cases}1.-.69\\2.-.01\\3.+.70\end{cases} \pm .07$$

predicted log pages swapped	replacement algorithm	deck (load sequence)	problem program	memory pages
	1. LRUV	1. GROUP	1. SMALL	1. 24
	2. FIFO	2. FREQY	2. MEDIUM	2. 20
	3. RAND	3. ALPHA	3. LARGE	3. 16

Eq. (1) shows that the four factors operate <u>multiplicatively</u> on the number of page swaps; the logarithm of the number of swaps (LPS) is described adequately by an additive term for each factor. There was a small persistent non-additivity - a two-factor interaction - between D and M but its inclusion only accounts for 4% of the total variation in LPS and so is included here in the error term. The latter, 0.07, means that on average the equation will give values for the 81 experimental conditions that are off by 16% (7 × 2.303). There are no outliers or other failures.

Eq. (1) conceals one important disparity between the replacement algorithms. The "random" algorithm has more (squared) random scatter (not accountable for by the factors) than the other two by a factor of 2.7. Also, the effects of the other two factors are slightly but not seriously different for that algorithm.

$$(2) \qquad \widehat{AC} \doteq 37 + A\begin{cases}1.\ +3\\2.\ -2\\3.\ -1\end{cases} + D\begin{cases}1.\\2.\\3.\end{cases}\overbrace{\begin{array}{|ccc|}\hline 1 & 5 & 1\\ 1 & 2 & -2\\ 2 & -4 & -6\\\hline\end{array}}^{\substack{M\\1.\ 2.\ 3.}} \pm .8$$

where AC is predicted activity count. Eq. (2) shows how the four factors affect AC. Here there is a much larger non-additivity of D and M, so a tiny table is given to show estimated AC for each D×M combination. No simple transformation will remove this anomaly.

Variation of the factor problem program (P) had no noticeable effect on AC, and this fact may well make this response more useful for the operating system when program cannot be controlled. Again, every variation in data that is not in the first three parts of the equation is pooled in the ±0.8 which represents, then, the average error of the equation in giving back the 81 "true" values. The average error in matching the observed values is $(0.8^2 + 2.1^2)^{1/2} = 5.05^{1/2} = 2.25$, since the average error (standard deviation) of single observations is 2.1. Again, there are no grossly defective vlaues and, as noted in the subject paper, only one value deviates noticeably from a normal distribution.

It is suggested that LPS is the more revealing response for understanding the system, even though the authors' immediate objective (to find a response that can be used under operating conditions) is better met by AC.

It is remarkable that a statistical design developed long ago for agricultural problems produces results of types entirely familiar in that field as well as in many engineering problems. Main effects dominate; one two-factor interaction appears; no higher-order terms are needed.

It should interest readers considering further experimentation on computer systems to know that the thoroughness of this pioneering effort need not be repeated. It is possible to take 27 observations (a "one-third replicate") scattered in a well-known way through the 81, that will permit the estimation of all main effects. The same third (each third from the 81 gives the same conclusions!) would clearly indicate just which new runs are required to identify and measure the D×M interaction for AC.

2: by L.A. Belady

The authors introduce a statistical notion called "interaction", a term often neglected while analyzing computer system behavior. I shall make some remarks on this; however, I would like to mention first another kind of interaction, equally infrequent, namely the one between individuals of different disciplines working together on the same problem.

Dr. Tsao, first of all a statistician, and myself worked together a few months back in Yorktown, motivated by some results reported in the papers under discussion now. In particular, we examined the value pattern of the IC-statistics and ultimately found a useful application of detecting localities of programs. (Locality - or working set - is a collection of memory sections, for example pages, the program needs in order to run, without demanding additional memory, for a specified execution-time interval.) A scheme was developed to measure on-line, i.e. in an operational environment, the most recent $(IC)_0$ values (derived from the originally conceived IC) for individual programs of a multiprogrammed mix. These values can then be meaningfully used to decide whether a program, at its page-demand time, should acquire an additional page at the expense of other programs or should be kept in its current memory partition. My point is that the particular way the statistician organized the data led to the development of a practical and potentially very useful operating system algorithm.

Let us turn our attention now to the authors' discovery that deck-arrangement and memory size display a strong interaction and that it is predominant among the six other pairwise interactions. In order to appreciate the significance of this statement one has only to recall that deck-arrangement can be interpreted as different ways of mapping the same program on the (virtual) memory space. Compilers, loaders and link editors are the components of operating systems having the potential role of intelligently executing this mapping such that, for example, the execution time be some minimum. The message of Tsao et al. is then clear: one cannot expect that a map thus constructed for a multiprogrammed paging machine be minimum in general; in fact, what is optimum for a given memory size could be detrimental for another. This result is but one instance of how statistical work can increase insight into system relationships during the process of interpreting measurements.

There are some clouds in the sky, though, and I am especially uncomfortable about the apparent arbitrariness with which levels of factors were chosen for the experiment. Conclusion 3 of the first paper, for example, should be taken critically. A somewhat different selection of problem programs could change the order and suggest another sequence of significance. The authors, of course, realize this, but there is still danger of becoming misunderstood. If, as we hope, we are about to enter an era in which statisticians and computer specialists fruitfully cooperate, then the former should never stop calling the attention of the latter to the data measured, the composition of the factors, in order to counteract the ever-existing urge to use canned results without really knowing their origin.

SESSION IV

INPUT ANALYSIS

CHAIRMAN: W. TIMLAKE

A COMPUTER PERFORMANCE MONITOR AND MARKOV ANALYSIS FOR MULTIPROCESSOR SYSTEM EVALUATION

John M. Feeley

1. INTRODUCTION

General Discussion

This study comprises the results of a preliminary analysis of the EXEC 8 multiprocessor used by the Naval Undersea Research and Development Center (NUC) as measured by the Computer Program Monitor (CPM) device in October 1970. The measured data have been placed into a form that can be used as input to a mathematical model describing the internal activity of the EXEC 8. By mathematically describing this internal activity, it can be observed that the activity within certain parts of the EXEC 8 accounts for the majority of the EXEC 8 system time used, but that this phenomenon is a function of user program, type of program and terminal demand (i.e. real time, batch, or the combination of both). It can also be seen that over short periods of time certain parts of the EXEC 8 are called much more frequently when there is a corresponding shift or change in the user program. The purpose of this analysis is to measure these changes within the EXEC 8 and to point out areas which need to be modified in order that the EXEC 8 might be used in a more optimum fashion both from the systems software and user program standpoints.

To accomplish this analysis, three basic steps were necessary:

- A method of measuring the EXEC 8 using the CPM under a variety of computer-user environments
- A method of recording and analyzing the gathered CPM data
- A method of mathematically modeling the EXEC 8 system that used the reduced CPM data as input.

Measuring method. To perform the CPM measurement, the EXEC 8 was considered in its modular form, where each module is a subroutine of the multiprocessor system. Since there are currently 211 modules or elements of the EXEC 8, measurements of only a few of the most commonly used were gathered. To extract these data, probes from the CPM were placed on the 1108 "p" register and the address within this register was compared with the known high and low core locations of the EXEC 8 modules. The data extracted contain information on the length of time each of the monitored EXEC 8 programs spends in execution and the number of times a module is entered per unit of time. These data form the raw data source upon which the analysis is based.

To understand how the EXEC 8 functions, and in particular how the EXEC 8 modules that were measured by the CPM interconnect, a block drawing was constructed and updated throughout the study as more information became available. The block diagram of the seven modules measured is found in figure 1-1. The lines between blocks (modules) represent the interconnections between EXEC 8 programs.

Since the interconnection paths between EXEC 8 modules remain fixed, the block diagram in figure 1-1 holds for all four of the environments measured. These environments were:

- Real-time environment with nothing else on the system;
- Benchmark environment with a selected set of programs that constitute the average-type program load, all placed on the system at once in a batch environment;
- Real-time and batched benchmark environment;
- Standard environment where the CPM was connected, but no knowledge of the input user programs was available.

The major CPM analysis effort concentrated on the real-time environment, primarily because of real-time systems that were slowing down the computations but, in addition, because the system was "dry" and there was more accurate knowledge of the user program specifically being executed.

<u>Reduction method.</u> The second step in the study was to reduce the raw data gathered by the CPM. Several programs were necessary to accomplish this since the raw CPM data were not gathered in a fashion amenable to direct statistical interpretation.

Because the data were gathered over irregular intervals of time (aperiodically), it was necessary to formulate an <u>interpolation program</u> to provide measurements over equal intervals of time. From these modified data, the sample mean and standard deviations were calculated, and some curve fitting routines were developed to determine if the data were of a particular distribution form.

In addition to the reduction and curve fitting routines, quick-look routines were used to examine trends in the raw data. These routines both listed the data and plotted them using the Calcomp plotter. Examples of these data are found in section 3.

Modeling method. Having measured the EXEC 8 using the CPM, and reduced the data using reduction techniques, the resulting data were then used as input for a mathematical model which was developed to describe the EXEC 8 interaction and throughput characteristics.

The mathematical model consists of a Markov process with a finite number of states or positions in which the phenomena being modeled can be as time increases. The EXEC 8, as modeled, can only be EXECUTING in one of the nine modules at any given time. Since the modules are interconnected and have multiple entry and exit points, the amount of time a given module is active over a specified delta of clock time is not constant but is rather considered to be a random variable. The Markov model developed will determine the expected time a module will be in an execute mode as well as the probability that the EXEC 8 will be in that particular module at any given time. This probability changes as the user program "flows through" the EXEC 8 system. The accumulation of time spent in any of the EXEC 8 modules, as well as the probability of being in these modules, is related directly to program throughput. As is pointed out below, when a great deal of time is spent in a module such as the dispatcher, the user program is not being executed; this results in extended throughput times.

Conclusions

In presenting a summary of the conclusions, the CPM analytical results conducted on the Test 6 data are presented first. It was these data that most completely measured the activity of the EXEC 8 under four different environments, and it was these data that were adapted for the Markov model of the EXEC 8 modularized system.

Figures 1-2 through 1-5 show the probabilities of being in one of the EXEC 8 modules at any given time, providing the system began operation by execution of the user program. Tables 1-1 through 1-4 list the corresponding model output probabilities. In all cases, they show that when the system stabilizes (to the right of the figure), the system operates in the EXEC 8 modules for the majority of the time. The parameters used to determine these curves were the average server rate within the modules, which was measured by the CPM, and the instantaneous transfer rates between the EXEC 8 modules, which were estimated based on data obtained by the CPM on the number of times each module was entered during the test.

It is to be noted, before examining each figure individually, that the dispatcher is the dominant module and the process is operating within this module between 58 and 95% of the time, depending on the selected operating environment. On the other hand, the user program operates over the same test environments between 18 and 1.7% of the time. This says that the system might be considered "EXEC 8 bound", depending on the functions that the EXEC 8 is performing directly and those it is performing for the user program.

EXEC 8 and the real-time simulation. Figure 1-2 shows the system operating under the real-time environment in which no other user programs are being introduced into the system. As might be expected, as the user program is executed, the probabilities of being in that module, $P_{up}(t)$, decrease until it reaches a steady-state value of 1.7%. On the other hand, the dispatcher probability, $P_{disp}(t)$, increases until its average operating time at steady state (after about 128×10^{-5} sec = 1,280 microsec) is 95%.

During the operation of the real-time simulation program there are no other user programs in the computer. If the user program for some reason requires a waiting time when operations are being carried out in the A/D and D/A converters, the 1108 continues to function within the dispatcher. With this in mind it is not surprising that the system operates within EXEC 8 98% of the time during the real-time simulation environment.

It was found that the 1108 system, when performing operations in the real-time environment, was not able to obtain certain of the required response times. This was undoubtedly true because of the amount of time being spent in the EXEC 8, and most probably within the dispatcher module of the EXEC 8. The CPM data for this environment were not amenable to step-by-step time slice analysis. Therefore, in times of CPM-recorded high-density response during the real-time simulation, exact analysis could not be performed when timing and range problems were apparent. This was for two reasons: first, the data were serially extracted from each EXEC 8 module so that time zero could not be lined up for all the modules, and second, the recording method was aperiodic so that exact results for identical time intervals could not be determined. Instead, the statistical results are based on averages over the entire measured interval for each of the modules, individually.

By using a CPM that was modified to examine all the modules at the same time, the time zero start problem could be eliminated and the information could be periodically recorded to enable exact analysis over certain critical time ranges such as the high-density compute ranges.

EXEC 8 and the real-time simulation plus benchmarks environment. While it was found that 98.3% of the time was spent in the EXEC 8 during the real-time simulation environment, the system was otherwise dry and the results might be expected. The next step was to analyze the system when the real-time simulation programs were time-sharing with a large number of standard user programs called "benchmarks".

In this environment, the EXEC 8 routines were executing at steady state (approximately 5120 microsec) 88% of the time and the dispatcher module, although still the dominant module, operated 70% of the time rather than 95%. This reduction of 25% of the clock time for the dispatcher indicates that it is in an idle state at least 25% of the time during the real-time simulation. With the addition of the benchmarks the amount of time the user program operates increases to 12% at steady state. As can be seen from figure 1-3, the output module, which contains all the EXEC 8 modules not directly measured by the CPM, increases to constitute 11% of the system clock time. This means that there is a considerable change in the I/O routines being called by the user program through Executive requests. These I/O routines are some of the EXEC 8 modules which were not measured by the CPM, but their existence was determined by the TRACE-EDIT program which was run on the benchmark programs. The EXEC 8 modules most frequently called by the user program were READ$ and PRINT$. These modules account for the additional activity in the output module due to the addition of the benchmark routines to the real-time environment.

The dispatcher is still in an idle mode some of the time and possibly a direct measure of the dispatcher idle time by the CPM would be in order, to get knowledge of the continual dominance of the dispatcher module in this environment.

EXEC 8 and the benchmark environment alone. When the benchmark programs are run alone in a batch environment, it is found that the User Program increases to a steady state of 18% of the operating time and that the dispatcher decreases from 70% in the previous environment to 58%. This means that the dispatcher spends an additional 12% of its time performing on the real-time simulation when it is included with the benchmark programs. This result is believed to occur because there is a great reduction of executive request calls to the dispatcher when the real-time simulation program is removed. The most common of these Executive requests is TWAIT$, which directly calls and ties up the EXEC 8 dispatcher module. These user program/EXEC 8 interaction results are made available from the TRACE-EDIT program, which was run on the real-time simulation program and is summarized in table 3-5.

In addition, during the benchmark environment, the output module increases from 11% in the real-time plus benchmark to 16% in the benchmark alone. This is due to the increased input and output that can be processed if the real-time TWAIT$ and other real-time slicing that takes priority is not holding up the system.

Of the environments measured, the benchmark environment was the most efficient from the standpoint of executing user program code that was being processed 18% of the time, while the output module, which includes the EXEC 8 control of the Read and Print routines, operated at a steady-state rate of 16%. Both of these modules have a definite effect on the system throughput and might be summed together to obtain an indicator, which in this case would be 34%.

As in the case of the real-time plus benchmarks environment, if an accurate measure of dispatcher idle time could be made it would be helpful in the analysis and could possibly be used to show that more user programs could be admitted into core at the same time to increase throughput.

EXEC 8 and the standard environment. The standard environment was measured to get an idea of how close it was to the selected benchmark environment. There was no control of programs flowing through the system, and all the activities were present in addition to the activity from the San Diego remote terminal.

It is obvious from the data that the idle time in the dispatcher module jumped considerably, since its operation increased from 58% in the previous environment to 77%. The user program time decreased from 18 to 9% and the output dropped from 16 to 8.7%. Because there was no control over this environment, the only value of measurements of this type would be as an indicator, and many more data would have to be taken to perform a statistical analysis on average user rates and module probabilities in a noncontrolled environment.

In general, the system convergence occurred around 5,120 microsec for all the modules in all the environments. It may also be noted that although convergence would not follow the same path had the initial state of the system been something other than operation within the user program, the final results would be the same at steady state. Thus, the steady-state probabilities are the most significant measure of average system performance. The steady-state probabilities for the four environments are listed in table 1-5 along with their earliest convergence time.

The results of the TRACE-EDIT program add more information to the benchmark environment. While the CPM could determine and measure activity that was purely user program and EXEC 8, it was unable to measure the portion of the EXEC 8 time that was called directly by and acting for the user program. Table 3-4 shows that 72% (126.559/175.375 × 100%) of the activity in the EXEC 8 was called for by the user program in the benchmark environment. During the benchmark environment, it can be seen from table 1-5 that the EXEC 8 is active on an average of 82% (100-18) of the time. The TRACE-EDIT data show that of this 82% active time 72×82 or 59% is User Program activity within the EXEC 8. Now, since the User Program spends 18% of the clock time outside the EXEC 8 and 59% of the clock time within EXEC 8, it is then responsible for executing 77% of the clock time. This leaves the remaining 23% of the time for the EXEC 8 to perform its multiprocessing functions. Because the TRACE-EDIT data in table 3-4 also show, in the column entitled

"Total ER time used", an amount of time that is actually larger than the run throughput time that was taken from the individual job accounting summaries, it is felt that the true accounting is not available to the TRACE-EDIT program and that there is actually a considerable amount of idle time in the EXEC 8 while operating in this environment. This could be further analyzed with additional CPM measurement.

An interesting result is obtained from the mathematical model when the 1108 system is considered to be operating initially within the dispatcher rather than the user program. In the steady-state solution, the probability of being in a particular module is a constant and is independent of the module the system was initially in. However, the convergence to the steady-state solutions is interesting. Closer examination of the dispatcher module shows that during idle times the 1108 spends most of its time there. This means that it is fairly likely that at time zero the system is in the dispatcher rather than the user programs. A plot was made (figure 1-6) of this phenomenon for the benchmark environment. The dispatcher now monotonically decreases to its steady-state value of 58% while the output module increases rapidly and then levels off slightly below its peak at a steady state of 16%. In all cases except the user program and the interrupt handle, the modules move upward to a high point using more and more CPU time and then the slope changes and they move gradually down to the steady-state solution. If the system is actually in the dispatcher at the initialization of the user program, then, based on the data and figure 1-6, it can be seen that the probability associated with operation within the user program is quite low and remains low for a significant length of time, while the other module probabilities move rapidly to their respective steady states.

Averages and program peaks. The EXEC 8 has been modeled using average data, which include certain peak transmission rates that could occur whenever another user program enters the system or whenever there is a change in user program coding that might, for example, switch the routine from a compute-bound user program into the EXEC 8 I/O routines. Even with these peaks, which can be seen in the plotted data of section 3 (figures 3-1 through 3-12), the model shows the system to converge in a time period that is much smaller than the occurrence of the peaks. This means that the system can effectively perform more of the peak-type operations, which means it will handle effectively more user programs without appreciable degradation. This general conclusion can be made based on the average 5,120-microsec convergence time, which is quite small relative to the peaking phenomena which occur once every 150 to 180 seconds.

Other tests. The test data are summarized in section 3 and are available in detail upon request. Of the tests examined, particular emphasis was placed on those dealing with the EXEC 8 software. While all the data are reduced and are available in section 3, the conclusions are based on Tests 2, 6 and 12. The data from Tests 2 and 12 filled in information that was necessary to complete the analysis of the Test 6 data using the Markov model. The data from Tests 2 and 12 show directly the amount of time the EXEC 8 spends within its modules. The times for the four environments are:

1. Real time 98.33%
2. Real time plus benchmarks 85.58%
3. Benchmarks 80% (estimate)
4. Standard 88.78% (Test 12)

The difference between these figures and 100% was the time spent by the 1108 system in the User Program. These values for the four environments are:

1. Real time 1.67%
2. Real time plus benchmarks 14.42%
3. Benchmarks 20%
4. Standard 11.22%

From the plots of the Test 6 real-time run data it was found that there were 21 different in-range and server rate shifts in the seven modules monitored. An attempt was made to consider these shifts by defining a new mathematical model matrix at the point of each shift. This, however, was impossible to do with the data available, as the time zero point at which to begin the process could not be exactly defined for all the modules since the rates had not been simultaneously recorded by the CPM. The mathematical model, however, has been modified to handle an indefinite number of rate shifts within the data.

In addition, it was obvious from the data plots that there were three distinct activities going on during the real-time and real-time plus benchmark environments. These activities had to do with the real-time simulation program that operates with three distinct call rates to the EXEC 8. These three transmission rates are obvious in each of the six modules plotted. In the case of the dispatcher (figure 3-5), the frequency with which it was called was inversely related to the other modules. It can be concluded that during large amounts of real-time activity there were fewer TWAIT$ directed to the dispatcher and the module did not perform as many of the usual EXEC 8 housekeeping chores. As an example (table 1-6), when the dispatcher is activated between 125 to 135 sec into the real-time run, the activation rate is 682 times/sec. The corresponding rate for the I/O module was 750 times/sec. However, at times 155 to 160 sec into the real-time run, the dispatcher rate was 2,989 per sec and the I/O dropped to 593 per sec. A complete data summary of these break points is found in table 1-6.

Conclusions regarding the model. The mathematical model and computer program for the analysis has been completed and coded in a form that will make any future analysis of this type easy to handle. The model is described in detail in section 4. It is felt that the model represents a significant breakthrough in the analysis of hardware-software systems that have been measured using the computer program monitoring device.

Assumptions

The following assumptions were made concerning the system and mathematical model:

- The service time for each of the EXEC 8 modules is exponentially distributed
- The seven EXEC 8 modules that were measured by the CPM, when combined with the dummy module labeled "User activity and output", constitute 100% of the EXEC 8 system. Either the 1108-CPU is operating within these eight modules or within the user program.
- Upon completion of execution of a particular set of instructions, the control is instantaneously transferred to another set of code within another EXEC 8 module.
- Associated with each instantaneous transfer there is a transfer probability. This probability is proportional to the number of times each of the adjacent EXEC 8 modules is activated.

2. GENERAL DISCUSSION

To collect and reduce the data extracted from the EXEC 8 eight basic steps were necessary.

1. Selection and measurement of the seven most critical EXEC 8 elements under all four environments.
2. Conversion of the CPM-extracted data to be compatible with the 1108.
3. Initial examination of the raw measured data.
4. Attempt to curve-fit the data to the Poisson distribution using various statistical tests.
5. Create a general-purpose program to adjust the data and find the mean and standard deviation.
6. Check for randomness of the selected and adjusted data by comparing with the calculated $\hat{\mu}$ and $\hat{\sigma}$ for the entire population of real-time module tests.
7. Determine the User Program interaction with the EXEC 8 by introducing both raw and summarized data from the TRACE-EDIT package for the real-time program and all benchmark routines.
8. Combine the data from various different tests to look for data compatibility.

The initial step in the data acquisition phase of the study was to establish which of the EXEC 8 elements were to be measured using the CPM. Although there were 13 tests conducted, the information collected on the seven selected EXEC 8 modules formed the best information for analytical purposes. The seven modules were:

1. Arbitrary device handler
2. COME
3. Dynamic allocator
4. Drum handler
5. Dispatcher
6. Interrupt handler
7. Input/output controller

Detailed Module Description

A block diagram of the seven executive elements is found in figure 1-1; these seven elements are described briefly by UNIVAC 1100 Series Systems Memorandum, Revision 10, Appendix A, as follows:

ADH: The ADH resident routine is the arbitrary device handler. It is used to perform I/O on devices driven by symbionts and maintenance as well as handling the user ERs, IOARB$, and IOAXI$.

COME: The COME resident I/O handler is part of the dispatcher. Its whole function is concerned with ESI activity. COME saves the circumstances of the EXEC when ESI interrupts occur, it queues up ESI interrupts, and handles ER for ESI interrupts. COME will either restore previous EXEC conditions or go to DISP to continue the run stream.

DA: This resident routine is the main portion of the dynamic allocator. It services requests for core space, release of core space, initial loading of programs, swap-in and swap-out of programs, loading of program overlays, and assignment of program control tables (PCTs). It calls on dynamic allocator core control (DACC) to accomplish I/O in conjunction with these functions.

DH: This resident routine is the FH432/FH1732 drum handler. It is called by the I/O routine to perform operations on the high-speed drums.

DISP: This resident routine is the main dispatcher. It contains ER control, subroutines for activity initiation (ACT) and deactivation (DEACT), and for obtaining the highest priority activity (NEXT), as well as routines to handle ER to FORK$ and TFOR$, SWAIT$, TWAIT$, and AWAIT$.

All switching (via switch lists) is done through DISP. A number of commonly used EXEC indicators and pointers are in its data bank.

IH: This resident routine handles all I/O interrupts (ISI and ESI). It queues or distributes control to I/O completion activities.

I/O: This resident element is the I/O controller, which handles all interface between the user and ISI hardware. All files are handled by exclusive block use with this routine. I/O handles ERs to IO$, IOWI$, IOW$, IOXI$, IOARB$, IOI$ and the @MODE control statement. It contains coding to check user access words and to establish interrupt activities on I/O request, as well as service routines to handle and log I/O error messages, and time all I/O channels. It depends on the configuration element, IOD.

Additional Elements

In addition to the EXEC 8 elements described, there is a user program which is thought of as the driver of the Executive activity. Since the user program does not reside in a fixed section of core, CPM measurement of this routine was not possible; therefore a TRACE-EDIT program was developed to determine measurements regarding these routines. The programs TRACE and EDIT are further discussed in this section.

The remaining item in figure 1-1 is labeled "USER ACTIVITY AND OUTPUT", and essentially accounts for all the rest of the EXEC 8 routines which, while they exist and are used, do not take up an appreciable amount of system time while a typical user program is being run.

Measurements for this dummy module were estimated on the basis of the CPM measurements of the surrounding modules. Once these elements were selected for measurement, the test logic for the CPM had to be determined, the CPM hardware had to be wired, and the probes had to be attached to the proper locations.

Since the CPM used had only one address comparator, only one EXEC 8 module at a time could be measured, and a new module could be measured only by rerunning the user program. Due to test logic, CPM hardware restrictions and the information deemed applicable for the study, the information measured was in almost all cases aperiodic, so there was no common delta clock time for each successive measurement recording.

The second step of the process was to make the data generated by the CPM compatible with the 1108. This was done using a short conversion routine that added a single dummy word to each record of the CPM recorded data.

The third step was to examine the raw data with the intention of determining trends. This was done by means of two programs, CPM1 and CPM6, developed by INFONET Division of CSC. These routines list the clock time, the length of the delta-time interval, and the CPM measurement within that delta interval of clock time. The CPM6 routine lists the cumulative data measured as the clock time progresses. The CPM1 routine lists the percentage of total clock time that the measured module was in operation.

A third method for examining the raw data was to display them using a Calcomp plot package which was developed for this study. This routine was used to plot all the raw data as they were recorded by the CPM and, in addition, to plot the cumulative data by adding the incremental deltas.

These three routines were quite valuable in establishing use trends within the EXEC 8 modules as the user program was executing. In the analysis, it was possible to split the real-time user program activity up into three distinct cycles and accurately pinpoint the times within each of the seven EXEC 8 modules when activities changed.

The fourth step in the data reduction was of a statistical nature. Due to CPM recording methods, many of the CPM measurements were taken over intervals of approximately 8 seconds. In most cases, these data were found to be representative of all the data extracted, and the remaining data, which were aperiodically recorded over much smaller intervals of variable length, were omitted.

The data from the 8-second interval were linearly adjusted to exactly 8 seconds and were then fit to a Poisson distribution using a chi square technique, which failed. Since the number of 8-second intervals was small, the chi square test, which required groups of 8-second intervals to build the data histogram, lost some of its validity. To avoid this problem, the Kolmogorov-Smirnov "d" test was used on the cumulative distribution, which used every data point (every 8-second interval) rather than group averages. Again, however, the data failed to be Poisson-distributed at any meaningful level of significance.

Finally, since the mean and variance of the Poisson distribution are equal, the sample mean $\hat{\mu}$ and the sample variance $\hat{\sigma}^2$ were compared with one another, and it was found that in most cases the sample mean was very close to the sample standard deviation $\hat{\sigma}$. It was then concluded that the data from any single test were not Poisson-distributed. It is possible that Poisson properties might be present if the data were grouped according to the information obtained from the plotted raw data which showed definite multimodal trends. The effort was discontinued here because of time constraints.

The fifth step taken was to develop a general method for handling and reducing all the CPM data. This was done by creating a program that would:

- Read in selected parts or all of the raw CPM data
- Select all information that came from time intervals of a specific size
- Adjust the information in these aperiodic time intervals

to a common time using linear interpolation
- Determine and print the selected maximum and minimum aperiodic time intervals
- Order and print the selected, adjusted data
- Determine the sample mean and standard deviation of the selected data

The sixth step was to test the results of the General Program with the results of the CPM6 cumulative values to see if the selected interval data were from the same distribution as the total overall data. There was very little difference in all cases, so a statistical test to determine if the data came from the same population was not felt necessary.

The seventh step was to introduce data from the user program into the analysis. This was done by means of a TRACE-EDIT program, which lists:

- The number of instructions executed between ER requests
- The amount of user-chargeable time (in ticks) the ER used
- The amount of time between ERs
- A summary of the above with a column tabulating the EXEC 8 module to which the ER was directed

The TRACE-EDIT program was run on the real-time routines and on all the standard benchmark routines. The information obtained from these runs is summarized in table 3-4 of section 3.

The eighth step in the data reduction effort was to examine all the test data gathered and to combine the results to look for data compatibility. Tests 2 and 6, for example, test the entire EXEC 8 and the individual EXEC 8 modules, respectively, under all four environments. The test information from these tests was compared for all four environments.

3. DATA SUMMARIZATION AND RESULTS

General Discussion

This section contains a partial summarization of the data from the CPM and the TRACE-EDIT package. The CPM data listed in this section contain the sample mean for each test and the TRACE-EDIT summary contains, for those routines which were traced, the number of ERs that occurred during the program, the time that was spent accomplishing the ER for which the user was charged, and the module in the EXEC 8 that the ER called directly.

The CM data are summarized in table 3-1, and the TRACE-EDIT data are summarized in table 3-4.

CPM Data Summary

The CPM data summarized here contain the average number of times the measured module was used and the average amount of time the module operated once it was entered. These averages or means are used in the Markov model of the "7-module" EXEC 8 system operating in real time which is diagrammed in figure 1-1. Complete data information can be supplied on request; this information contains not only the mean but the standard deviation, skewness, sample size and information regarding the size of the window and interpolation used.

Comparison of "Window" Data and Total Average Data

Since the data were recorded aperiodically, a common unit of time had to be designated in order to describe the transition and server rates of the various EXEC 8 modules. A considerable amount of the data collected in Test 6, the "Module execution events and timing" test, had measurements that were taken over intervals of approximately 8 seconds. Therefore, in the case of the Test 6 data, the samples were taken from these windows (intervals) and the data were adjusted to correspond to exactly 8 seconds. It was felt that these data would be representative of the entire sample but, in order to check this, a comparison was made of the window data event rate from Test 6 and the total cumulative Test 6 average event rate. Table 3-2 shows the comparison, which points out a few discrepancies. This error is felt to occur from the linear adjustment of the data to fit the exact 8-second window. In cases where a large variation exists, the information from the total cumulative average was used, as being more representative of the data.

In addition to the mean, standard deviation and skewness being available for the CPM data, the adjusted information was also ordered and is available for curve fitting and other statistical techniques. However, the statistic will be only as good as the specially selected adjusted data.

Interrelationships Between Tests 2 and 6

While Test 2 measures the entire EXEC 8 activity and records the total time spent executing EXEC 8 programs, Test 6 measures seven of the individual modules of the EXEC 8 package. The seven modules described in sections 1 and 2 of this paper were selected because it was thought that the majority of the time spent in the EXEC 8 mode would be spent in the seven elements or modules. The time spent in each of the seven modules is then compared with the total time the EXEC 8 was in operation in each of the four environments. The results are presented in table 3-3A.

To determine the amount of time actually spent in the user program, the Test 2 results were subtracted from unity. To determine the amount of time spent in the EXEC 8 modules which were not measured by the CPM the Test 6 total was subtracted from the Test 2 total; this result was assigned to the dummy module "Output and user activity". Table 3-3B lists the user program times and the OUTPUT module times.

CPM Raw Data Plots

In order to determine the form of the CPM data, plots were made of the unadjusted raw data. These plots point out:

- anomalies in the data
- range of the data
- breakpoints and cycles in the data
- rough probability distribution of the data

For example, from the plots of the EXEC 8 module data from Test 6 in the real-time mode (figures 3-1 through 3-12), it can be seen that there are three distinct record levels of data for each of the modules. The low record rate for most of the modules indicates a low-order activity in the real-time program. On the other hand, the high record rate indicates the high activity which occurs during the later phase of the real-time simulation program.

Certain anomalies may be found in the data, such as the record rate of the Dispatcher (figure 3-5), which is highest when other module rates are low. This indicates that the dispatcher module is responding to EXEC 8 system functions rather than user program functions and is performing housekeeping tasks while waiting for a user program directed executive request (ER).

Another anomaly that is easily seen from the data is the Dynamic Allocator record rate (figure 3-4), which is every 10 seconds as opposed to the aperiodic record rates of the other modules.

From each of the plots, the high and low range of the data and the data rate changes are apparent and affect the system operating time. From the range, a scheme for setting up a histogram of the data can be arrived at, and from the rate changes, the data can be divided into a possible multiple of distributions, each one describing a particular section of data. Rough characteristics of the distribution are also apparent since the relative number of high and low record rates is immediately observable and tells whether the distribution is relatively uniform or is skewed left or right.

TRACE-EDIT Data Summary

The TRACE-EDIT program is designed to trace the interactions between the user program and the EXEC 8 modules. The EXEC 8 modules are called by the user program by means of ERs, which switch control from the user program to the programs contained in the executive. The TRACE program follows all the ERs and the EDIT program prints them out with other data pertaining to the number and type of ER. The EDIT program also produces, at the end of its individual ER printout, a summary of the ER activity.

The TRACE-EDIT programs were successfully run on most of the benchmark programs, and with partial success on the real-time simulation programs.

The total time taken by the ERs was determined by multiplying the number of ERs selected by the TRACE-EDIT routine by the mean number of ticks taken by these ERs. This number is then converted to seconds. For example: in table 3-4, there were 937 ERs selected by the TRACE-EDIT program in the benchmark AERDYN. The mean number of ticks per ER was 7.04 and each tick is 200 microseconds. Therefore, 7.04 × 937 × 0.000200 = 1.3192 seconds. It is to be noted, however, that a certain number of ERs could not be counted by the program and were <u>not</u> included in the calculation. This was done later in the column entitled "Total Time Estimated for all ERs", which used the total number of ERs as the result. It should also be noted that the results based on the total number of ERs differ considerably from the calculations that used only the selected ERs. This difference occurred because of the inaccessibility of certain ER accounting words when the TRACE-EDIT program was being run. More accurate information could be made available if the EXEC 8 were modified slightly to allow the TRACE routine access to these important accounting areas.

Although without exact information regarding the interrelationship between the user program and the EXEC 8 modules that was to be obtained through the TRACE-EDIT program, the ER flow information was obtained. A summary of the results is presented in table 3-4.

For the real-time simulation program analysis, it was desirable to know the flow rate of ERs from the user program to specific modules within the EXEC 8. This information cannot be obtained from the TRACE-EDIT program due to the inaccessibility of the wall clock time within the EXEC 8. However, the number of ERs directed by the user program to the EXEC 8 is available and is presented in table 3-5.

Other problems in tracing the real-time simulation program through the analog/digital converter were caused by an excessive number of TWAIT$ Executive requests. This problem occurred in all the simulation TRACE-EDIT runs independent of a large number of user program changes. However, information regarding the type of interaction with the EXEC 8 modules is present in table 3-5 even though the flow rates (interaction rates) are not.

4. MATHEMATICAL MODEL OF THE EXEC 8 NINE-MODULE SYSTEM

The analysis of the EXEC 8 multiprocessor system lends itself to a Markov model which describes the interactions of the system modules and the user program. Before going directly into the model itself, some definitions and tools need to be explained and derived.

The Poisson Process

The Poisson process is a statistical method that deals with the number of changes that occur in nature in a given time interval. The process states that whatever the number of changes that a system undergoes in the time interval $(0,t)$, the conditional probability that a change occurs in $[t,t+\Delta t]$ is $\lambda\Delta t + o\Delta t$, and the probability of more than one change occurring in $[t,t+\Delta t]$ is $o\Delta t$.

For this case, the parameter $\lambda(t)$ is the instantaneous server rate of a module and is defined as follows:

$$\lambda(t) = \lim_{h\to 0} \frac{P(t<T<t+h \mid T>t)}{h}$$

$$= \lim_{h\to 0} \frac{F(t+h)-F(t)}{h} \cdot \frac{1}{[1-F(t)]}$$

$$= \frac{f(t)}{1-F(t)} \qquad (1)$$

where h = small interval of time, f(t) = the server density function, F(t) = the server distribution function, and T = service time.

When an exponential service time is assumed, $\lambda(t)$ becomes independent of time (homogeneous) and can be expressed as the constant λ. The remaining term in the Poisson process defined above is $o\Delta t$, which is called "little oh of t" and is a function f(t) such that:

$$\lim_{t\to 0} \frac{f(t)}{t} = 0 \qquad (2)$$

From the above definitions, a set of n+1 difference equations can be obtained which describe the probability of being in any one of the n+1 states at time $t+\Delta t$.

$$P_0(t+\Delta t) = (1-\phi_0(\lambda)\Delta t)\ P_0(t) + \sum_{\substack{i=0\\ i\neq 0}}^{n} \phi_{i,0}(\lambda)\Delta t P_i(t)$$

$$P_1(t+\Delta t) = (1-\phi_1(\lambda)\Delta t)\ P_1(t) + \sum_{\substack{i=0\\ i\neq 1}}^{n} \phi_{i,1}(\lambda)\Delta t P_i(t)$$

$$\vdots \qquad\qquad \vdots \qquad\qquad \vdots$$

$$P_n(t+\Delta t) = (1-\phi_1(\lambda)\Delta t)\ P_n(t) + \sum_{\substack{i=0\\ i\neq n}}^{n} \phi_{i,n}(\lambda)\Delta t P_i(t)\ , \qquad (3)$$

where $P_i(t)$ is the probability of being in state i at time t, $\phi_i(\lambda)\Delta t+o\Delta t$ is the probability of leaving state i in Δt, $\phi_{i,j}(\lambda)\Delta t+o\Delta t$ is the probability of moving from state i to state j in Δt given a transition did occur.

By rearranging these difference equations and taking the limit as $\Delta t \to 0$, the following set of differential equations is obtained:

$$P_0'(t) = -\phi(\lambda)\, P_0(t) + \sum_{\substack{i=0\\ i\neq 0}}^{n} \phi_{i,0}(\lambda)\, P_i(t)$$

$$P_1'(t) = -\phi_1(\lambda) P_1(t) + \sum_{\substack{i=0\\ i\neq 1}}^{n} \phi_{i,1}(\lambda)\, P_i(t)$$

$$P_n'(t) = -\phi_n(\lambda) P_n(t) + \sum_{\substack{i=0\\ i\neq n}}^{n} \phi_{i,n}(\lambda)\, P_n(t) \qquad (4)$$

This series of equations can be written in matrix notation as

$$P'(t) = P(t)A \qquad (5)$$

where

$$A = \begin{matrix} -\phi_0(\lambda) & \phi_{0,1}(\lambda) \;\cdots & \phi_{0,n}(\lambda) \\ \phi_{1,0}(\lambda) & -\phi_1(\lambda) & \vdots \\ \phi_{n,0}(\lambda) & \phi_{n,1}(\lambda) & -\phi_n(\lambda) \end{matrix} \qquad (6)$$

Since the number of states is finite, the solution of Eq. (6) subject to the initial condition $P(0) = (1,0,0,\ldots 0)$ is given in the following form:

$$P(t) = P(0)\exp(A\cdot t) \qquad (7)$$

The process described above can be expressed in terms of a Markov chain. By definition, a Markov process is a stochastic process $\{X(t), t<T\}$ for which the conditional probability

$$P\{\underline{X}(t)\varepsilon E \,|\, \underline{X}(\tau) = \xi$$

has the property that

$$P\{\underline{X}(t)\varepsilon E|\underline{X}(t) = \xi\}$$

$$= P\ \underline{X}(t)\varepsilon E|\underline{X}_1(t_1)=\xi_1,\ \underline{X}_2(t_2)=\xi_2,\ldots,\underline{X}_n(t_n)=\xi_n,\ \underline{X}(\tau)=\xi\}$$

for all $t_1<t_2\ldots t_n < \tau$ and for all possible values of the random variable $\xi_1,\xi_2,\ldots,\xi_n$, ξ; i.e., $\underline{X}(\tau)$ is independent of the history of the process for $t < t$.

The Poisson process is a simple case of a Markov process and when Eq. (4) is described in matrix notation

$$P_k'(t) = -\phi_k(\lambda)\ P_k(t) + \sum_{j\neq k} \phi_{j,k}(\lambda)\ P_j(t) \tag{8}$$

it is equivalent to the Chapman-Kolmogorov equation for a homogeneous Markov process with a denumerable state space [1]

$$dP_k(t)/dt = -c_kP_k(t) + \sum_{j\neq k} c_jp_{jk}P_j(t) \tag{9}$$

where $k=0,1,2,\ldots$, $c_k = \phi_k(\lambda)$ and $c_jp_{jk} = \phi_{jk}(\lambda)$.

The terms $\phi_k(\lambda)$ and $\phi_{jk}(\lambda)$ from Eq. (8) are, respectively:

$$\phi_k(\lambda) = \lim_{\Delta t\to 0} \frac{1-P_{k,k}(\Delta t)}{\Delta t},\quad \phi_{jk}(\lambda) = \lim_{\Delta t\to 0} \frac{P_{k,j}(\Delta t)}{\Delta t} \tag{10}$$

Module Service

Before showing a detailed example of the EXEC 8 nine-module system model, the following example of a three-module system with different server rates, λ, and instantaneous transmission probabilities, p_{ij}, is examined.

Example. Consider the following three modules which service information and then transfer control to another module to continue the process. Consider also that the service time of each of the three modules is exponentially distributed and that the system continues to "cycle" indefinitely. This means that there are no "absorbing barriers" in the system, and the processing does not all terminate in a single module with probability 1.0 but rather continues to cycle. The modules are depicted in figure 4-1.

There are three states in which the system can be operating, namely: "A" executes, "B" executes or "C" executes. These states are denoted symbolically by placing a bar above the executing module symbol, so that the three states are: $\overline{A}BC$, $A\overline{B}C$ and $AB\overline{C}$.

Let the mean service times for modules A, B and C be as follows:

$$\mu_A = 0.26666, \quad \mu_B = 0.48, \quad \mu_C = 0.253334$$

It is to be noted that $\Sigma\mu_i = 1.0$ since the system is in one of the three states 100% of the time. Also, since the service time is exponential, the service rates are $\lambda_A = 1/\mu_A$, $\lambda_B = 1/\mu_B$, and $\lambda_C = 1/\mu_C$.

Let the transition rates be defined by the "in range" rates r of the modules (the "in range" rates were measured for the EXEC 8 using the CPM) as follows:

$$r_A = 10, \quad r_B = 9, \quad r_C = 9.5$$

and let the interaction rates, r_{ij}, between the modules be:

$$r_{AB} = 9 \quad r_{BA} = 0.5 \quad r_{CA} = 9.5$$

$$R_{AC} = 1 \quad r_{BC} = 8.5 \quad r_{CB} = 0$$

(These interaction rates were estimated in the case of the CPM data.)

The probabilities associated with the interactions are then:

$$p_{AB} = \frac{r_{AB}}{r_{AB} + r_{AC}} = \frac{r_{ij}}{\sum_j r_{ij}} = 9/10 = 0.9$$

$$p_{AC} = 0.1$$

$$p_{BA} = 0.5/9.0 = 0.0556$$

$$p_{BC} = 8.5/9.0 = 0.9444$$

$$p_{CA} = 1.0$$

$$p_{CB} = 0$$

The inputs are now defined for the Markov model of this three-module system and the following difference equations can be defined:

$$P_0(t+\Delta t) = (1-\phi_0(\lambda)\Delta t)P_0(t) + \sum_{\substack{i=1 \\ i\neq 0}}^{2} \phi_{i,0}(\lambda)\Delta t\ P_i(t)$$

$$= (1-\lambda_A\Delta t)P_0(t)+\lambda_B\Delta t p_{BA} \cdot P_1(t)+\lambda_C\Delta t p_{CA} \cdot P_2(t)$$

$$P_1(t+\Delta t) = \lambda_A\Delta t p_{AB} \cdot P_0(t) + (1-\lambda_B\Delta t)\ P_1(t) + 0$$

$$P_2(t+\Delta t) = \lambda_A\Delta t p_{AC} \cdot P_0(t) + \lambda_B\ t p_{BC} \cdot P_1(t)+(1-\lambda_C\Delta t)P_2(t)$$

Dividing by Δt and taking the limit of the above three equations gives:

$$\lim_{\Delta t\to 0} \frac{P_0(t+\Delta t)}{\Delta t} = \frac{(1-\lambda_A\Delta t)P_0(t)}{\Delta t} + \frac{\lambda_B\Delta t p_{BA}\cdot P_1(t)}{\Delta t} + \frac{\lambda_C\Delta t p_{BA}\cdot P_2(t)}{\Delta t}$$

$$\lim_{\Delta t\to 0} \frac{P_1(t+\Delta t)}{\Delta t} = \frac{\lambda_A\Delta t p_{AB}P_0(t)}{\Delta t} + \frac{(1-\lambda_B\ t)P_1(t)}{\Delta t} + 0$$

$$\lim_{\Delta t\to 0} \frac{P_2(t+\Delta t)}{\Delta t} = \frac{\lambda_A\Delta t p_{AC}P_0(t)}{\Delta t} + \frac{\lambda_B\Delta t p_{BC}\cdot P_1(t)}{\Delta t} + \frac{(1-\lambda_C\Delta t)P_2(t)}{\Delta t}$$

These equations lead to the following set of three first-order differential equations:

$$\lim_{\Delta t\to 0} \frac{P_0(t+\Delta t)-P_0(t)}{\Delta t} = P_0'(t) = -\lambda_A P_0(t)+\lambda_B p_{BA}\cdot P_1(t)+\lambda_C p_{CA}\cdot P_2(t)$$

$$P_1'(t) = \lambda_A p_{AB}\cdot P_0(t)-\lambda_B\cdot P_1(t)$$

$$P_2'(t) = \lambda_A p_{AC}\cdot P_0(t)+\lambda_B p_{BC}\cdot P_1(t)-\lambda_C P_2(t)$$

Written in matrix notation, these equations become:

$$P'(t) = P(t) \cdot A$$

where $P'(t) = (P_0'(t),P_1'(t),P_2'(t))$, $P(t) = (P_0(t),P_1(t),P_2(t))$, and

$$A = \begin{matrix} -\lambda_A & \lambda_A p_{AB} & \lambda_A p_{AC} \\ \lambda_B p_{BA} & -\lambda_B & \lambda_B p_{BC} \\ \lambda_C p_{CA} & 0 & -\lambda_C \end{matrix}$$

The matrix equation is solved using a matrix function solution

$$P(t) = P(0) \exp A \cdot t$$

where P(0) is the initial condition vector for t = 0 and in this case (as in the case of the EXEC 8 model) was

$$P(t=0) = (P_0(0), P_1(0), P_2(0))$$
$$= (1,0,0)$$

where the probability of being in the first state (e.g. $\overline{A}BC$) at time zero is 1.0.

The values filling the "A" matrix for this example are

$$A = \begin{matrix} -0.3750009-1 & 0.3375008-1 & 0.3750009-2 \\ 1.041667-2 & -0.1875000-1 & 0.1770833-1 \\ 0.3749990-1 & 0.0 & -0.3749990-1 \end{matrix}$$

and the results of the matrix function are

t_1	P_A	P_B	P_C
1	0.963	0.033	0.004
2	0.928	0.064	0.008
4	0.862	0.121	0.017
8	0.746	0.216	0.037
16	0.571	0.349	0.080
32	0.374	0.470	0.115
64	0.268	0.501	0.232
128	0.265	0.482	0.253
256	0.267	0.480	0.253
512	0.266666	0.480	0.253334

These results are plotted in figure 4-2.

As can be seen from the data and the figure, Module A which "drives" the function (since $P_A(t=0) = 1.0$) moves from its dominant position over the interval [t=0,t=250] to its steady-state position of $P_A = 0.266666$. Module C, on the other hand, continually rises to a final steady state of 0.253334. Module B moves up using more time to about t = 64 where it diminishes again to a steady state of $P_B(t) = 0.4800$.

It should be quite clear from this example that when dealing with a relatively complicated network without a mathematical model it is difficult to predict the network interactions. It is for this reason that the mathematical model of the EXEC 8 system was chosen.

Also, it is of interest to note that the mathematical model provides probabilistic results for a system whose initial state begins in modules B or C. The plot above is for the system beginning in module A, but figures 4-3 and 4-4 show the interactions of the modules had the system "started" in module B or C.

As can be seen, the modules reach the same steady-state solution which is independent of the starting point. The expected time spent in any of the modules is $t \cdot p(\infty)$, which has meaning at steady state, for large t, since the system, once it reaches steady state, remains there indefinitely. This means that on an average one could expect modules A, B or C to be in operation 26.66 sec, 48.0 sec and 25.34 sec, respectively, in any average hundred-second interval. However, during the time the modules are reaching steady state the time spent in any interval Δt is found directly by $\Delta t P(t+\Delta t/2)$ for any value of t.

<u>The EXEC 8 nine-module system model</u>. The general model that fits the EXEC 8 system is presented here in shortened form by making use of the equations of the example.

Skipping directly to the differential equations, it is found that the EXEC 8 model considered here is described by:

$$P_0'(t) = -\lambda_{UG}(p_{UA}+p_{UB}+p_{UC}+p_{UD}+p_{UG})P_0(t)+p_{HU}\lambda_G P_7(t)$$

$$P_1'(t) = p_{UA}\lambda_U P_0(t)-\lambda_A(p_{AE}+p_{AH})P_1(t)+p_{FA}\lambda_F P_6(t)+p_{GA}\lambda_G P_7(t)$$

$$P_2'(t) = p_{UB}\lambda_U P_0-\lambda_B(p_{BE}+p_{BH})P_2(t)+p_{FB}\lambda_F P_6(t)+p_{GB}\lambda_G P_7(t)$$

$$P_3'(t) = p_{UC}\lambda_U P_0(t)-\lambda_C(p_{CE})P_3(t)+p_{FC}\lambda_F P_6(t)$$

$$P_4'(t) = p_{UD}\lambda_U P_0(t)-\lambda_D(p_{DE}+p_{DH})P_4(t)+p_{FD}\lambda_F P_6(t)$$

$$P_5'(t) = p_{AE}\lambda_A P_1(t)+p_{BE}\lambda_B P_2(t)+p_{CE}\lambda_C P_3(t)+p_{DE}\lambda_D P_4(t)-p_{EF}\lambda_E P_5(t)$$

$$P_6'(t) = p_{EF}\lambda_E P_1(t)-\lambda_F(p_{FA}+p_{FB}+p_{FC}+p_{FD})P_6(t)+p_{GF}\lambda_G P_7(t)$$
$$+ p_{HF}\lambda_H P_8(t)$$

$$P_7'(t) = p_{UG}\lambda_U P_0(t) - \lambda_G(p_{GA}+p_{GB}+p_{GF}+p_{GH})P_7(t)$$

$$P_8'(t) = p_{AH}\lambda_A P_1(t) + p_{BH}\lambda_B P_2(t) + p_{DH}\lambda_D P_4(t) + p_{GH}\lambda_G P_7(t)$$
$$- \lambda_H(p_{HU}+p_{HF})P_8(t)$$

The subscripts on the constants refer to the module block designators A through G and U, and the module interactions are indicated by a double subscript on the corresponding constant (see figure 1-1).

These differential equations have for their solution (expressed in matrix notation)

$$P(t) = P(0)\{\exp A \cdot t\}$$

where $P(t) = (P_1(t), P_2(t), \ldots P_9(t))$, $P(0) = (1,0,0,\ldots 0)$, and

	state at t	State at t+Δt								
		U	A	B	C	D	E	F	G	H
	U	$-\Sigma_U$	$p_{UA}\lambda_U$	$p_{UB}\lambda_U$	$p_{UC}\lambda_U$	$p_{UD}\lambda_U$	0	0	$p_{UG}\lambda_G$	0
	A	0	$-\Sigma_A$	0	0	0	$p_{AE}\lambda_A$	0	0	$p_{AH}\lambda_A$
	B	0	0	$-\Sigma_B$	0	0	$p_{BE}\lambda_B$	0	0	$p_{BH}\lambda_B$
	C	0	0	0	$-\Sigma_C$	0	$p_{CE}\lambda_C$	0	0	0
A =	D	0	0	0	0	$-\Sigma_D$	$p_{DE}\lambda_D$	0	0	p_{DH} D
	E	0	0	0	0	0	$-\Sigma_E$	$p_{EF}\lambda_E$	0	0
	F	0	$p_{FA}\lambda_F$	$p_{FB}\lambda_F$	$p_{FC}\lambda_F$	$p_{FD}\lambda_F$	0	$-\Sigma_F$	0	0
	G	0	$p_{GA}\lambda_G$	$p_{GB}\lambda_G$	0	0	0	$p_{GF}\lambda_G$	$-\Sigma_G$	$p_{GH}\lambda_G$
	H	$p_{HU}\lambda_H$	0	0	0	0	0	$p_{HF}\lambda_H$	0	$-\Sigma_H$

The terms "$-\Sigma$" indicate that this element is filled with the negative of the off-diagonal terms in the respective row (e.g. $-\Sigma_A = -(p_{AE}\lambda_A + p_{AH}\lambda_A)$).

This equation is solved using a Markov model computer program and the probabilistic results are found in section 1.

Model Flexibility

Matrix size. Currently, the computer program is written to handle a state space up to 37. The size of the allowable state space is limited only by computational expenses, which increase rapidly as the size of the matrix and the number of matrix operations increase. The state space of 9 requires approximately 7 seconds of CPU time, while the example having a state space of 3 required 2 seconds.

Scale factor. Due to the expansion properties of the exponential function

$$\text{Exp } A \cdot t = I + At + \frac{(At)^2}{2!} + \dots ,$$

computational problems arise when either the values in the "A" matrix are large or when a large value of t is selected. This is because the factorial in the denominator and the power series in the numerator form a fraction which does not converge for many terms. Because the "A" matrix has the property that the diagonal elements are negative while all others are positive, it can be shown that exp $A \cdot t$ always converges, but it may not converge for a great number of terms with very large magnitude that limit the computer accuracy. For this reason a scale factor is introduced in the matrix. The equation is solved, then (exp $A \cdot t \cdot s$) or a new value of t is selected and

$$t > t \cdot s$$

It is then possible to solve the matrix equation using expansion techniques, but the output values are limited to values of t (or $t \cdot s$) which fall in the progression $t = 1,2,4,\dots,2^N$ ($t \cdot s = 1,2,4,\dots,2^N$). If $s = 1.0$ then the values of t for which the matrix equation is solved are $1,2,4,\dots,2^N$ while if $s = 0.10$ the values of t for which the equation is solved are $0.1,0.2,0.4,0.8,1.6,\dots$. If the two runs of the matrix equation are combined, the resulting values are for

$$t = 0.1,0.2,0.4,0.8,1.0,1.6,2.0,3.2,4.0,\dots .$$

Other intermediate values can be found by proper selection of the scale factor. Thus, even though for computational reasons only certain values of the equation are available for a given computer run, other values can be found by selecting the proper scale factors.

Constant matrix shifting in time. One of the most valuable features of the model from a flexibility standpoint is its ability to handle a multiple of matrix element changes in time.

Due to the Poisson assumptions defined earlier, the "A" matrix must be composed of elements which are constants. If, in fact, the process is not Poisson, the elements in the matrix and also in the Chapman-Kolmogorov equation (Eq.(9)) are not constants. To solve this problem, Gantmacher [4] has suggested that the process be examined as "piece-wise" continuous, and that over certain Δt's the matrix be taken as a constant, adjusting the matrix elements at the end of each Δt. This method, of course, approaches the true solution as $\lim \Delta t \to 0$, but the computations with a digital computer become long and tedious.

If there are fairly regular trends in the data as they are recorded, the size of the Δt's selected can be manageable and the computational time needed to solve the matrix exponential equations is then tenable.

In the case of the CPM data, from both the plots and the adjusted data, certain trends and changes were visible at 21 points. These time periods were isolated and recorded for each of the modules measured, but the attempt to align them at a starting time was not possible due to the CPM recording procedure used. Had the data been chronologically recorded from each module at the same time, the multiple matrix shift method could have been applied and, if new data are gathered, this technique and model are available for application to the study.

<u>Multiple active elements</u>. Another feature of the Markov model is that it will handle a situation where there are multiple active modules at any given time. In the 1108 system this implies that there are more than one CPU present simultaneously acting on the system. Take the case of a simple three-module example (figure 4-1): the state space might, with multiple-state execution capability, be extended from

$$\begin{matrix} \bar{A}BC \\ A\bar{B}C \\ AB\bar{C} \end{matrix} \qquad \text{to} \qquad \begin{matrix} \bar{A}BC \\ A\bar{B}C \\ AB\bar{C} \\ \overline{AB}C \\ A\overline{BC} \\ \bar{A}B\bar{C} \end{matrix}$$

Certainly, even with a single CPU, multiple events could be modeled, such as the input/output channels which move block data with very little interconnection with the CPU. A benefit from a model of this would be more rapid throughput if it were found that the modules could be re-arranged or modified to enable less waiting time between operations.

The difference and differential equations used to solve the multiple-active-state problem are determined and solved in the same manner as the single-state problem. In the case of multiple states with equal probabilities, certain steps can be taken to group the states, thereby reducing the state space and computation time [8].

REFERENCES

1. The advanced theory of statistics, Kendall and Stuart, 3d edition, volumes 1, 2 and 3, Hafner 1967, 1968, 1969.
2. Stochastic processes, Parzen, Holden-Day, 1965.
3. An introduction to probability theory and its application, Feller, volume 1, 2d edition, Wiley, 1960.
4. Matrix theory, Gantmacher, volume 2, Chelsea, 1960.
5. Elements of the theory of Markov processes and their applications, Bharucha-Reid, McGraw-Hill, 1960.
6. Introduction to probability and statistics, Lindgren, Macmillan, 1957.
7. Stochastic processes, Takacs, Methuen, 1962.
8. "A method of predicting and apportioning systems effectiveness estimators", J. Feeley, 10th Annual West Coast Reliability Symposium, p. 135, 1969.
9. "Computer program monitor program summary report (CPM1) and counter profile report (CPM6)", Computer Sciences Corp., M. German and D. James, 1970.
10. Dynamic programming and Markov processes, Howard, MIT Press, 1966.

TABLE 1-1. Summary of Markov Run Real-Time Environment

Time x 10^{-5}	Module Probability User Program	I/O	ADH	COME	DH	DA	DISP	IH	OUTPUT
1	0.95	0.011	0.0044	0.0	0.00059	0.000017	0.0042	0.21	0.0047
2	0.90	0.021	0.0088	0.0	0.00077	0.000059	0.016	0.039	0.013
4	0.81	0.035	0.016	0.0	0.00079	0.00018	0.060	0.045	0.028
8	0.67	0.044	0.023	0.0	0.00070	0.00038	0.18	0.040	0.039
16	0.47	0.038	0.024	0.0	0.00058	0.00047	0.41	0.028	0.034
32	0.23	0.025	0.016	0.0	0.00045	0.00033	0.69	0.014	0.021
64	0.065	0.016	0.0098	0.0	0.00037	0.00020	0.89	0.0038	0.011
128	0.019	0.013	0.0080	0.0	0.00035	0.00016	0.95	0.0011	0.0087
256	0.017	0.013	0.0079	0.0	0.00034	0.00016	0.95	0.00094	0.0086
512	0.017	0.013	0.0079	0.0	0.00034	0.00016	0.95	0.00094	0.0086
1024	0.017	0.013	0.0079	0.0	0.00034	0.00016	0.95	0.00094	0.0086
2048	0.017	0.013	0.0079	0.0	0.00034	0.00016	0.95	0.00094	0.0086
10	0.61	0.044	0.024	0.0	0.00066	0.00044	0.24	0.037	0.039
20	0.39	0.034	0.022	0.0	0.00054	0.00044	0.50	0.023	0.030
40	0.16	0.021	0.014	0.0	0.00042	0.00028	0.77	0.0098	0.017
80	0.039	0.014	0.0088	0.0	0.00036	0.00018	0.92	0.0023	0.00998
160	0.017	0.013	0.0079	0.0	0.00034	0.00016	0.95	0.00097	0.0086
320	0.017	0.013	0.0079	0.0	0.00034	0.00016	0.95	0.00094	0.0086
640	0.017	0.013	0.0079	0.0	0.00034	0.00016	0.95	0.00094	0.0086

TABLE 1-2. Summary of Markov Run Real Time Plus Benchmarks Environment

Time x 10^{-5}	Module Probability User Program	I/O	ADH	COME	DH	DA	DISP	IH	OUTPUT
1	0.99	0.0022	0.00037	0.21–4	0.00023	0.43–4	0.00092	0.0073	0.0012
2	0.97	0.0046	0.00080	0.40–4	0.00035	0.00016	0.0032	0.011	0.0042
4	0.95	0.0085	0.0016	0.75–4	0.00046	0.00055	0.010	0.014	0.013
8	0.91	0.013	0.0027	0.00013	0.00053	0.0014	0.031	0.015	0.030
16	0.82	0.016	0.0037	0.00023	0.00068	0.0026	0.086	0.013	0.053
32	0.69	0.019	0.0047	0.00041	0.0010	0.0036	0.20	0.011	0.071
64	0.49	0.025	0.0062	0.00070	0.0016	0.0050	0.38	0.0080	0.086
128	0.28	0.031	0.0078	0.0010	0.0022	0.0066	0.57	0.0045	0.10
256	0.15	0.035	0.0087	0.0012	0.0025	0.0075	0.68	0.0024	0.11
512	0.13	0.036	0.0089	0.0013	0.0026	0.0077	0.70	0.0020	0.11
1024	0.12	0.036	0.0089	0.0013	0.0026	0.0077	0.70	0.0020	0.11
2048	0.12	0.036	0.0089	0.0013	0.0026	0.0077	0.70	0.0020	0.11
10	0.88	0.014	0.0030	0.00016	0.00056	0.0018	0.044	0.014	0.037
20	0.79	0.016	0.0040	0.00027	0.00077	0.0029	0.011	0.013	0.060
40	0.63	0.021	0.0051	0.00049	0.0012	0.0040	0.025	0.010	0.076
80	0.42	0.027	0.0067	0.00081	0.0018	0.0055	0.44	0.0068	0.091
160	0.22	0.033	0.0082	0.0011	0.0023	0.0069	0.61	0.0036	0.10
320	0.14	0.035	0.0089	0.0013	0.0026	0.0076	0.69	0.0022	0.11
640	0.12	0.036	0.0089	0.0013	0.0026	0.0077	0.70	0.0020	0.11
1280	0.12	0.036	0.0089	0.0013	0.0026	0.0077	0.70	0.0020	0.11

TABLE 1-3. Summary of Markov Run Benchmarks Environment

Time x 10^{-5}	Module Probability User Program	I/O	ADH	COME	DH	DA	DISP	IH	OUTPUT
1	0.99	0.0039	0.00050	0.47-4	0.00038	0.29-4	0.00034	0.0027	0.0010
2	0.98	0.0070	0.00094	0.91-4	0.00059	0.00010	0.0013	0.0042	0.0035
4	0.96	0.011	0.0016	0.00017	0.00078	0.00030	0.0047	0.0056	0.011
8	0.93	0.015	0.0025	0.00029	0.00087	0.00070	0.017	0.0061	0.026
16	0.87	0.018	0.0034	0.00046	0.00099	0.0012	0.052	0.0057	0.048
32	0.76	0.023	0.0044	0.00071	0.0013	0.0017	0.13	0.0050	0.070
64	0.59	0.031	0.0061	0.0011	0.0018	0.0024	0.26	0.0039	0.097
128	0.39	0.042	0.00082	0.0015	0.0024	0.0034	0.42	0.0026	0.13
256	0.23	0.049	0.0098	0.0019	0.0029	0.0041	0.54	0.0015	0.15
512	0.18	0.052	0.010	0.0020	0.0031	0.0043	0.58	0.0012	0.16
1024	0.18	0.052	0.010	0.0020	0.0031	0.0043	0.58	0.0012	0.16
2048	0.18	0.052	0.010	0.0020	0.0031	0.0043	0.58	0.0012	0.16
10	0.91	0.016	0.0028	0.00034	0.00090	0.00086	0.024	0.0060	0.033
20	0.84	0.019	0.0037	0.00053	0.0010	0.0013	0.072	0.0055	0.055
40	0.71	0.025	0.0049	0.00082	0.0014	0.0019	0.017	0.0047	0.078
80	0.53	0.035	0.0068	0.0012	0.0020	0.0027	0.31	0.0035	0.11
160	0.33	0.045	0.0088	0.0017	0.0026	0.0036	0.46	0.0022	0.14
320	0.21	0.051	0.010	0.0019	0.0030	0.0042	0.56	0.0013	0.16
640	0.18	0.052	0.010	0.0020	0.0031	0.0043	0.58	0.0012	0.16
1280	0.18	0.052	0.010	0.0020	0.0031	0.0043	0.58	0.0012	0.16

TABLE 1-4. Summary of Markov Run Standard Environment

Time x 10^{-5}	Module Probability User Program	I/O	ADH	COME	DH	DA	DISP	IH	OUTPUT
1	0.98	0.0013	0.00063	0.88-4	0.00054	0.38-4	0.0039	0.010	0.0042
2	0.96	0.0027	0.0013	0.00017	0.00076	0.0013	0.012	0.013	0.012
4	0.92	0.0052	0.0024	0.00032	0.00092	0.00038	0.031	0.013	0.027
8	0.85	0.0084	0.0039	0.00060	0.0011	0.00098	0.074	0.012	0.051
16	0.72	0.012	0.0053	0.0011	0.0014	0.0020	0.17	0.010	0.076
32	0.54	0.015	0.0070	0.0020	0.0021	0.0035	0.33	0.0078	0.087
64	0.32	0.020	0.0092	0.0032	0.0030	0.0054	0.55	0.0046	0.088
128	0.15	0.024	0.011	0.0043	0.0036	0.0070	0.71	0.0021	0.087
256	0.093	0.025	0.011	0.0046	0.0038	0.0075	0.76	0.0013	0.087
512	0.090	0.025	0.011	0.0046	0.0039	0.0075	0.77	0.0013	0.087
1024	0.090	0.025	0.011	0.0046	0.0039	0.0075	0.77	0.0013	0.087
2048	0.090	0.025	0.011	0.0046	0.0039	0.0075	0.77	0.0013	0.087
10	0.81	0.0094	0.0043	0.0072	0.0012	0.0013	0.097	0.012	0.060
20	0.67	0.0013	0.0058	0.0013	0.0016	0.0025	0.21	0.0097	0.081
40	0.47	0.017	0.0077	0.0023	0.0024	0.0041	0.40	0.0068	0.088
80	0.25	0.022	0.0098	0.0036	0.0032	0.0060	0.61	0.0036	0.088
160	0.12	0.025	0.011	0.0045	0.0037	0.0023	0.74	0.0017	0.087
320	0.091	0.025	0.011	0.046	0.0039	0.0075	0.77	0.0013	0.087
640	0.090	0.025	0.011	0.046	0.0038	0.0075	0.77	0.0013	0.087
1280	0.090	0.025	0.011	0.046	0.0039	0.0075	0.77	0.0013	0.087

TABLE 1-5. Steady-State Probabilities

User Program (%)	I/O (%)	ADH (%)	COME (%)	DH (%)	DA (%)	DISP (%)	I/H (%)	Output (%)
Real-Time Environment — Convergence Time 2,560 microseconds								
1.7	1.3	0.79	0.0	0.034	0.016	95	0.094	0.36
Real Time + Benchmarks — Convergence Time 10,240 microseconds								
12	3.6	0.89	0.13	0.26	0.77	70	0.20	11
Benchmarks — Convergence Time 5,120 microseconds								
18	5.2	1.0	2.0	3.1	4.3	58	1.2	16
Standard — Convergence Time 5,120 microseconds								
9.0	2.5	1.1	0.46	0.39	0.75	77	0.13	8.7

TABLE 1-6. Ordered Module Intervals and Corresponding Server Time/Second and In Range Rate/Second for a Single Cycle of the Simulation Program

Δt	UP	IO	ADH	DH	DA	DISP	IH	OUT
0-15	0.0075 1,619.718	0.0205 648.80	0.019069 368.425	0.0000845 10.775	0.00005257 1.75	0.9385 307.45	0.0001635 20.207	
15-20						0.9910 95.35		
20-30					0.00007726 2.3			
30-40				0.001674 226.945				
40-45				0.0013778 180.113			0.0005740 39.643	
45-50						0.9355 717.20		
50-55			0.03756 732.05		0.00006179 1.9			
55-75						0.9305 673.65		
75-80		0.0722 2,300.65						
80-100					0.00007667 2.25			
100-110				0.0015514 209.48			0.001279 84.269	
110-120			0.038515 750.365					
120-125					0.00005273 1.75			
125-135						0.9625 682.845		
135-140		0.0615 1,960.90						
140-150				0.0010 134.125	0.00008716 2.367		0.002835 202.895	
150-155			0.03078 593.15					
155-160						0.7395 2,989.05		
160-170							0.0024733 124.093	
170-175				0.0016035 63.885	0.000552 14.30			
175-180		0.04905 1,544.20						

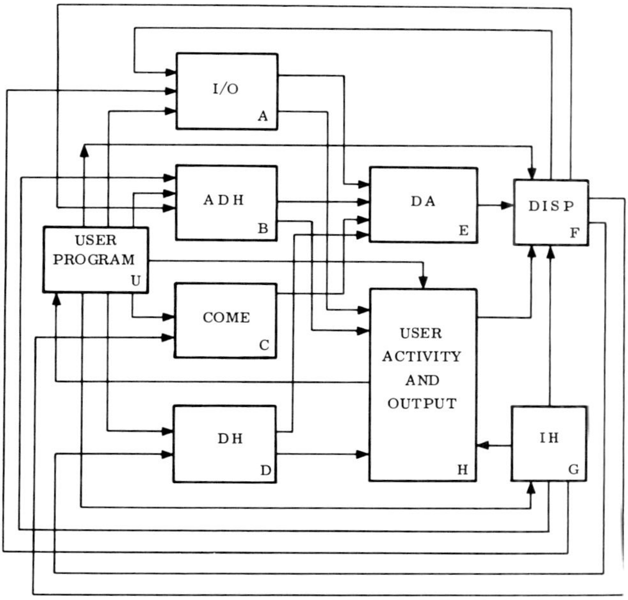

Fig. 1-1. System block diagram.

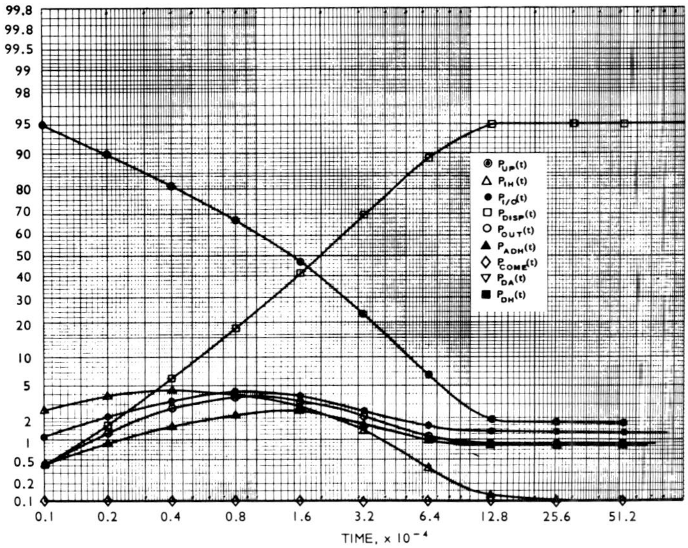

Fig. 1-2. Test 6 real time.

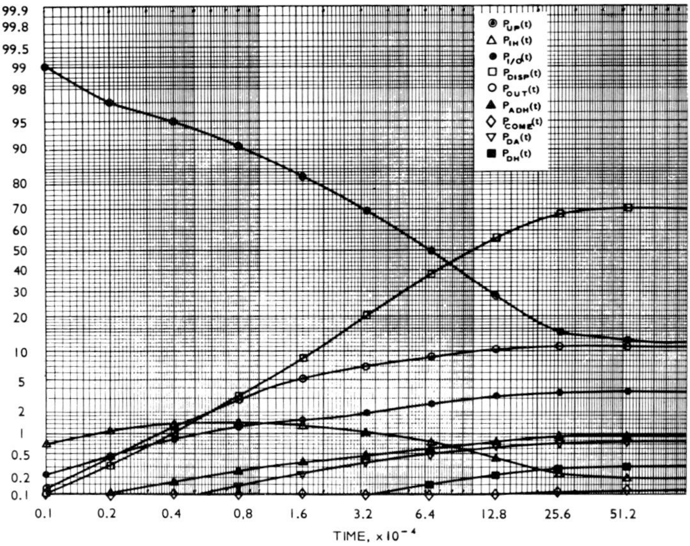

Fig. 1-3. Test 6 real time plus benchmarks.

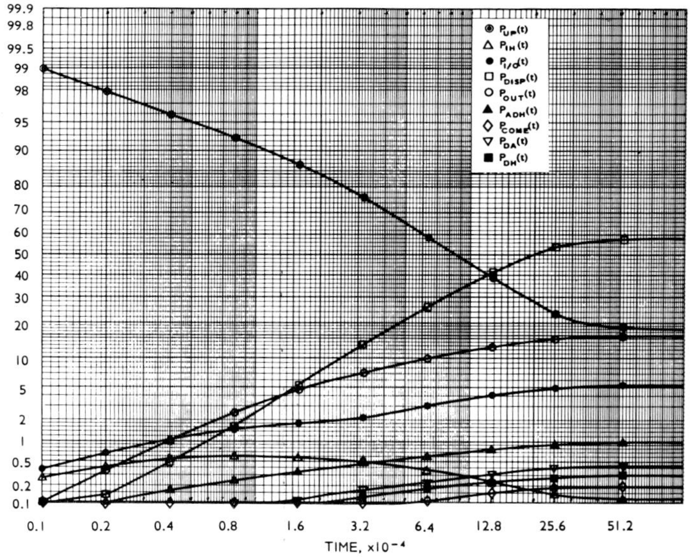

Fig. 1-4. Test 6 benchmarks.

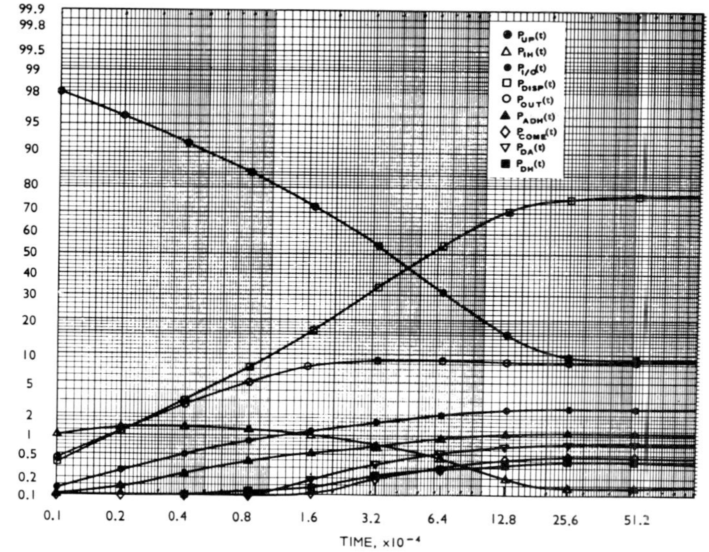

Fig. 1-5. Test 6 standard.

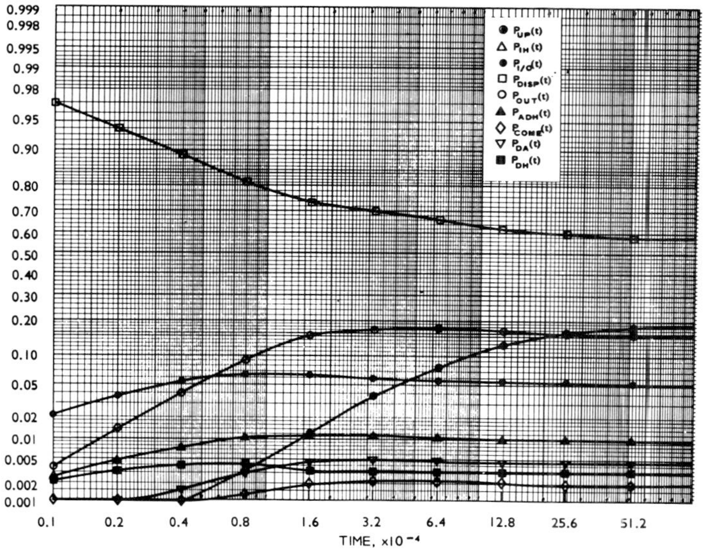

Fig. 1-6. Test 6 benchmarks.

TABLE 3-1. CPM Measured Test Data

Test 1 Channel Connect Test

Real-Time Environment (Tape No. 3179)	Mean Recorded Time/Sec	Real-Time Environment (Tape No. 4147)	Mean Recorded Time/Sec
Drum Controller	0.0689	Drum Controller	0.0041
Analog/Digital	0.0013	Analog/Digital	0.0004
Digital/Analog	0.0006	Digital/Analog	0.0004
FASTRAND II	0.4948	FASTRAND II	0.0373
CATS	0.0	CATS	0.0
VIII C Control	0.1346	VIII C Control	0.0
CTMC	0.9999	CTMC	0.9999
Console	0.9994	Console	0.9999
Benchmark Environment (Tape No. 4134)	**Mean Recorded Time/Sec**	**Standard Environment (Tape No. 3633)**	**Mean Recorded Time/Sec**
Drum Controller	0.1818	Drum Controller	0.0372
Analog/Digital	0.0	Analog/Digital	0.0
Digital/Analog	0.0	Digital/Analog	0.0
FASTRAND II	0.9056	FASTRAND II	0.9214
CATS	0.0	CATS	0.0
VIII C Control	0.6400	VIII C Control	0.0054
CTMC	0.9999	CTMC	0.9999
Console	0.9999	Console	0.9997
Real-Time Plus Benchmarks (Tape No. 4083)	**Mean Recorded Time/Sec**		
Drum Controller	0.1427		
Analog/Digital	0.0006		
Digital/Analog	0.0004		
FASTRAND II	0.5671		
CATS	0.0		
VIII C Control	0.2488		
CTMC	0.9999		
Console	0.9989		

TABLE 3-1. CPM Measured Test Data (Cont'd)

Test 2 EXEC 8 Utilization	
Real-Time Mean Value	= 98.3261% (Test Tape No. 1302)
Real-Time Plus Benchmarks Mean Value	= 85.5828% (Test Tape No. 3965)
Benchmarks Alone Mean Value	= 7.9193% (Test Tape No. 4005)
Standard Environment Mean Value	= 79.3173% (Test Tape No. 721
Standard Environment (Not Guard Mode Plus EXEC 8) Mean Value	= 10.9286% (Test Tape No. 3847)

Test 3 Software Interrupt Timing	
Environment	Mean Time Interrupts are Prevented from Execution
Standard (Tape No. 3599)	0.1618 — 03/sec
Standard (Tape No. 3950)	0.2141 — 02/sec
Benchmark (Tape No. 4160)	0.22758 — 02/sec
Real Time Plus Benchmarks (Tape No. 108)	0.38424 — 02/sec
Standard (Tape No. 1610)	0.19463 — 02/sec

Test 4 Instruction Execution Rate/B Register Conflict	Mean Events/Sec
Benchmark Plus Real-Time Environment (Tape No. 840)	
Instruction Execution Events	1,331,755
B-Register Conflict/System Program	20,466
B-Register Conflict/User Program	1,538
Guard Mode Faults	0.001639
Benchmark (Tape No. 4123)	
Instruction Execution Events	1,401,056
B-Register Conflict/System Program	15,840
B-Register Conflict/User Program	1,153
Guard Mode Faults	0.0

TABLE 3-1. CPM Measured Test Data (Cont'd)

Test 4 Instruction Execution Rate/B Register Conflict (Cont'd)	
	Mean Events/Sec
Standard (Tape No. 2353)	
Instruction Execution Events	1,304,625
B-Register Conflict/System Program	18,288
B-Register Conflict/User Program	2,619
Guard Mode Faults	0.001668
Standard Environment (Tape No. 2908)	
Instruction Execution Events	1,188,510
B-Register Conflict/System Program	14,097
B-Register Conflict/User Program	1,351
Guard Mode Faults	0.005838
Standard Environment (Tape No. 4108)	
Instruction Execution Events	824,937
B-Register Conflict/System Program	8,403
B-Register Conflict/User Program	2,787
Guard Mode Faults	0.0
Benchmark (Tape No. 4198)	
Instruction Execution Events	1,332,016
B-Register Conflict/System Program	18,597
B-Register Conflict/User Program	1,758
Guard Mode Faults	0.001196
Real-Time (Tape No. 1375)	
Instruction Execution Events	1,503,019
B-Register Conflict/System Program	14,960
B-Register Conflict/User Program	156.6
Guard Mode Faults	0.0

Test 5 Real-Time Module Operation Test	
	Mean Recorded Time/Sec
Real-Time Plus Benchmarks Environment (Tape No. 3463)	
Real-Time Clock Interrupt Counter	0.2936
ISI External Interrupt	0.2223
ISI Input Monitor	0.1454
ISI Output Monitor	8.5440

TABLE 3-1. CPM Measured Test Data (Cont'd)

Test 5 Real-Time Module Operation Test (Cont'd)	Mean Recorded Time/Sec
Function Monitor*	0.0
ESI Output Monitor*	0.0
ESI Input Monitor*	0.0
Time Within ADH	0.01599
ISI Channel Plus External Interrupt	0.08458-07
Real-Time (Tape No. 1462)	
Real-Time Clock Interrupt Counter	0.2019
ISI External Interrupt*	0.0
ISI Input Monitor	0.04429
ISI Output Monitor	32.262
Function Monitor*	0.0
ESI Output Monitor*	0.0
ESI Input Monitor*	0.0
Time Within ADH	0.05089
ISI Channel Plus External Interrupt	0.3249-07
Real-Time Environment No. 1 (Tape No. 4131)	
Real-Time Clock Interrupt Count	0.4270
External Interrupt (ISI)	0.2365
Input Monitor (ISI)	0.1379
Output Monitor (ISI)	14.8190
Function Monitor*	0.0
Output Monitor (ESI)*	0.0
Input Monitor (ESI)*	0.0
Time Within ADH	0.011211
Channel 4 External Interrupt (ISI)	0.0841

Test 6 Module Execution Events and Timing	Per Second
Real-Time Environment No. 1 (Tape No. 3879)	
Interrupt Handler (Module No. 1)	
Cumulative Time in Module	0.0009
Cumulative Events in Module	60.6213

*Not recorded.

TABLE 3-1. CPM Measured Test Data (Cont'd)

Test 6 Module Execution Events and Timing (Cont'd)	Per Second
Arbitrary Device Handler (Module No. 2)	
Cumulative Time in Module	0.0082
Cumulative Events in Module	158.9045
Drum Handler (Module No. 3)	
Cumulative Time in Module	0.0003
Cumulative Events in Module	40.3307
Dynamic Allocator (Module No. 4)	
Cumulative Time in Module	0.0001
Cumulative Events in Module	4.7797
Dispatcher (Module No. 5)	
Cumulative Time in Module	0.9499
Cumulative Events in Module	542.3999
Input/Output (Module No. 6)	
Cumulative Time in Module	0.0131
Cumulative Events in Module	414.1569
Benchmark Plus Real-Time Environment No. 2	
Module No. 1 Interrupt Handler (Tape No. 1381)	
Cumulative Time in Module	0.0023
Cumulative Events in Module	140.3449
Module No. 2 Arbitrary Device Handler (Tape No. 3854)	
Cumulative Time in Module	0.0084
Cumulative Events in Module	162.7852
Module No. 2 Arbitrary Device Handler (Tape No. 1381)	
Cumulative Time in Module	0.0076
Cumulative Events in Module	145.7729
Module No. 3 Drum Handler (Tape No. 1381)	
Cumulative Time in Module	0.0023
Cumulative Events in Module	146.4145
Module No. 4 Dynamic Allocator (Tape No. 1381)	
Cumulative Time in Module	0.0072
Cumulative Events in Module	145.1533

TABLE 3-1. CPM Measured Test Data (Cont'd)

Test 6 Module Execution Events and Timing (Cont'd)	
	Per Second
Module No. 5 Dispatcher (Tape No. 1381)	
Cumulative Time in Module	0.6770
Cumulative Events in Module	1,308.9721
Module No. 6 Input/Output (Tape No. 1381)	
Cumulative Time in Module	0.0309
Cumulative Events in Module	920.4343
Benchmark Environment No. 3	
Module No. 4 Dynamic Allocator (Tape No. 3777)	
Cumulative Time in Module	0.0036
Cumulative Events in Module	74.8051
Module No. 1 Interrupt Handler (Tape No. 4141)	
Cumulative Time in Module	0.0013
Cumulative Events in Module	69.0908
Module No. 2 Arbitrary Device Handler (Tape No. 3775)	
Cumulative Time in Module	0.0061
Cumulative Events in Module	109.4068
Module No. 5 Dispatcher (Tape No. 4110)	
Cumulative Time in Module	0.5880
Cumulative Events in Module	1,832.0674
Standard Environment No. 4	
Module No. 1 Interrupt Handler (Tape No. 401)	
Cumulative Time in Module	0.0009
Cumulative Events in Module	39.6138
Module No. 2 Arbitrary Device Handler	
Cumulative Time in Module	0.0029
Cumulative Events in Module	54.7442
Module No. 3 Drum Handler (Tape No. 4140) (177 Observations)	
Cumulative Time in Module	0.0108
Cumulative Events in Module	108.6327

TABLE 3-1. CPM Measured Test Data (Cont'd)

Test 6 Module Execution Events and Timing (Cont'd)	
	Per Second
Module No. 3 Drum Handler (Tape No. 4171)	
(179 Observations	
Cumulative Time in Module	0.0027
Cumulative Events in Module	92.1636
Module No. 4 Dynamic Allocator (Tape No. 839)	
Cumulative Time in Module	0.0050
Cumulative Events in Module	94.7150
Module No. 4 Dynamic Allocator	
Cumulative Time in Module	0.0062
Cumulative Events in Module	110.2767
Module No. 5 Dispatcher (Tape No. 4043)	
Cumulative Time in Module	0.7724
Cumulative Events in Module	1,038.4131
Module No. 6 Input/Output (Tape No. 3988)	
Cumulative Time in Module	0.0049
Cumulative Events in Module	97.0644
Module No. 5 Dispatcher (Tape No. 3995)	
Cumulative Time in Module	0.5220
Cumulative Events in Module	442.3917
Module No. 1 Interrupt Handler (Tape No. 4095)	
Cumulative Time in Module	0.0019
Cumulative Events in Module	104.8765

Test 8 Module Execution Events and Timing	
	Per Second
Standard Environment No. 4 (Tape No. 4162)	
COME (Module No. 7)	
Cumulative Time in Module	0.0012
Cumulative Events in Module	10.20

TABLE 3-1. CPM Measured Test Data (Cont'd)

Test 9 Hardware Interrupt Timing	
	Mean Recorded Time/Sec
Standard Environment (Tape No. 796)	
Cumulative Time Within Interrupt	0.000125
Benchmark Environment (Tape No. 4294)	
Cumulative Time Within Interrupt	0.0001121

Test 11 Real-Time Clock Interrupt Timing Test	
	Mean Recorded Time/Sec
Standard Environment (Tape No. 4304)	
Cumulative Time Between RTC Interrupt	0.012300
Cumulative Time Between RTC Interrupt	0.013037
Cumulative Time Between RTC Interrupt	0.013204
Cumulative Time Between RTC Interrupt	0.013849
Number of Real-Time Clock Interrupts	0.61636
Benchmark Environment (Tape No. 4304)	
Cumulative Time Between RTC Interrupt	0.012300
Cumulative Time Between RTC Interrupt	0.013037
Cumulative Time Between RTC Interrupt	0.013204
Cumulative Time Between RTC Interrupt	0.013849
Cumulative Time Between RTC Interrupt	0.014500

Test 12 EXEC 8 Utilization Test (No Exec Control Register)	
Standard Environment (Tape No. 1529)	
Time in EXEC 8	Mean = 0.9512
Standard Environment (Tape No. 4064)	
Time in EXEC 8	Mean = 0.8878

TABLE 3-2.

Module Name	Real Time		Real Time Plus Benchmarks		Benchmark		Standard	
	Events/Sec		Events/Sec		Events/Sec		Events/Sec	
	Window	Total	Window	Total	Window	Total	Window	Total
IH	60.62	62.11	140.34	142.51	86.36	69.78	131.10	212.73
ADH	158.90	159.87	145.77	146.25	136.76	109.67	54.74	---
DH	40.33	40.35	146.42	147.65	87.24	104.06	92.16	92.60
DA	3.82	4.45	145.15	155.69	74.81	75.35	94.71	94.51
DISP	458.24	550.98	1,309.00	1,312.92	1,832.07	1,847.39	1,038.41	1,038.67
I/O	365.04	419.20	920.43	915.83	915.83*	915.83*	97.07	133.47
COME	0.0	0.0	---	10.32	---	10.32	---	10.32

Module Name	Real Time		Real Time Plus Benchmarks		Benchmarks		Standard	
	Operation Time/Sec		Operation Time/Sec		Operation Time/Sec		Operation Time/Sec	
	Window	Total	Window	Total	Window	Total	Window	Total
IH	0.0009	0.93647-3	0.0023	0.23258-2	0.0013	0.12714-2	0.0019	0.15676-2
ADH	0.0082	0.81904-2	0.0076	0.76677-2	0.0061	0.61153-2	0.0029	0.28617-2
DH	0.0003	0.36660-3	0.0023	0.23477-2	---	0.18294-2	0.0027	0.10112-2
DA	0.0001	0.16468-3	0.0072	0.77481-2	0.0036	0.36311-2	0.0050	0.59692-2
DISP	0.9499	0.95178	0.6770	0.68852	0.5880	0.59057	0.7724	0.77532
I/O	0.0131	0.13284-1	0.0309	0.30701-1	---	0.30701-1	0.0049	0.63997-2
COME	---	0.0	---	0.12151-2	---	0.12151-2	0.0012	0.12151-2

*Estimate

TABLE 3-3A. Test 2/Test 6 Data Relationship

		Real Time	Real Time Plus Benchmarks	Benchmarks	Standard
Test 2		0.9833	0.8558	0.8000*	0.8878**
Test 6	IH	0.00093647	0.00232579	0.00127137	0.00156758
	ADH	0.00819038	0.00766766	0.00611530	0.00286170
	DH	0.00036660	0.00234767	0.00182942	0.00101121
	DA	0.00016468	0.00774809	0.00363107	0.00596924
	DISP	0.95177694	0.68851806	0.59057307	0.7753225
	I/O	0.01328392	0.03070174	0.03070174	0.00639970
	COME	0.0	0.00121508*	0.00121508*	0.00121508
Total		0.97471899	0.74052409	0.63533729	0.79434701

*Estimate

**Results from Test 12

TABLE 3-3B.

	Real Time	Real Time Plus Benchmarks	Benchmarks	Standard
All EXEC 8 Modules	0.9833	0.8558	0.8000*	0.8878**
USER Program	0.0167	0.1442	0.2000*	0.1122**
EXEC 8 Modules	0.9747	0.7405	0.6353	0.7943
OUTPUT Modules	0.0086	0.1153	0.1647	0.0935
Total	1.0000	1.0000	1.0000	1.0000

*Estimate

**Based on data from Test 12

TABLE 3-4. TRACE-EDIT Program Summary

Program Name	Total ER Time (Sec)	Number of ERs Used by Edit	Total No. ERs	Mean Time/ER (Ticks)	Total Time Estimate for all ERs	Total* CPU Time
AERDYN	1.3192	937	1,050	7.04	1.4784	3.0880
BEEF	0.0032	3	20	5.33	0.02132	1.2160
DOCPRO	0.2928	177	275	8.27	0.4549	0.4590
ECAP	46.3620	16,101	31,702	14.40	91.3218	76.0000
FFRM	2.5936	1,104	2,162	11.75	5.0807	5.0820
GRKSYM	0.0024	3	42	4.0	0.0336	28.4810
MATHPK	0.3724	198	217	9.40	0.4080	7.2820
MIMIC	2.3114	605	855	19.10	3.2629	12.8960
PIN	66.1894	250,800	251,799	1.32	66.4759	20.1960
POLY	6.8530	5,325	11,275	6.43	14.4996	18.0000
STATPK	0.2602	171	225	6.81	0.3065	2.6750
Totals	126.5596	275,444	299,622	93.85	181.141	175.375
REAL TIME SIMULATION	44.2688	168	122	--	--	--

*This column of figures was found by averaging the CPU times from 3 runs on each Benchmark program.

TABLE 3-5. Edit of Real Time Simulation Program

ER Summary for Real Time Simulation Program

ER	Number Occur	Number Used	CPU Time		Wall Clock		EXEC Module
			Mean	S.D.	Mean	S.D.	
EXIT$	6	0	0.00	0.00	0.00	0.00	TERM
FORK$	5	3	5.00	1.00	0.00	0.00	DISP
PRINT$	2	2	5.00	2.83	0.00	0.00	PRINT$
IOARB$	2	2	5.50	0.71	0.00	0.00	IO
DATE$	2	2	1.50	0.71	0.00	0.00	TIME
TIME$	1	1	1.00	0.00	0.00	0.00	TIME
TWAIT$	393	154	1,436.97	705.29	0.00	0.00	DISP
RT$	5	4	2.50	0.58	0.00	0.00	EXERR
NAME$	2	0	0.00	0.00	0.00	0.00	NAME
ACT$	2	0	0.00	0.00	0.00	0.00	DISP
DACT$	2	0	0.00	0.00	0.00	0.00	DISP

Time	Total No. ER, 5	No. Used	Total ER Time	Mean	S.D.
CPU	422	168	00:44.2658	1,317.52	783.34
WALL	422	416	00:00.0000	0.00	0.00

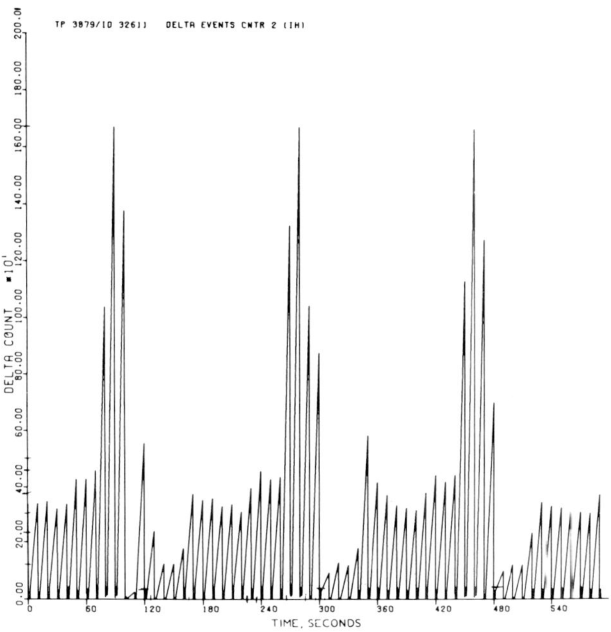

Fig. 3-1. Interrupt handler.

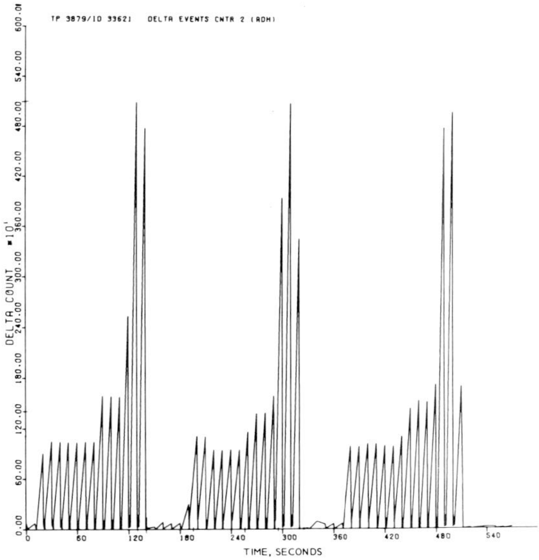

Fig. 3-2. Arbirary device handler.

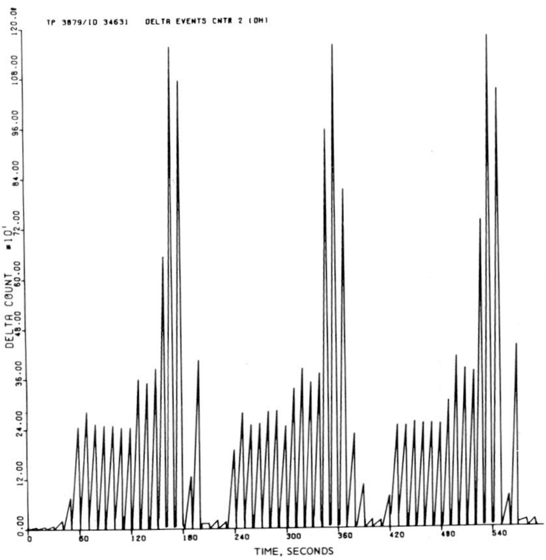

Fig. 3-3. Device handler.

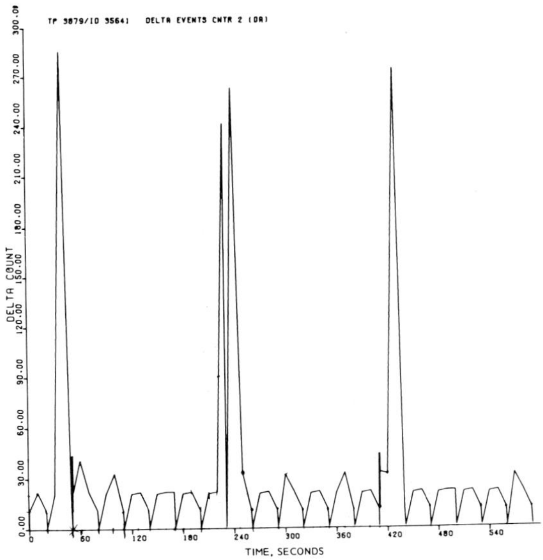

Fig. 3-4. Dynamic allocator.

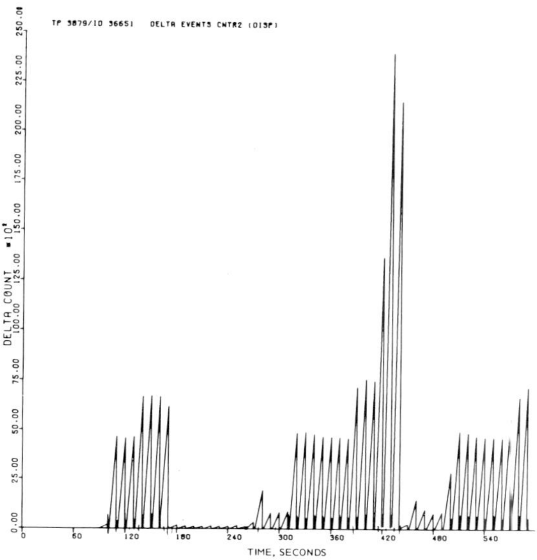

Fig. 3-5. Dispatcher.

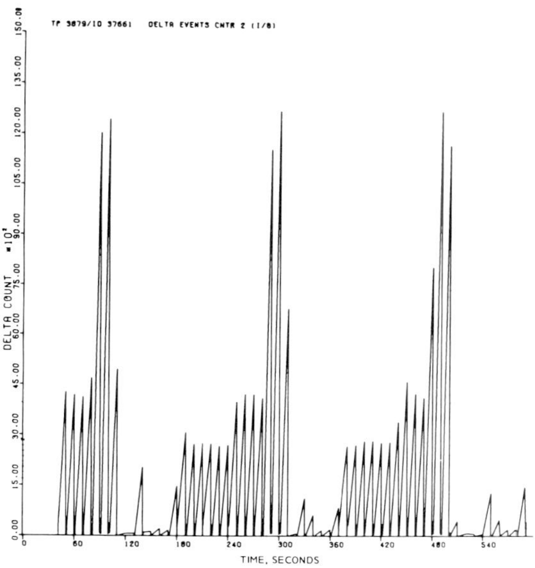

Fig. 3-6. Input/output.

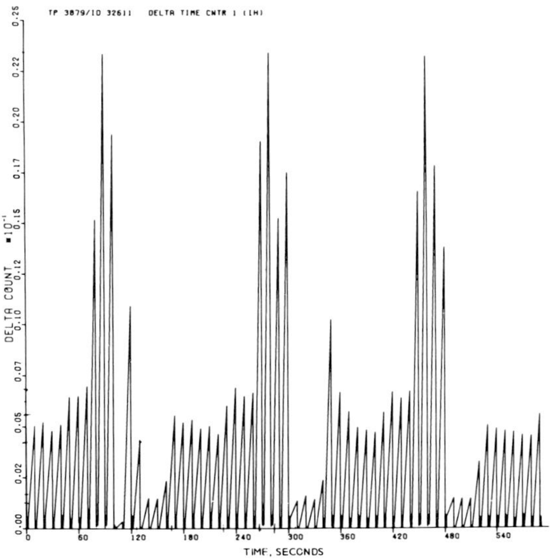

Fig. 3-7. Interrupt handler.

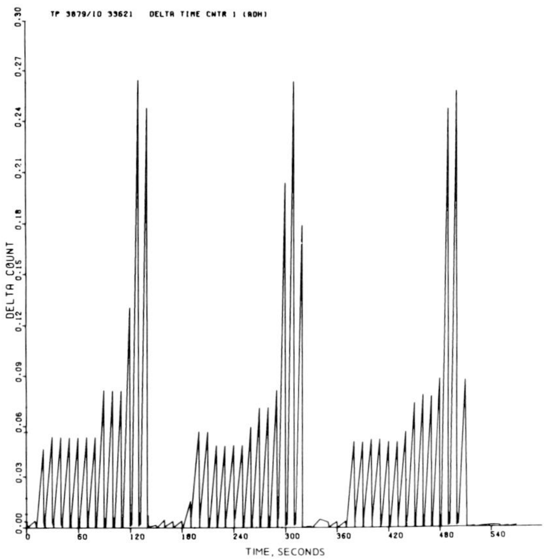

Fig. 3-8. Arbitrary device handler.

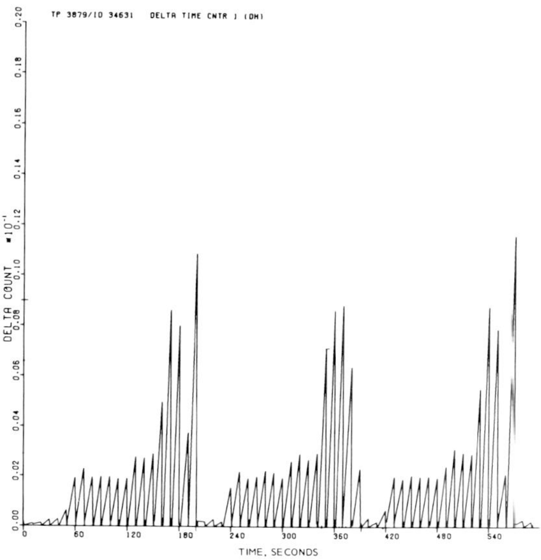

Fig. 3-9. Drum handler.

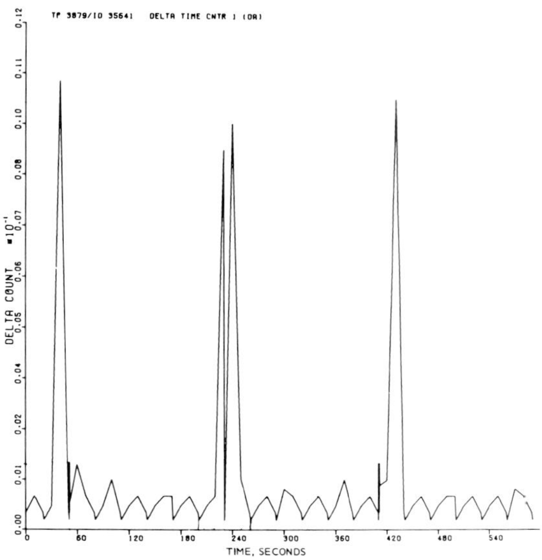

Fig. 3-10. Dynamic allocator

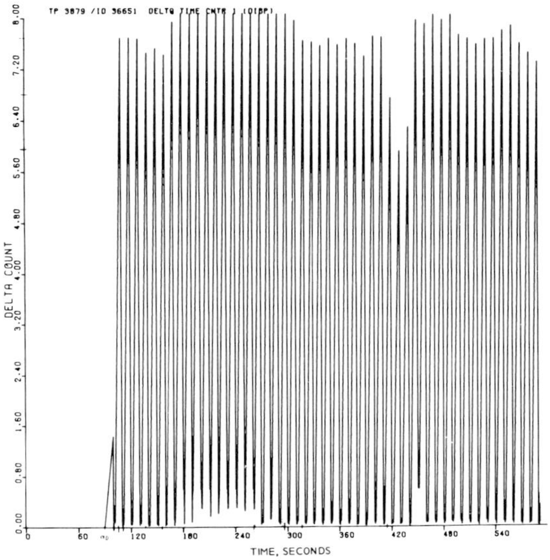

Fig. 3-11. Dispatcher.

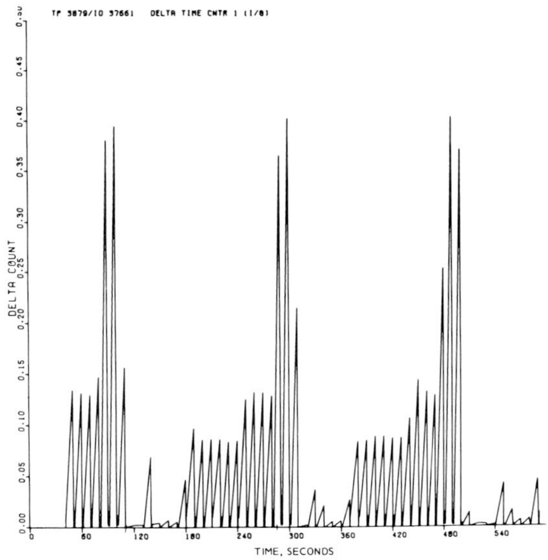

Fig. 3-12. Input/output.

TABLE 4-1. Constants Used in the Markov Model for all Four Environments

Symbol	Environment			
	Real Time	RT + BM	BM	STD
p_{UA}	0.1988	0.15634	0.49198	0.04755
p_{UB}	0.0758	0.02497	0.05891	0.02294
p_{UC}	0.0	0.00176	0.00554	0.00432
p_{UD}	0.0191	0.02521	0.05590	0.03881
p_{UG}	0.7063	0.79172	0.38767	0.88638
p_{AE}	0.00736	0.103	0.038	0.079
p_{AH}	0.99264	0.897	0.962	0.921
p_{BE}	0.00736	0.103	0.038	0.079
p_{BH}	0.99264	0.897	0.962	0.921
p_{CE}	1.0	1.0	1.0	1.0
p_{DE}	0.00736	0.103	0.038	0.079
p_{DH}	0.99264	0.897	0.962	0.921
p_{EF}	1.0	1.0	1.0	1.0
p_{FA}	0.67676	0.751	0.803	0.419
p_{FB}	0.25809	0.120	0.096	0.202
p_{FC}	0.0	0.008	0.009	0.038
p_{FD}	0.06514	0.121	0.091	0.341
p_{GF}	0.31848	0.362	0.387	0.450
p_{GH}	0.34681	0.372	0.398	0.478
p_{GA}	0.24231	0.253	0.192	0.049
p_{GB}	0.09241	0.040	0.023	0.023
p_{HU}	0.14305	0.120	0.086	0.188
p_{HF}	0.85695	0.880	0.914	0.812
λ_U	5,265	1,248	900	2,139
λ_A	31,556	29,831	29,831	17,729
λ_B	19,519	18,993	182,783	18,375
λ_C	0.0	8,600	8,600	8,600
λ_D	110,047	64,195	57,811	15,433
λ_E	27,022	20,219	20,930	16,018
λ_F	578	1,906	3,127	1,339
λ_G	66,323	61,960	53,676	132,956
λ_H	71,474	11,708	11,536	11,764

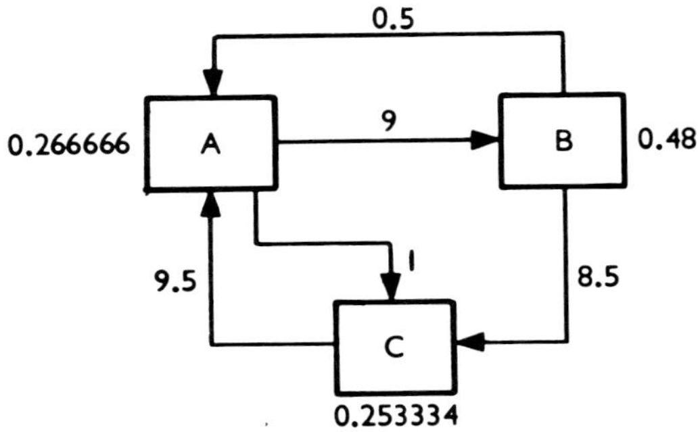

Figure 4-1. Simple three-module example.

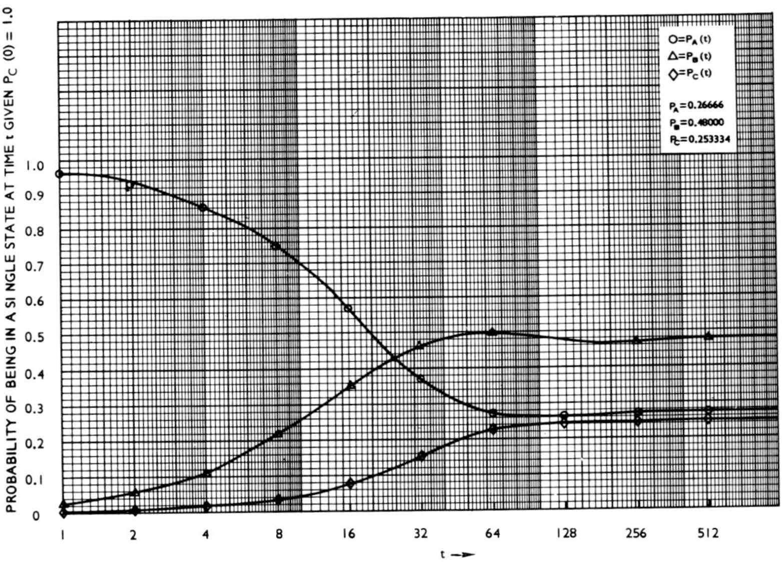

Fig. 4-2. Probability of being in a particular module given system began in module A.

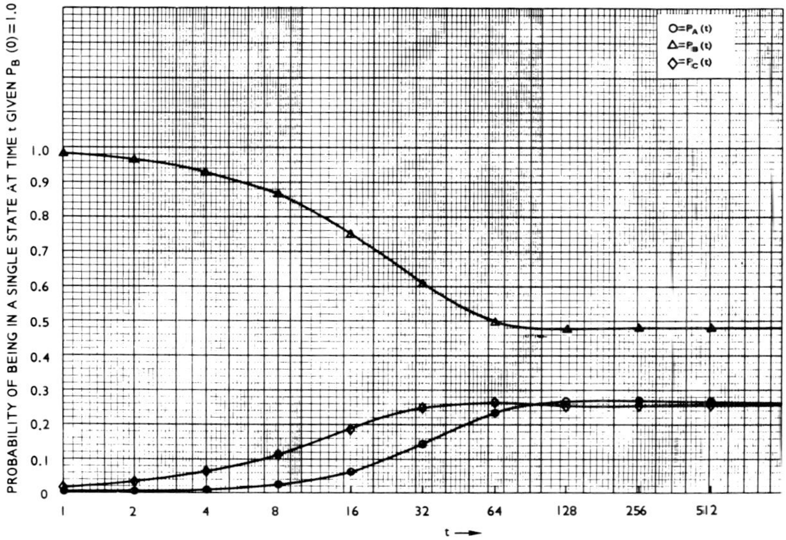

Fig. 4-3. Probability of being in a particular module given system began in module B.

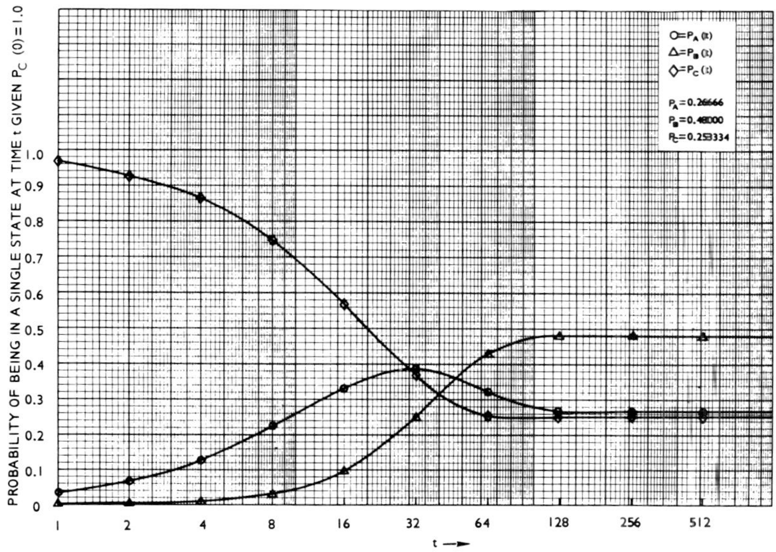

Fig. 4-4. Probability of being in a particular module given system began in module C.

STATISTICAL QUANTIFICATION OF INSTRUCTION AND OPERAND TRACES

Sant R. Arora and Frank L. Wu

ABSTRACT. Performance of a high-speed buffer of a particular size, besides depending upon the size of the backing storage, critically depends upon the statistical characteristics of the trace of memory addresses referenced for obtaining instructions and operands. Simulation techniques are usually employed in such evaluation studies. In order to carry out the simulation, it is essential that the sequence of memory addresses referenced for instructions and operands resembling real environments is generated. Two possibilities exist; one of them is that the trace of memory addresses is captured by instrumentation techniques when benchmarks are running in a real system and the second possibility is that these traces are synthetically generated.

Instrumentation techniques have been very popular because of their ability to capture the sequence of operation with sufficient accuracy, but these techniques are limited to the existing systems. Whenever we are faced with the problem of evaluating environments different from the ones currently available, synthetic generation of traces becomes essential. For example, we may like to study different levels of multiprogramming and multiprocessing for which real data are not available. Before any synthetic traces can be generated, it is essential that the statistical characteristics of the traces are properly identified and quantified.

This paper attempts to identify different segments in memory, to study jump behavior among different segments and to study the autocorrelations and cross-correlations of the request of addresses with each segment. The motivation behind the study is to enhance the ability to extrapolate environments which do not currently exist and to obtain more realistic input data for simulation studies.

INTRODUCTION

Over the past several years there has been an increased demand for larger and faster computer memories. A properly designed hierarchical structure in memory systems has been recognized as one of the most effective ways for meeting these demands while maintaining good cost/performance ratios. The basis for the suitability of the hierarchical structure lies in the inherent heterogeneity of the various data elements with respect to their rate of reference [1,2]. It is recognized that the rate of reference for a particular data element does not remain static. It varies over time, being high in certain periods and low in others; this makes schemes which include the possibility of dynamically moving the data within various levels of hierarchy superior to those that are restricted to static loading. Furthermore, it is also recognized that the task of identifying changes in the rate of data reference is not an easy one, and any schemes to move the data within the hierarchy should be automatic, controlled either by some sort of self-learning type algorithm or based upon some internally collected statistics.

Among the several technologies that have been developed in the hierarchical memory design, the one that has received a great deal of attention in the literature during recent years is that which provides a high-speed buffer integrated into the central processing unit (CPU), which has been commonly called cache memory [3,5,7]. This cache memory works in conjunction with another memory called the backing storage. The cache memory and the backing storage are both partitioned into equal-size blocks and these blocks are organized into columns (Fig. 1). When the CPU issues a request for memory reference, an initial check is made to find whether the block containing the address referenced is in cache or not. If the block is already in cache, execution by the CPU proceeds without any reference to the backing storage, and no movement of blocks is involved; otherwise the required block is loaded from the backing storage into the corresponding column of the cache memory, and if this column in the cache is already full, the block which has not been demanded for the longest time in that column is pushed out of cache. This replacement algorithm is commonly known in the literature as LRU* (which implies that the Least

*Several other replacement algorithms have been suggested in the literature.

Recently Used block is pushed out). The contents of the cache are dynamically adjusted as the execution of the program proceeds.

In evaluating the cost performance of the system, the performance may be measured in terms of the average reference access time (ARAT), which may be denoted by T. It is given by

$$T = t_1 p + t_2(1 - p) \tag{1}$$

where t_1 = average time required to make the reference available to CPU from the cache memory, t_2 = average time required to make the reference available to CPU from the backing storage, and p = probability of finding the required reference from cache. This last will be referred to as the probability of hit. The probability of hit provides an interim parameter by which the performance may be measured.

One of the important parameters which determines the comparative suitability of one hierarchical system over another is the nature of the users' environments. To evaluate the performance of memory systems with cache design, one phase of the task will be to test performance in several typical users' environments, and the second phase of the task, which is perhaps more difficult, will to be identify worse conditions where cache design may be a poor performer.

A number of studies have been made [3,4,6] to evaluate the performance of cache memory system. The usual approach followed in these studies has been to use recorded traces of memory references of some representative test program and feed them into a simulator which monitors the loadings and unloadings into and out of the cache memory in accordance with the demand sequence in the given trace and computes the probability of hit. Whenever we are faced with the problem of evaluating environments different from the one currently available, synthetic generation of traces becomes essential. Before any synthetic traces can be generated, it is essential that the statistical characteristics of traces are properly identified and quantified.

The purpose of this paper is to define essential statistical characteristics of a trace and to define probability of hit as a function of these trace characteristics. This approach of linking the probability of hit with the statistical characteristics of the trace will give a better understanding of the factors which adversely affect the probability of hit. The task of extrapolating the values of the probability of hit with different levels of multiprogramming and multiprocessing will become easier and perhaps more reliable. More importantly, if the worse conditions for the cache system have to be identified, it is very essential that as a first step we should define measures which will quantify those essential characteristics of a trace which are important from the point of view of performance.

The design parameters which play an important role in the performance of cache systems are the cache size, block size, backing storage size and number of columns. If we assume that b denotes the number of words per block, J denotes the total number of columns in the backing storage and cache, and I denotes the total number of rows in the backing storage, then addresses from $(kJ + j-1)b$ to $(kJ + j-1)b + (b-1)$ for $k = 0,1,2,\ldots,I$ will belong to column j. In fully associative memories, both the cache and the backing storage are organized in a single column, whereas in set associative there are J columns where $J > 1$.

STATISTICAL ANALYSIS OF A TRACE

Blocks in any one column of the backing storage commonly share the space in the corresponding column of the cache. On first reference to a block, the block will have to be initially loaded in the cache. On subsequent references to this block, whether the block will be in cache or not will be determined by the number of distinctive blocks that have been referenced from the same column during this time. If the number of distinctive blocks referenced from the same column is greater than or equal to m, where m denotes the number of blocks per column in the cache memory, the block on its second reference would already be out of the cache memory. The portion of the trace between two consecutive references to a block will be called a loop. A trace is essentially a number of loops with smaller loops embedded in larger loops. We will define the size of a loop originating from a block in column i by the number of distinctive blocks visited between the two consecutive references to the same block in column i and will denote this by s_i. If we further decompose s_i into the number of distinctive blocks over the various columns, we have

$$s_i = \sum_{j=1}^{J} s_{ij} \qquad (2)$$

where s_{ij} = number of distinctive blocks from column j in the loop originating from a block in column i. Each s_{ij} and s_i are random variables.

Probability of hit gets built up from two accounts. First, a number of consecutive references may be made from a block before we leave that block, and second, when we leave that block it is still in cache when it is re-referenced. We will define the probability of block hit as the fraction of times a block is found in cache on its re-reference. Probability of block hit for the various columns and for the overall system are given by

$$\text{Prob. of block hit in column j} = U_j = \text{Prob}[s_{jj}<m] \qquad (3)$$

$$\text{Overall prob. of block hit} = U = \sum_{j=1}^{J} a_j u_j \qquad (4)$$

where a_j = fraction of memory references from column j.

It is true that the various distinctive blocks within a loop are referenced for varying numbers of times and the actual sequence of references within the loop is much more complex, but all these details are unnecessary as far as the probability of block hit is concerned.

Probability of hit can be easily computed once we know the probability of block hit, the average number of references made to a block before we leave that block, and fractions of the first time and referenced block demands.

$$p = \frac{v - q(1-U) - (1-q)}{v} \qquad (5)$$

where v = average number of references made to a block before we leave that block, and q = fraction of re-referenced block demand.

It may be pointed out that v is influenced more by the block size, whereas U is influenced more by the remaining design parameters, i.e. cache size, backing storage size and number of columns. In order to compute the probability of hit we need to know the values of v, a_j and distribution of s_{jj} for various j. Joint distributions of s_{ij} and $s_{ij'}$ will be useful in estimating the probability of hit for various levels of multiprogramming and multiprocessing when the statistical characteristics in uniprogramming mode are known.

EXTRAPOLATION TO MULTIPROGRAMMING AND MULTIPROCESSING MODES FROM UNIPROGRAMMING MODE

Once the statistical characteristics of a trace in uniprogramming mode, measured in terms of s_{ij}, are available, it will be possible to obtain estimates for the probability of hit in multiprogramming and multiprocessing modes. While passing from a uniprogramming to a multiprogramming and multiprocessing mode, what essentially happens to the loop originating from a block in column i is that this loop may be augmented by a number of other fractional or full loops originating from some other columns. For example, if in the loop originating from a block in column i, another loop originating from a block in column j is inserted with probability w_j, and if the new parameters of the augmented loop are denoted by primes, then we will have

$$s'_{ii} = s_{ii} + \sum_{j=1}^{J} s_{ji} w_j \tag{6}$$

Knowledge of s'_{ii} will enable us to obtain estimates for the new probability of block hit U'. If some criterion could be established which would give the number of loops to be inserted as a function of the level of multiprogramming and multiprocessing, extrapolation of the values of the probability of block hit would be possible under these modes.

ILLUSTRATION OF STATISTICS FROM A SAMPLE DATA STREAM

We have analyzed data from three different environments, namely scientific, business and program development. The UNIVAC 1108 hardware instrumentation device was used to collect the data. The device captures the addresses of each operand referenced by instructions, the jump-to addresses plus the number of instructions advanced since the last jump. A data reduction program was used to process the raw data and generate the actual sequence of memory addresses requested for instructions, read operands and write operands. We give sample statistics from a mix of six scientific programs executed in a batch mode as an illustration.

The following observations may also be made regarding the statistical characteristics of address trace in general:

1. Distribution of a_j over the various values of j, when j's are numbered in the monotonic decreasing order of the values of a_j, follows a negative exponential distribution.* This points out that the distribution of the number of memory references over the various columns is not uniform. For the observed sample, the distribution of a_j is shown in figure 2.

2. Distributions of s_{ii} over the various i were observed to follow very sharp negatively exponential distributions.* Distributions of s_{ii} for the sample program lie between the two curves shown in figure 3. It may be seen that on the average approximately 35% of block hits are captured by the cache with one block per column. An additional 46% of block hits are captured by the cache with 4 blocks per column. When the cache has 8 blocks per column, an additional 17% of block hits will be captured. This points out the decreasing rate of marginal utility of additional blocks per column in the cache.

The trace of memory references is composed of the instruction stream and the data stream. It has been observed in general that the instruction stream is markedly more sequential than the data stream. This highly sequential nature of the instruction stream is what makes the instruction part of the loop lie mostly in the blocks neighboring the originating block, and this is what perhaps makes the distribution of s_{ii} quite steeply negatively exponential. In designs where the degree of set association is quite high, the instruction stream will rarely introduce distinctive block from column i within the loop. The distinctive blocks within the loop from the data stream are more randomly distributed over all the columns and thus do not add to many distinctive blocks to s_{ii}. In multiprogramming and multiprocessing modes, when loops get augmented, the contribution to s_{ii} is increased significantly because of the concentrated behavior of the instruction streams within the augmented loops.

The decreasing level of the concentration of loops over the various columns, as we move away from the originating column, is shown in figure 4 by plotting Prob(s_{ij}>C) against

---

*Continuous negative exponential distributions have been fitted to the discrete distributions of a_j and s_{ii}.

d, where d is defined below and measures the distance in the forward direction of any column j from the originating column i:

$$d = \begin{cases} j-i & \text{if } j = i+1,\ldots,J \\ J-i+j & \text{if } j = 1,\ldots,i-1 \end{cases}$$

$$\text{and } i = 1,\ldots,J$$

It may be observed that the curve flattens very quickly and stays flat over a large number of columns, rising again a little towards the end of the cycle. The difference in heights between the beginning and the ending parts of the curve points out the relatively higher degree of concentration of the loop in the forward direction as compared to the backward direction.

SUMMARY

Knowledge of the distribution of s_{ij} can be very useful in determining the optimal degree of set association. As the level of set association is decreased, a stage is reached when decreasing the level any further will add very little to the probability of block hit and will not justify the additional expenditure due to longer columns in cache. This knowledge can also be used in extrapolating performance of the given environment to a different level of multiprogramming and multiprocessing.

The problem of identifying worse conditions of high-speed buffer performance is very difficult. Statistics s_{ii} and s_{ij} have been observed for a limited number of environments. To extrapolate these statistics to other environments without determining the s_{ii} and s_{ij} values from observations is very risky. On the other hand, a_j is a parameter which can be easily monitored. Thus a great deal of data is available on the a_j parameter. Our efforts, therefore, are devoted to defining an estimate for s_{jj} as a function of a_j, from which the probability of hit can be easily evaluated. The parameter a_j appears to be useful in extrapolating to determine worse-case conditions.

REFERENCES

1. S.R. Arora and A. Gallo: Optimal sizing, loading and re-loading in a multi-level memory hierarchy system, UNIVAC Tech. Memo. 01EN 108.

2. C.J. Conti: Concepts for buffer storage, Computer Group News 9 (March 1969).
3. J.S. Liptay: Structural aspects of the system/360 model 85. II. The cache, IBM Syst. J. 7, no. 1 (1968).
4. R.L. Mattson, J. Gecsei, D.R. Slutz, and I.L. Trager: Evaluation techniques for storage hierarchies, IBM Syst. J. 9, no. 2 (1970).
5. G.G. Scarrott: The efficient use of multi-level storage, Proc. IFIPS Congress 1965, Spartan Books.
6. S.M. Sisson and M.J. Flynn: Addressing patterns and memory-handling algorithms, FJCC, 1968.
7. M.V. Wilkes: Slave memories and dynamic storage allocation, IEEE Trans. Elec. Comp., April 1965, 270-71.

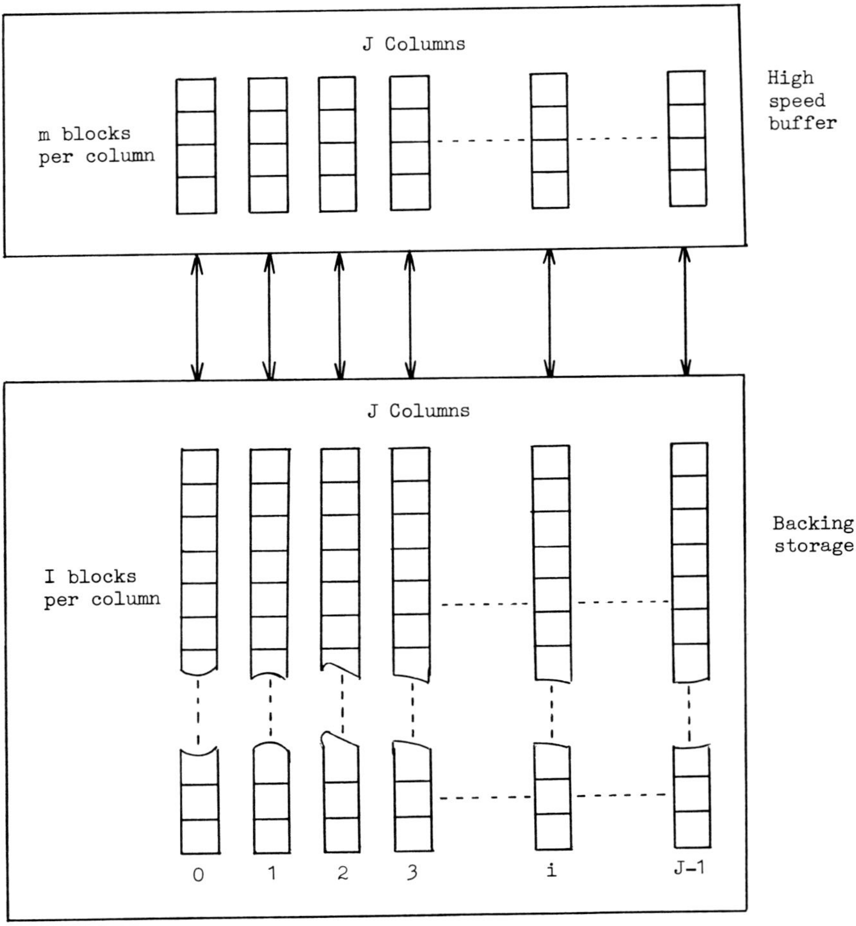

Fig. 1. Layout of a set associative memory.

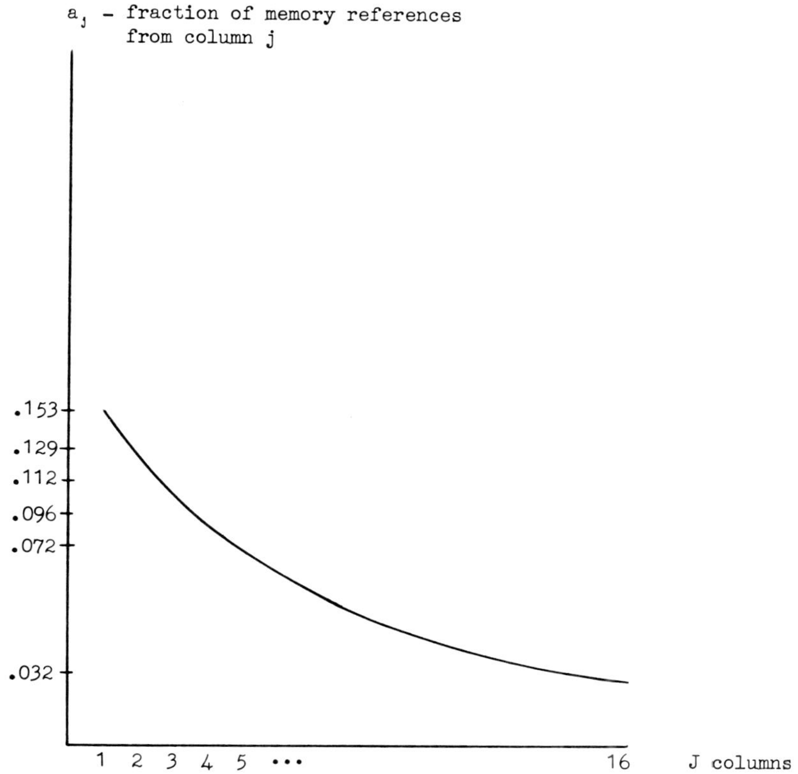

Fig. 2. Frequency distribution of memory references over J columns, where J = 16 in this case. (Columns are renumbered according to the decreasing order of the relative frequencies of memory references.)

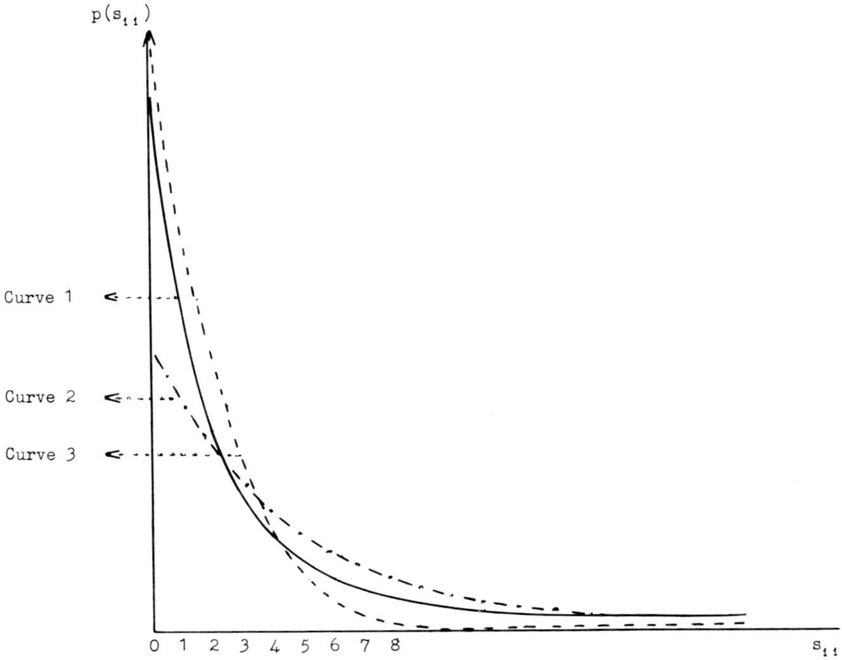

Fig. 3. Distribution of the number of distinctive blocks from column i in a loop originating from column i. Curve 1 shows the average behavior over J column. Curve 2 shows the column with the lowest probability of block hit. Curve 3 shows the column with the highest probability of block hit.

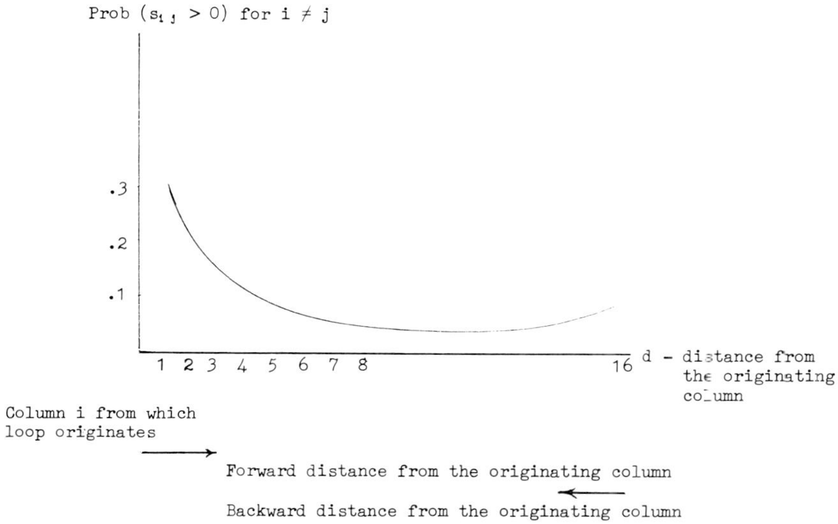

Fig. 4. The higher level of concentration of the loops in the neighborhood of the originating column.

MEASUREMENT AND IMPROVEMENT OF PROGRAM BEHAVIOR UNDER PAGING SYSTEMS

J.L. Baer and G.R. Sager

I. INTRODUCTION

Paging systems are now a reality in a growing number of multiprogramming and time-sharing installations. Much has been said about the inefficiency of such systems. Most efforts to remedy this situation have focused on possible improvements of the supervisors; almost none have been directed to the processor or application program levels. Since now the concept of virtual memory [10] is well entrenched in the architecture of large computing systems, it should be taken into account not only when designing the resource allocation and scheduling algorithms of the operating systems, but also when implementing compilers and writing application programs. It is mainly on this latter aspect that we concentrate our effort in this paper.

We first define the parameters of importance which govern the performance of a paging system as well as the measurements needed to monitor the behavior of programs. Then the tool we implemented in order to make these measurements available, namely WIPS (Washington Interpretive Paging System) will be described. Results of experiments on different versions of users' programs will be reported and will show how simple changes in a program can yield interesting improvements in its behavior while running under a paging system. Finally, we conclude this paper with suggestions for processor implementation and a wide range of experiments.

II. PARAMETERS OF A PAGING SYSTEM. APPROPRIATE MEASUREMENTS

A. Virtual Memory Concept. Definitions - Paging System

The set of addresses that a programmer can use will be called the <u>(virtual) address space</u> V, while the set of physical memory locations in main memory will be the <u>memory space</u> M. When at execution time reference to some address is done in the object code, this will be a reference to a virtual address. Therefore there must be a scheme for mapping the address space into the memory space. Such a function f is called an address map and is defined as follows:

$$f: V \rightarrow M \text{ such that if } x \varepsilon V$$
$$f(x) = \{y \text{ if } x \text{ is in } M \text{ at location } y\}$$
$$\{\text{undefined otherwise}\}.$$

Generally, $|V| >> |M|$.

Figure 1 represents an address map.

In the case where f(x) is undefined, we have a missing-item fault. That is, the item we want to reference is in a higher level of the memory hierarchy and must be brought into core. In order to do this, the following decisions must be made:

-- If the memory is full, which item of M must be removed from M. This is the replacement rule.
-- When to put the missing item in M, i.e. the loading rule.
-- Where to put the missing item in M, i.e. the placement rule.

Before going further, one has to define the missing item. Here we consider these items to be pages, that is blocks of fixed size, and we call our virtual memory system a paging system.

In paging systems, the address field of an instruction referring to a memory address is divided into two parts: the page number and the displacement. The page number - in fact a virtual page number - corresponds to the argument x of the address map function. The displacement, also called a line number, gives the address within the page. The transformation of a virtual address to a physical address is done through a page table mechanism as shown in figure 2.

B. Paging System Parameters

The main parameters directing the use of a paging system are:

Replacement rule: There has been an extensive study of replacement algorithms and measurements of their behavior. We shall look into four of them, namely: FIFO (first in, first out): LAM (look-aside memory), an improved version of FIFO [15]; LRU (least recently used); OPT (optimal). We then concentrate on LRU for reasons which will become apparent in section III. We limit ourselves to these four because the first three may be implemented in hardware with increasing degrees of difficulty while the last one - not realizable in practice since it implies the knowledge of the total string of memory references - gives an upper bound on the efficiency one can expect.

Loading rule: If one waits to load a given page until it is requested, we then have a demand paging loading rule. It can be shown [17] that this is an optimal rule whenever the time to load k pages is k times the time to load one page. If this could be the case for large-core-storage type of second-level memories, it is not true for rotating devices such as drums or disks. However, we consider only demand paging in this report.

Placement rule: In the case of a paging system, there is no parameter associated with the placement rule since all blocks to be placed are of the same size and fit exactly in the 'holes' freed by the replacement rule.

Page size: The page size p is an important parameter for the overhead associated with the loading of pages in core. It also is a direct factor in the efficiency of the replacement rules.

Physical memory size: Evidently the amount of real core, i.e. r resident pages, allocated to a program is of primary importance. A large amount of core is always welcomed but the virtual memory concept has for its principal objective circumventing the lack of real core.

Style of programming: Although the user's programming style cannot be thought of as a parameter of the paging system *per se*, it has a strong influence on the behavior of the whole system. For example, if we contrast the following segments of FORTRAN programs:

```
   DO  10  I = 1, 1000          DO  10  I = 1, 1000
   A(I) = I                  10 A(I) = I
   B(I) = I + 1                 DO  20  I = 1, 1000
10 C(I) = 2 * I              20 B(I) = I + 1
                                DO  30  I = 1, 1000
                             30 C(I) = 2 * I
```

and if A, B, C are arrays of 1000 words with page size of 1000, we can see that for the program on the left, two more pages will be necessary for efficient execution of the DO-loop than for the program on the right. We shall return to this problem in section IV.

C. Measurement on Paged Programs

In a paging system, one reduces the space that a program may occupy by paying the cost of having to transfer pages from one memory level to another; that is, by spending extra I/O time. Measurements of programs running under a paging system will then mainly involve recording the number of page faults f versus the main memory allotted, with a number of parameters (outlined above) influencing this ratio, mainly the page size p, the primary memory allotted r, and the replacement rule (say A).

The following measurements are of interest.

1. Measurements depending only on the program and the page size. In a conventional system without overlay, one has to reserve the amount of memory needed in all possible cases, that is for all possible sets of data. For a given execution only a portion of this memory space will be addressed due, for example, to conditional branches, excessive dimensioning of arrays, etc. In a paging system, since only blocks in which at least one word is accessed are transferred, the total amount of memory needed M(p) (in supposing an infinite amount of primary memory, i.e. no replacement) should be decreasing with decreasing page size [3,18].

2. Influence of the primary memory size allocated to one program. Given a replacement algorithm (cf. #5 below) and p, one can measure f for different r. Results showing the decrease of f when increasing r are abundant [1-3,5,7, 11-13,18,19] (for exceptions, see #5 below). An equivalent measure is the average number of instructions or the average number of memory accesses between page faults.

3. Influence of the page size given a primary memory allotment. For a given memory size, i.e. a fixed r × p, we can record f with varying p. Known measurements have shown that in general the improvements due to smaller p are greater than the savings accomplished when increasing the allotment of primary memory for a given page size [1,2,7,13].

4. Utilization of paged words. It is important to know the percentage of paged words which have been accessed. This can be done on a gross level by measurements from (2)

above. That is, one can define the ratio n_t of total number of words transferred from second to first level over n_p, total number of references. In fact, this measure is equivalent to those already defined in (2) [1]. More important could be what has been called superfluity by Kuck [14] and which can be defined as follows. For each residency of a given page, let s_i be the number of words referenced in that page. S, the superfluity, is then defined by the expression

$$S = \left(\sum_{i=1}^{f} s_i\right)/(f \times p)$$

To our knowledge, we are the first to report measurements of superfluity.

5. Replacement algorithms. In [17] one can find a classification of replacement algorithms in two categories: stack algorithms which have the property of being such that f decreases when r increases (e.g., Belady's optimal replacement algorithm OPT [3] and the least-recently-used (LRU) policy which pushes, when necessary, to the secondary level the page in core which has not been referenced for the longest time); and those which may present the possible anomaly of f increasing with r (e.g. the first-in-first-out policy where pages are replaced in their order of arrival, and the "look-aside memory" [15]). Experiments of type (2) and (3) above have been done with different replacement algorithms.

6. Style of programming. We have seen a trivial example of how some obvious changes in a program could improve its behavior when running under a paging system. Techniques have been developed for list processing languages [4,8] as well as algorithms to partition matrices efficiently [16]. More recently, comparative experiments have been performed [5,6,20] on some test programs showing how rather simple changes can improve the running characteristics of the programs.

These measurements can be recorded thanks to a tool which we now describe.

III. A SOFTWARE PAGING SYSTEM

The Washington Interpretive Paging System (WIPS) is a software package designed to emulate the execution of a subject program on an XDS Sigma 5 computer [21]. By duplicating much of the hardwired actions of the Sigma 5 in software,

WIPS is able to intercept and analyze actions of the subject program. In particular, WIPS intercepts memory references and does the necessary swapping to insure that the desired information is in core, thereby enabling the subject program to run as though the full 128K words of storage were available even though the CSTL Sigma 5 we used has only 24K words of storage.

In addition to increasing the address space of the computer, WIPS could serve as a powerful analytic device. For example, simple additions to the existing code could make it into an interactive debugging tool or cause it to accumulate statistics on instruction mixes. More involved modifications could simulate a multiprocessor computer or a processor with a different instruction set. Of prime importance to this study is the fact that WIPS can output a record of the memory references it intercepts which can later be used to study the effects of various memory allocation techniques (hardware and software) on the performance of the subject program.

WIPS is divided into three main modules: the CPU, the Memory Manager, and the CAL Interpreter. A detailed description of WIPS may be found in [22]. A possible option in the use of WIPS is the recording on tape of all memory references as well as the type of operation (instruction or data fetch) which caused that reference. We describe now the instrumentation process.

The Simulated Runs

As indicated in figure 3, experiments are broken into a two-phase operation. In Phase I, the user program is executed by WIPS, which generates a magnetic tape containing a record of all memory references made by that program. The format of the trace tape is given in figure 4. In Phase II, particular values of page size, real core size, and replacement algorithm (p,r,A) are input to a "simulation" along with the trace tape for the run to be simulated. From this simulated run we get such statistics as a map showing which locations were referenced, the number of pulls required, the number of pushes required, the superfluity and the rate of page demand.

It would be possible to combine Phase I and Phase II, but this would require re-interpreting the user program for each (p,r,A) triple of interest and would prevent the user program from using the peripherals required by the simulation program (i.e. card reader, line printer). As it stands, the user program is only prevented from usage of one tape unit, while the simulation program is free to use any peripheral. In addition, generating a trace tape has the advantage of solidifying the subject program, as it will no longer be sensitive to changes in the library between experiments.

Under this two-phase operation, we may conduct the two experiments of interest as follows. First, we may study the effects of programming style on paging behavior by operating Phase I once for each of two or more programs which accomplish the same algorithms by different styles and then operating Phase II once on each trace tape for each triple (p,r,A) of interest. One may then compare the results when the same value of (p,r,A) is used in conjunction with trace tapes representing different styles. Second, we may study the effects of p, r and A on paging behavior by operating Phase I once, then comparing the Phase II outputs when different values of (p,r,A) are applied to the resulting tapes.

As we have seen, Phase I of the experimentation process involves the generation of data (namely, a chronological sequence of memory references made by a subject program) whereas Phase II utilizes these data to simulate the run on a hypothetical paging system. It should be obvious that the simulation phase is very unsophisticated. It is not necessary actually to swap pages between primary and secondary memory. We need only keep a record of what pages <u>would</u> be in real core in an actual run; and, if we want to observe the number of pushes required, which of these pages has been written into (such information is available from the ST field of the trace word). Variation of the replacement algorithm (A) is achieved by running the trace tape through a separate Phase II processor for each of the algorithms of interest.

The page size (p) and real core size (r) are input to these processors. Typically, a simulated run of 400,000 references requires five to seven minutes of computer time for each triple (p,r,A). In the case of the LRU algorithm, however, there is a method [17] which can improve this figure significantly.

Algorithms having the stack property (cf. section II) are distinguished by the fact that at any time (where time is measured by what memory reference is being processed -

the ith reference in the reference stream occurs at time i) the pages residing in a real core of size r will include all those pages which would have resided in real core had its size been r-1. Thus, if a program references C+1 distinct pages during a run, it is possible to order a table of pages residing in real core such that at any time the first r-1 entries ($0 < r \leq C+1$) correspond to the pages which would have been in real core had its size been **r**. In the case of the LRU algorithm, this ordering is achieved by looking up the page just referenced in the table, pushing the entries above it down one position, and placing it in the topmost position. On the initial reference to a page, it will not be found, so all existing entries are pushed down one position and an entry made for the newly referenced page in the topmost position of the table. This method is illustrated by the SYMBOL code in figure 5, which implements a technique known as "look-down" because of the fact that it combines the operation of table lookup with pushdown required by the LRU ordering.

As can be seen, the code in figure 5 counts how many times each position is referenced in the table lookup process. Since the absence of a page from positions 0 through r-1 indicates that that page would not have been in real core of size r and necessitates looking at the rth position of the table, we see that the rth position of the COUNT array will indicate how many page faults would have occurred for a real core of size r. Hence, a single pass through the reference stream will yield the number of faults for each possible core size for a given page size. Using this method on the aforementioned trace of 400,000 references, the "fast LRU" (FLRU) can process 130 (p,r,LRU) triples in about twelve minutes, yielding data for all possible core sizes possible for page sizes between 64 and 1024 words. Since the shape of the curves for various paging characteristics tends to vary little from one algorithm to another (cf. section IV), the FLRU processor is a valuable "quick look" tool which indicates the (p,r,A) triples which are interesting enough to run on the slower processors.

We now give results pertaining to our experiments.

IV. EXPERIMENTS

A. Measurements on a Sample Program

The measurements described in section II were first applied to a sample program, a syntax-graph analyzer [3].

We obtained the following results:

1. M(p) versus p. Under the normal operating system, the program required 6490 words of which 2177 only were affected. The maximum amount of memory needed M(p) for a given page size p is plotted in figure 6. For $64 \leq p \leq 512$, one can see that M(p) is almost linear with p, confirming previous results.

2. f as a function of p, r and A. Table 1 shows the number of page faults for p varying from 64 to 512, a primary memory allotment from 2K to 6K, and the replacement algorithm cited in the previous sections.

As could be expected, f decreases with an increase in r for a given p for all four possible As (cf. figure 7). In all cases, LRU behaves the best of the non-optimal algorithms while in all but one case (p=64, r=48) LAM is better than FIFO. That is, investing in the hardware to implement more sophisticated algorithms will pay at run time.

The effect of page size p when one keeps a constant $r \times p$ (i.e. constant primary memory allotment) is shown in figures 8 and 9. As can be seen, increasing r and decreasing p has a large effect on f. However, there is a point where f may not decrease any more even for an optimal algorithm. In the case of a rotating paging device, one could investigate the trade-off between the number of seek-times (related to f) and the transfer times (related to $f \times p$).

3. Rate of page demand and superfluity. The rate of page demand, i.e. the cumulative number of page faults versus the number of instructions executed, is plotted in figure 10. One can see that after an initial flurry for the first few hundred instructions, the page demand rate is almost linear when one considers intervals of 10,000 instructions. We have found this to be true for all triplets (p,r,A) for this particular program, and it would be interesting to see if this is a general property or a peculiarity of the monitored program.

The superfluity S for different triplets (p,r,A) is given in table 2. The only trend noticeable is that S increases with decreasing p, but there is no correlation between A and S or even S and $r \times p$.

B. Improvements due to Changes in the High-Level Code

In order to see the effects of changes in the high-level code on the behavior of programs, we made some slight modifications to the above program. A second example (part of a Monte-Carlo simulation for scheduling parallel processes) was selected because first, like the previous one, it had been written and used on a conventional system, and second, the style of programming of the two programmers (the authors of this report) was widely different (one favoring small subroutines and numerous labelled commons, the other using blank common and larger subroutines with a fair number of arguments). We now summarize our results.

Example 1. The original version was referencing five vectors of 500 words each corresponding to the fields of 500 possible nodes of the syntax graph. The modified program converted this structure into a two-dimensional (5 by 500) array and it was stored starting on a page boundary. The only other modification was to change the relative locations of two small common blocks of two and three words respectively. In a second modification to this program, we generated the compiler output in such a way that code and data were separated. Each subroutine generated a code module, a simple variable and temporary module as well as a possible array module. Then these data modules, along with the common blocks, all ordered by size, were loaded into pages following a first-fit method. This second version had for its main purpose to have as many "clean" pages as possible.

Let f_{mod} and f_{ori} be the number of page faults under the modified and original versions. We define I_{jK} as the ratio f_{mod}/f_{ori} for a given $r \times p = jK$. Thus I_{jK} shows the improvement obtained for different page sizes and a constant primary memory allotment of jK words.

Figures 11(a,b,c,d) show I_{2K}, I_{3K}, $M(p)$ versus p, and f versus r for p = 256 for both the modified and the compiler versions. As can be seen, definite improvements are obtained mostly for smaller page size. For larger p the gain is not as good mostly because the program needs only 6500 words at worst.*

Example 2. A run of 300,000 references was recorded. The modifications were a change in the order of dimensioning in one array and decomposition of DO-loops, as shown in

*Some statistics for f_{ori} may look different from those obtained in measurements of the above paragraphs. It is because we fixed the library code on a page boundary for all experiments, thus reducing its side effects.

section II. No "compiler" version was attempted, nor was the prevention of crossing page boundaries by often-referenced arrays. However, these simple changes again gave some interesting improvements, as shown in figures 12(a,b,c,d) which record the same measures as figure 11.

V. CONCLUSION

In this paper we have described WIPS, an interpreter used for instrumenting and running programs under a paging system. We have defined measurements of interest for the behavior of programs and reported results on sample programs. Some examples have shown how rather trivial changes in the source code can decrease significantly the number of page faults. It is to be investigated how much of these changes could be done - or may be suggested - by the compiler. Some optional compiler directives such as PAGE could be introduced to align specific arrays on page boundaries or prevent them from crossing page boundaries. It seems that these applications could be of importance for small page size, as in "cache" memories. More experiments of this type are planned, as well as the use of WIPS for study of multiprocessors and hardware-software mapping devices in a multiprogramming environment.

Acknowledgments. We thank H. Golde, E. Hunt, E. Strasbourger, and M. Zosel for their help in implementing Phase I of WIPS.

REFERENCES

1. Anacker, W., and C.P. Wang, "Performance evaluation of computing systems with memory hierarchies", IEETC 16, 765-773 (1967).
2. Baylis, M.H.J., D.G. Fletcher and D.J. Howarth, "Paging studies made on the I.C.T. Atlas computer", IFIP Congress Proceedings D113-118, 1968.
3. Belady, L.A., "A study of replacement algorithms for a virtual storage computer", IBM Syst. J. 5, 78-101 (1966).
4. Borrow, D.G., and D.L. Murphy, "Structure of a LISP-system using two-level storage", CACM 10, 155-159 (1967).
5. Brawn, B.S., and F. Gustavson, "Program behavior in a paging environment", AFIPS FJCC Proceedings 33, part 2, 1019-1032, 1968.
6. Brawn, B.S., F. Gustavson and E.S. Mankin, "Sorting in a paging environment", CACM 13, 483-494 (1970).
7. Coffman, E.G., and L.C. Varian, "Further experimental

data on the behavior of programs in a paging environment", CACM 11, 471-474 (1968).
8. Cohen, J.A., "Use of fast and slow memories in list processing languages", CACM 10, 82-86 (1967).
9. Cohen, D.J. and C.C. Gotlieb, "A list structure form of grammars for syntactic analysis", Comp. Surveys 2, 65-82 (1970).
10. Denning, P.J., "Virtual memory", Comp. Surveys 2, 153-189 (1970).
11. Fine, G.H., C.W. Jackson and P.V. McIsaac, "Dynamic program behavior under paging", Proc. 21st Nat. ACM Conf., 223-229, 1966.
12. Freibergs, I.F., "The dynamic behavior of programs", AFIPS FJCC Proceedings 33, part 2, 1163-1168, 1968.
13. Joseph, M., "An analysis of paging and program behavior", Comp. J. 13, 48-54 (1970).
14. Kuck, D.J., and D.H. Lawrie, The use and performance of memory hierarchies, Report 363, Dept. of Computer Science, University of Illinois, 1969.
15. Lee, E.F., "Study of look-aside memory", IEEE Trans. Computers C-18, 1062-1064 (1969).
16. McKellar, A., and E.G. Coffman, "The organization of matrices and matrix operations in a paged multiprogramming environment", CACM 12, 153-165 (1969).
17. Mattson, R.L., J. Gecsei, D.R. Slutz and I.L. Traiger, "Evaluation techniques for storage hierarchies", IBM Syst. J. 9, 78-117 (1970).
18. O'Neill, R.W., "Experience using a time-sharing multiprogramming system with dynamic address relocation hardware", AFIPS SJCC Proceedings 30, 611-621, 1967.
19. Pinkerton, T., Program behavior and control in virtual storage computer systems, CONCOMP Report 4, University of Michigan, 1968.
20. Sayre, D., "Is automatic folding of programs efficient enough to displace manual?", CACM 12, 656-660 (1969).
21. XDS Sigma 5 Computer Reference Manual, 1968.
22. Baer, J.L. and G.R. Sager, Measurements and improvements of programs running under a paging system, Technical Report 71-07-04, Computer Science Group, University of Washington, 1971.

TABLE 1

f vs r for four different replacement algorithms and different p

a) p = 512

r \ f	OPT	LRU	LAM	FIFO
4	6,275	8,126	10,770	11,599
5	2,360	3,423	4,747	5,299
6	968	1,537	1,760	2,146
8	482	775	999	1,239
10	234	306	535	578
12	65	118	335	349

b) p = 256

r \ f	OPT	LRU	LAM	FIFO
4	10,869	14,462	18,209	21,317
6	4,059	6,208	7,777	9,352
8	1,653	2,794	3,515	4,101
10	923	1,533	1,866	1,885
12	638	819	860	992
16	282	527	823	892

c) p = 128

r \ f	OPT	LRU	LAM	FIFO
8	5,968	9,832	11,306	14,058
12	2,191	3,906	4,309	5,104
16	1,220	1,546	1,652	1,985
20	844	1,225	1,372	1,496
24	485	1,030	1,321	1,421
32	46	52	113	139

d) p = 64

r \ f	OPT	LRU	LAM	FIFO
32	1,436	2,221	2,356	2,465
48	258	924	1,323	1,143

TABLE 2

SUPERFLUITY RESULTS

A / p/r	OPT	LRU	LAM	FIFO
512/6	.173	.177	.163	.153
512/4	.035	.026	.027	.026
256/12	.214	.203	.250	.252
256/8	.138	.079	.077	.076
64/48	.264	.297	.342	.375
64/32	.340	.349	.353	.355

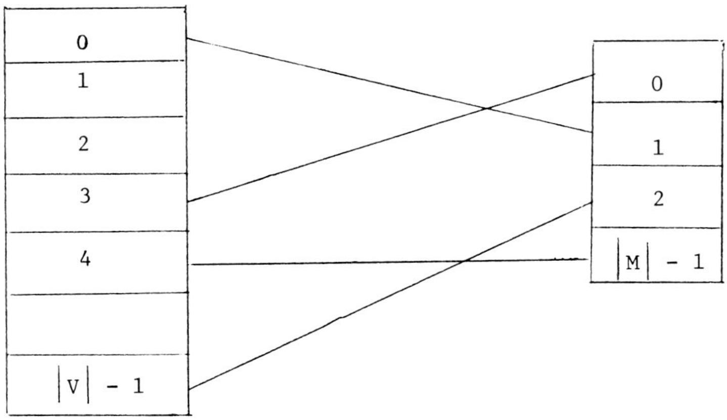

Fig. 1. Address map example.

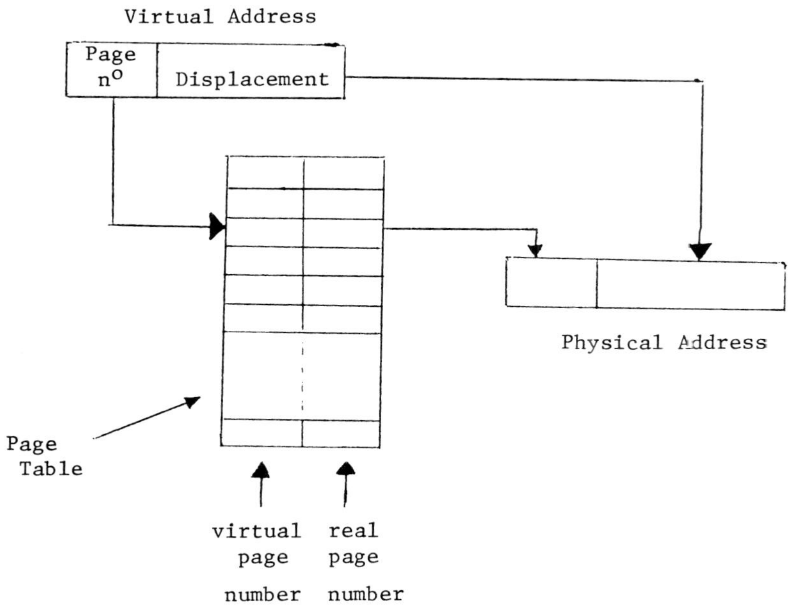

Fig. 2. Address transformation in a paging system.

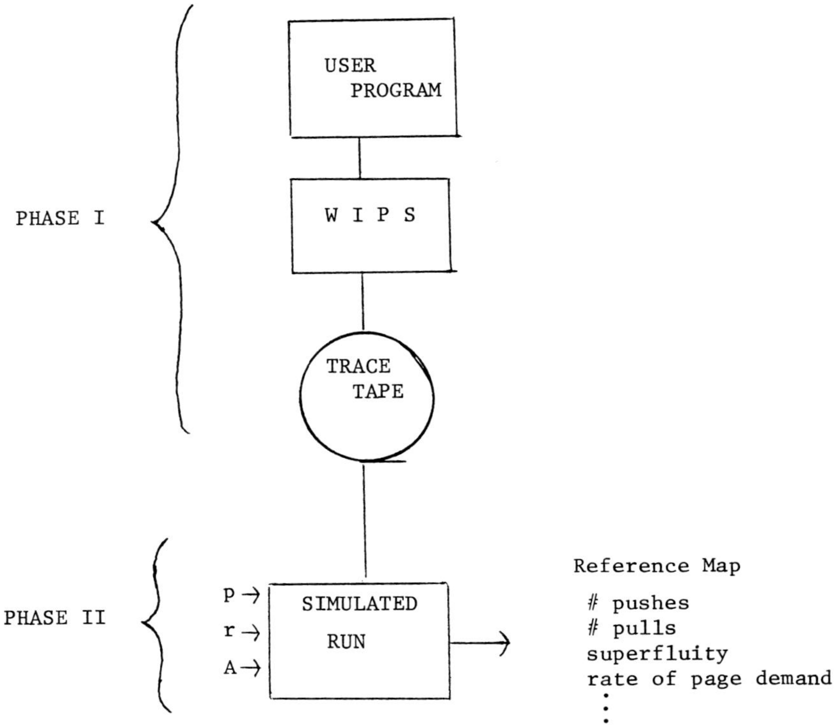

Fig. 3. Overall system.

0 1	2	3	4	5 14	15 31
	SV	ST	IO	LENGTH	ADDRESS

SV: save bit

ST: store bit (to distinguish between 'clean' and 'dirty'pages)

IO: Instruction or Operand bit

LENGTH: Number of contiguous words referenced

ADDRESS: The virtual address referenced

(512 trace words per record)

Fig. 4. Trace word format.

```
START      LW,3     C           get table size
           LI,2     0           initialize index
           LW,5     PAGTAB,2    get first entry from table
LOOKDOWN   MTW,1    COUNT,2     count a fault
           CW,5     REQPAG      compare entry with requested page
           BE       FOUND       equal?
           AI,2     1           no, continue the loop
           XW,5     PAGTAB,2    lookdown step
           BDR,3    LOOKDOWN
           MTW,1    C           not found, increase table size
           LW,5     REQPAG      make entry for page
FOUND      STW,5    PAGTAB      page just referenced goes on top
```

Fig. 5. The lookdown technique for LRU.

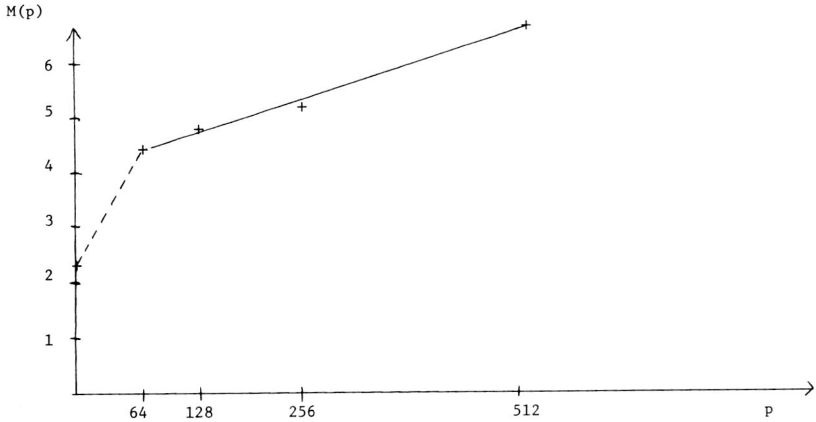

Fig. 6. M(p) versus p.

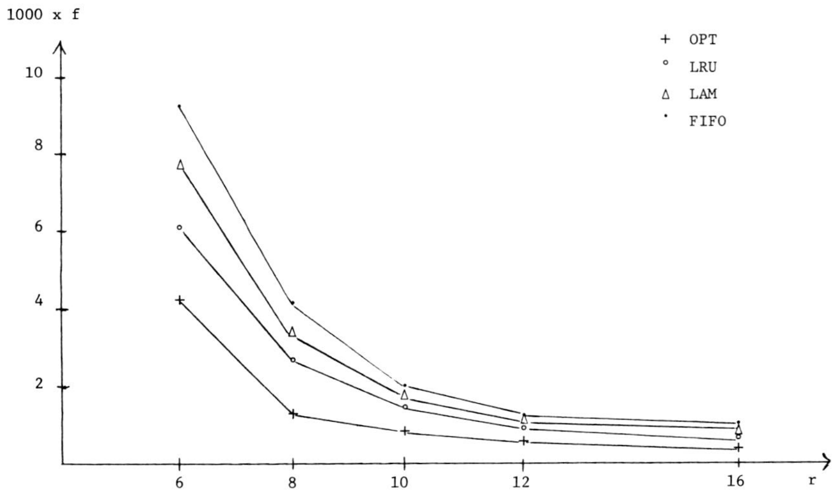

Fig. 7. f versus r for p = 256.

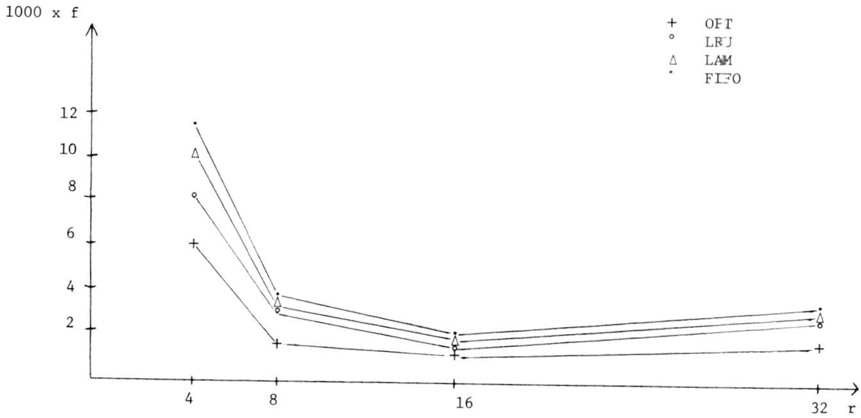

Fig. 8. f versus r for r × p = 2K.

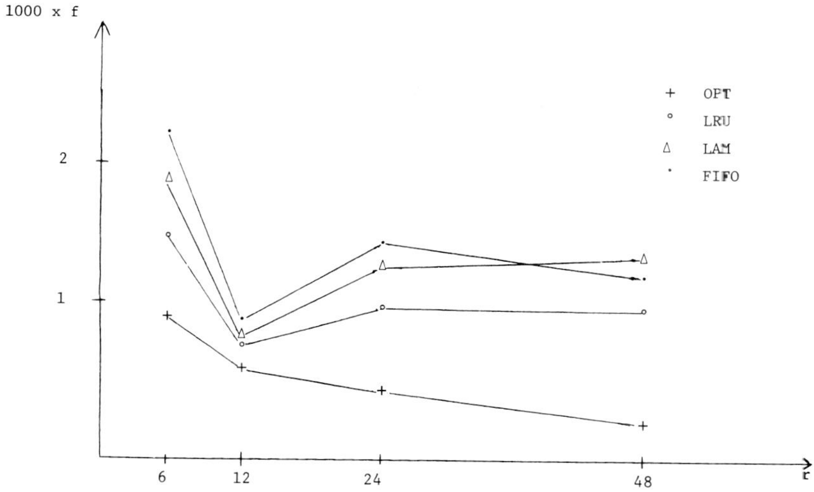

Fig. 9. f versus r for r × p = 3K.

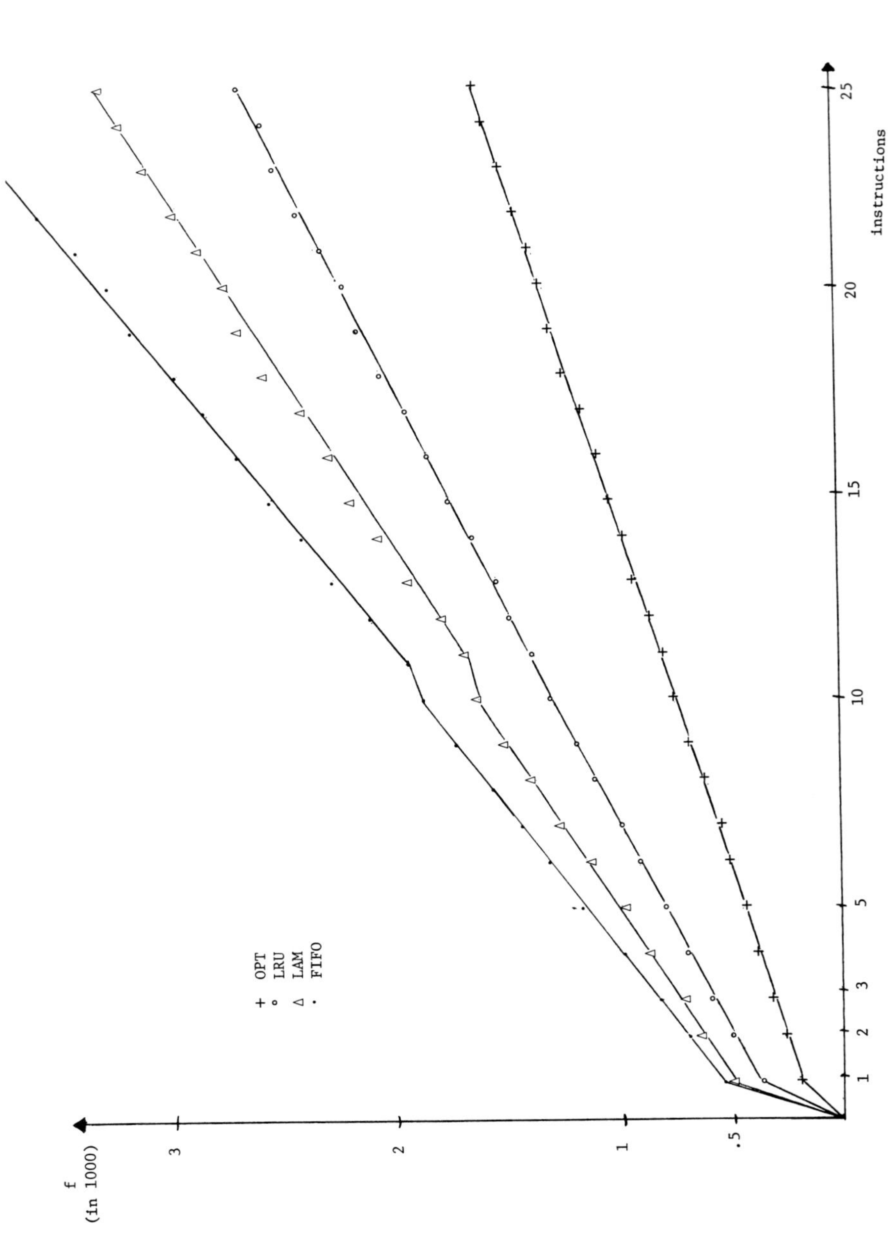

Fig. 10. Rate of page demand. r = 8, p = 256.

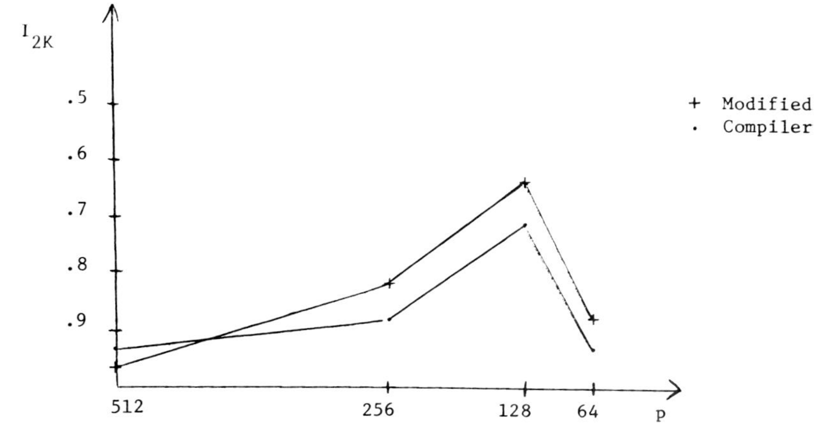

FIGURE 11(a). I_{2K} vs p

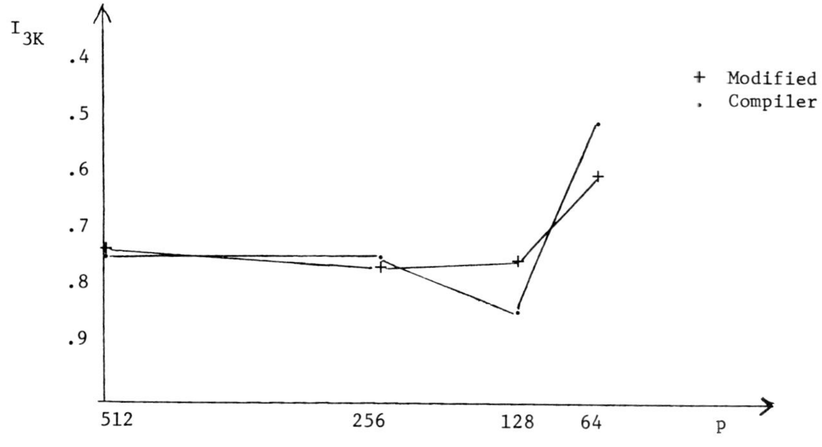

FIGURE 11(b). I_{3K} vs p

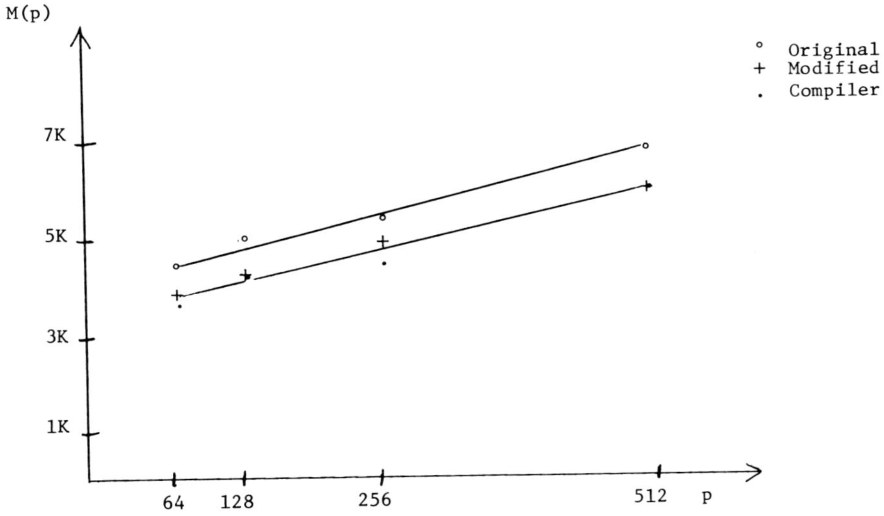

FIGURE 11(c). M(p) vs p

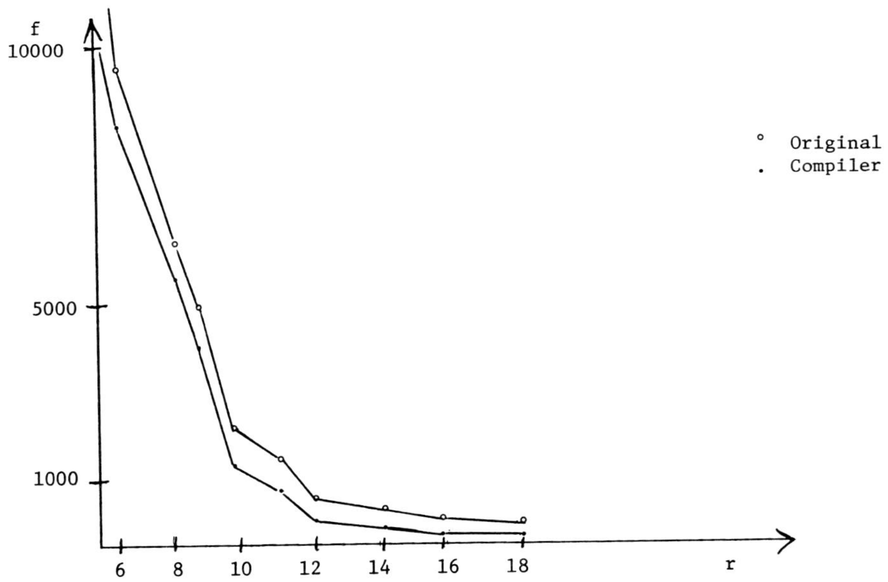

FIGURE 11(d). f vs r for p = 256

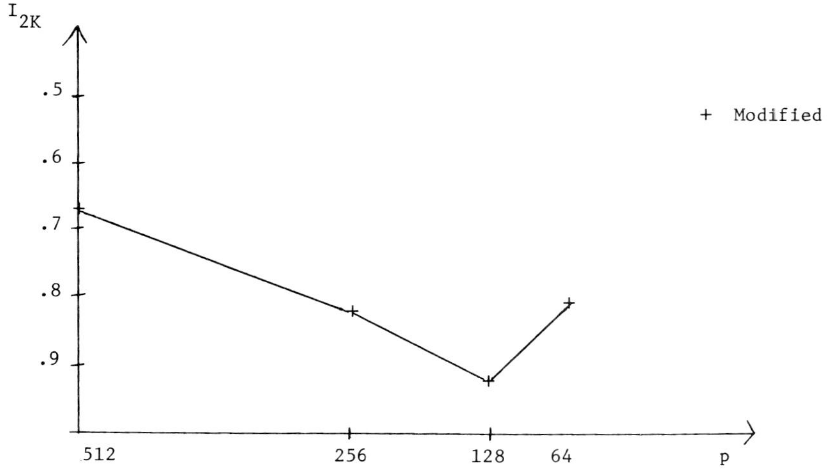

FIGURE 12(a). I_{2K} vs p

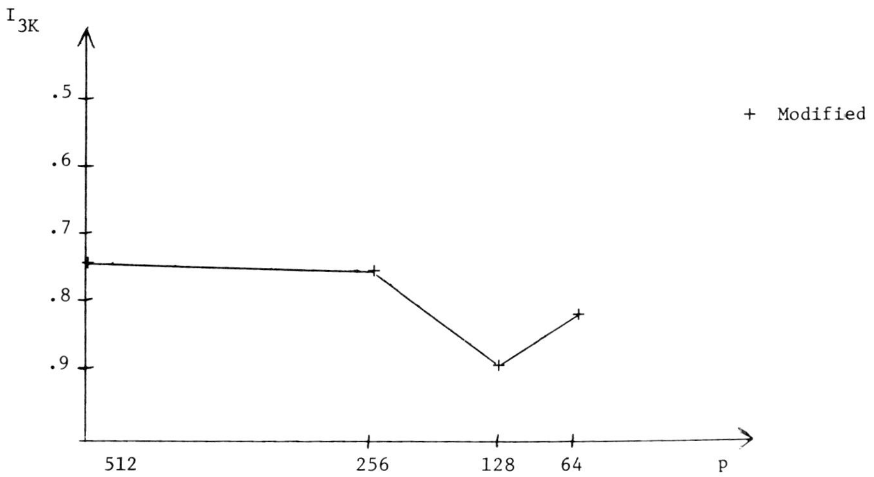

FIGURE 12(b). I_{3K} vs p

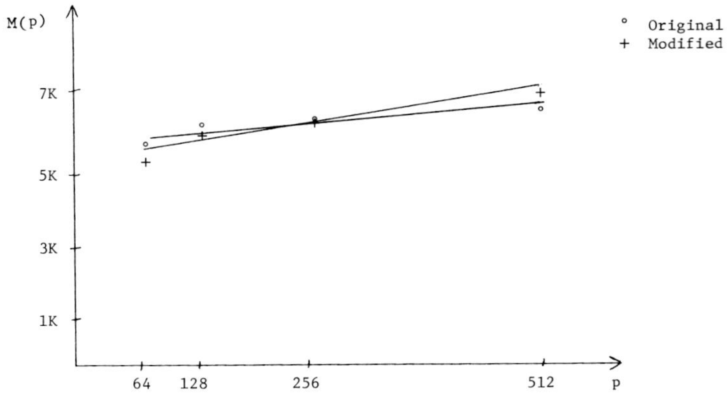

FIGURE 12(c). M(p) vs p

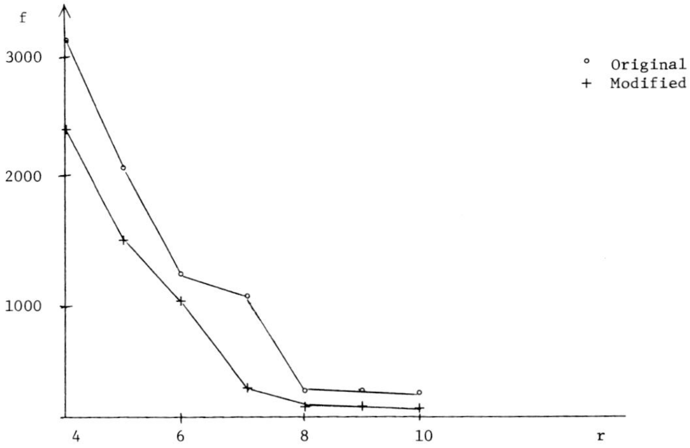

FIGURE 12(d). f vs r for p = 256

STATISTICAL ANALYSIS OF SERIES OF EVENTS IN COMPUTER SYSTEMS

P.A.W. Lewis and P.C. Yue

SUMMARY. SASE IV is a computer program which has been designed to implement the methodology for the statistical analysis of series of events (stochastic point processes) described in a recent monograph by Cox and Lewis. Various subroutines perform, among other things, tests for monotone and cyclic trends, tests for renewal and Poisson processes and two different types of spectral analysis. The program can also be used to analyze any series of positive random variables, and has been programmed both in FORTRAN and APL.

An application of the program to the analysis of a series of message arrivals at a teleprocessing information retrieval system is described, as well as some very recent developments in the statistical analysis of point processes.

I. INTRODUCTION: THE SASE IV PROGRAM

Series of events occurring randomly in time or space arise in many technological and scientific contexts. In statistics these event processes are called "stochastic point processes". The events may be the failures of a computer or the arrivals of nerve impulses at a synapse; the emissions of particles by a radioactive source or the errors occurring during transmission of binary data. An example is a study (Lewis [1964]) of computer failure patterns in time. Detailed statistical analysis (Lewis [1964]) assessed how one event type (maintenance) affected another (computer failure). It became evident even then that computer "reliability" would have to be broadened to include not only physical failures but also "congestion" failures.

Here, as an example of empirical event series in computer systems, we consider message arrivals at a teleprocessing information retrieval system.

A computer program, SASE IV (Statistical Analysis of Series of Events)*, has been developed by P.A.W. Lewis,

*SASE IV is available from IBM Program Information Department, Hawthorne, N.Y. 10532, as Program No. 360G130001.

with the assistance of A.M. Katcher, for the analysis of point processes and essentially implements methodology for such an analysis given in Cox and Lewis [1966].

SASE IV is an outgrowth of the earlier SASE I and SASE II programs which were executed on the IBM 7094 computer and described by Lewis [1966]. As implemented on the IBM system/360, the program can analyze series consisting of as many as 2000 events - twice the number which could be handled by its predecessor. The new program also provides a plotting capability (developed by A. Weis) which is executed on a standard printer, thus allowing visual output for SASE IV on machines lacking special graphics equipment.

To understand the type of analysis performed by SASE IV, it is necessary to consider series of events in some detail. Such a series is shown schematically in figure 1, events being represented by dots on the time axis.

In SASE IV, only univariate series of events are considered. The events can be distinguished only by the times at which they occur; other quantitative or qualitative data, such as the type of the event, are ignored for the purpose of analysis. Consequently the times-to-events t_i, with $0<t_1<t_2<t_3<\ldots$, or the times-between-events, x_i, can characterize the process.

There is, however, another equivalent way in which a series of events may be characterized. This is in terms of the counting process $N(t)$, the number of events occurring from the onset of measurement to time t. The representations of the series of events in terms of the times between events $\{t_i\}$ and the number of events occurring in a given interval $N(t)$ are related by the following fundamental relationships, implicit in figure 1:

$$N(t) < n \text{ if and only if } t_n = \sum_{j=1}^{n} x_i > t \qquad (1)$$

and

$$\text{prob}\{N(t) < n\} = \text{prob}\{T_n > t\}$$

for $n = 1,2,\ldots$.

These relationships seem almost trivially obvious; yet they have important consequences. For example, a statistical analysis based on estimated correlation or spectral properties of the counting process, $\{N(t)\}$, is generally not equivalent to an analysis based on correlation or spectral properties of the interval process, $\{X_i\}$. Thus, if one is to explore a point stochastic process in detail, analysis should be made both of the number of events occurring over time and of the sequence of intervals separating one event from another.

The SASE IV program, consisting of a main program and ten subroutines, carries out the computations for these two correlational analyses as well as some more particular types of analysis; for details, see Lewis [1966] and Lewis and Katcher [1969]. The overall analysis differs from an ordinary time-series analysis in two chief respects: (1) the $\{X_i\}$ (interval) process is a time series in which the variables take on only positive values and do not have a Gaussian (normal) distribution, and (2) the analysis of the $\{N(t)\}$ (counting) process has no direct counterpart in standard time-series analysis. With regard to (1), a central role is played by the exponential distribution. Moreover, SASE IV can be used to analyze sequences of positive random variables such as the successive response times in a computer system.

Further distinctions between the analysis of series of events and ordinary time-series analysis become evident in considering trends in a series of events. There are those which are best specified in terms of time, while others are most appropriately considered as functions of the event parameter i. Thus, computer failure-susceptibility might increase smoothly with time, affecting the intervals between events (failures) indirectly and in a complex fashion. On the other hand, if the events are the occasions when a rat presses a bar for food in a conditioning experiment, the increasing rate of events will be primarily a function of the number of previous bar pressings.

SASE IV includes a capability for detection of several types of trends. Thus, the TREND subroutine tests to determine if in a Poisson process the rate changes over time. (A Poisson process is an event process in which the number of events in non-overlapping time intervals are independent and have Poisson distributions.) Other subroutines (see figure 2) specify further details concerning the nature and properties of the trend. Subroutine TREND also calculates quantities which can be used to test for trends using least-squares regression methods. These methods are flexible and allow various types of trends to be tested, working reasonably well under fairly weak assumptions as to the detailed structure of the series under study.

Where TREND does not indicate the presence of trends, stationary processes are involved. SASE subroutines (figure 2) then test for fixed-rate Poisson processes; for correlation between intervals, i.e. for "renewal processes", in which the probabilistic structure of events and intervals "starts from scratch" after an event occurs. (A fixed-rate Poisson process is one particular instance of a renewal process, but so are others in which the values of the variable are independent and have the same distributions.)

SASE IV subroutines also carry out spectral analysis for both the $\{N_t\}$ and $\{X_i\}$ processes.

SASE IV has been widely used on a variety of computers with no apparent problems. This portability has been achieved at the cost of considerable sloppiness or looseness of programming.

Several drawbacks in SASE IV include the need to load the whole program (approximately 250K bytes of core are needed), and the limitation to 2000 events. It is hoped to rectify these problems in a later version of the program. Also, several new developments in the statistical analysis of series of events which are described in the next section of this paper will be incorporated.

An APL version of parts of SASE IV has been written. It offers several advantages over the FORTRAN version but at present is limited by workspace size and input-output problems.

II. RECENT DEVELOPMENTS

Recent developments in the statistical analysis of point processes have been summarized by Lewis [1972]; see also Cox [1972] and Brown [1972]. Here we highlight some which are of interest in analyzing data from computer systems.

1. Multivariate point processes. A drawback in SASE IV is that it analyzes only univariate point processes. Multivariate point processes are also of great importance and are of two kinds. The first kind arises in the situation in which events in a point process are typed, e.g. one

may be analyzing interrupts in a computer and they may be classified by some qualitative information. The second kind is the input-output situation such as the two series of events consisting of times at which jobs arrive at a queue and depart from the queue.

Clearly this is an important area of analysis. The main impetus for its study has come from neurophysiology, where inputs and outputs to nerve cells consist of spike trains which can be characterized by the times at which the spikes occur. Relevant work is that of Cox and Lewis [1966, ch. 3] and Perkel et al. [1967], and the recent study by Cox and Lewis [1972]. Much work remains to be done.

2. Detrending point processes. In many fields of application, it has become increasingly apparent that stationary point processes are at best a convenient mathematical fiction. Most data exhibit fairly subtle trends and methods for testing for these trends are known (Cox and Lewis [1966, ch. 3]); other data, however, exhibit gross trends, e.g. time-of-day effects in the series of arrivals at a queue, and techniques for the analysis and characterization of such data are only now beginning to be developed.

The situation is analogous to that in ordinary regression analysis and time-series analysis where one wants, for example, to estimate parameters in an assumed (linear) function for the mean, test the model for the mean function and then examine the model which is assumed for the residuals. The latter could include examining the residuals to test for independence, estimating the spectrum of the residuals and testing the assumed normality of the residuals. Techniques for these problems in the linear normal model are known (see e.g. Hannan [1970]).

By comparison, in point processes one might want to (i) estimate the rate function $\lambda(t)$, using either specific functional models or smoothing techniques; (ii) test specific functional models for $\lambda(t)$; (iii) detrend the point process, examine the 'residual' process and test the usual hypothesis that the events are generated by a homogeneous Poisson process.

When dealing with nonhomogeneous Poisson processes, the most appropriate detrending technique (Lewis [1970], [1972]) seems to be to transform the time-scale so that the ith event occurring at time t_i now occurs at time

$$\tau_i = \int_0^{t_i} \hat{\lambda}(u)du,$$

where $\hat{\lambda}(u)$ is some estimate of the rate function $\lambda(t)$. Note that if $\lambda(t)$ is known, the τ_i's are a Poisson process. When $\lambda(t)$ is estimated from the data, the distributional problems associated with determining properties of the $\{\tau_i\}$ processes are difficult.

Fairly regular point processes are probably best dealt with by using log transforms of the intervals between events and then using ordinary time-series methods (Cox and Lewis [1966, ch. 3]).

3. Spectral analysis of point processes. By spectral analysis of a point process (Bartlett [1963]) we mean the second-order spectrum of the counting function $N(t)$. Brillinger [1972] has put this spectral analysis on a firm footing in the context of a general spectral theory for stationary interval functions such as $N(t)$. He also proposed the use of higher-order spectra.

The spectral analysis may also be thought of as an ordinary spectral analysis of a function $dN(t)$ which is a series of delta functions occurring at random times $\{t_i\}$ (see Lewis [1970] for a heuristic interpretation). Note that it is not a spectral analysis of the intervals between events $x_i = t_i - t_{i-1}$. The latter spectrum is useful for differentiating between renewal processes (for which it is flat) and non-renewal point processes. It may in fact be preferable to higher-order spectra of counts (Brillinger [1972]) in that it can be expected to exhibit fewer sampling fluctuations.

Note that a renewal process is completely specified by its second-order spectrum. The spectrum is in fact essentially the Fourier transform of the renewal density or the intensity function.

One drawback to the spectral analysis of point processes is the large amount of time required for computation of spectral estimates. Only recently have French and Holden [1971], in an important paper, found a way to use the fast Fourier transform (FFT) in this problem. To illustrate their technique, consider the case when the series is observed for a fixed time t_0 and n events occur at times $t_1 < t_2 < \ldots < t_n$, where $t_n < t_0$. Note that n is the observed value of the random variable $N(t_0)$.

Empirical spectral analysis of the point process is usually based on the finite Fourier-Stieltjes transform of N(t),

$$H_{t_0}(\omega) = (\pi t_0)^{-1/2} \int_{t=0}^{t_0} e^{it\omega} dN(t) \tag{3}$$

$$= (\pi t_0)^{-1/2} \sum_{j=1}^{n} \exp(it_j\omega) \tag{4}$$

$$= (\pi t_0)^{-1/2} \{ \sum_{j=1}^{n} \cos(t_j\omega) + i \sum_{j=1}^{n} \sin(t_j\omega)\}, \tag{5}$$

and the periodogram

$$I_{t_0}(\omega) = |H_{t_0}(\omega)|^2 \tag{6}$$

The sum in (4) is not in the form required for the fast Fourier transform FFT algorithm (see, e.g., Cooley, Lewis and Welch [1970]), so that for each frequency for which the spectrum is required the number of operations required is of the order n. The periodogram is usually computed at frequencies ω_p such that $t_0\omega_p = 2\pi p$, where $p=1,\ldots,P$, since the estimates are "well-behaved" at these points (Cox and Lewis [1966, ch. 5]). Since P is of the order of n, the total number of computations is of order n^2.

Note that a point process can be sampled completely through knowledge of the event times, and the spectrum is not band-limited. However, French and Holden [1971] considered the realistic case where an upper frequency of interest $\omega_p = W$ was given, and considered the continuous time process

$$Y(t) = \int_0^{t_0} \frac{\sin W(t-\tau)}{2\pi(t-\tau)} dN(t) \tag{7}$$

$$= \sum_{j=1}^{n} \frac{\sin W(t-t_j)}{2\pi(t-t_j)} \tag{8}$$

The Stieltjes convolution of $N(t)$ with the $\sin(Wt)/(2\pi t)$ function is equivalent to multiplying the spectrum by a frequency window which is flat to frequency W and zero thereafter. By sampling $Y(t)$ at intervals of width $\Delta t = \pi/W$, no frequency information is lost (aliasing) in the spectrum of $Y(t)$, which is the same as that of the point process up to frequency ω. The FFT can be applied to transform the sampled values of $Y(t)$; <u>the main point in the technique, however, is that the sine functions have the same value for all t of the form $k\Delta t$, where k is an integer</u>. Thus, computation of the $Y(k\Delta t)$'s is relatively simple and a considerable gain in computing speed should be obtained.

There are some problems with this technique, e.g. it is not bias-free, as asserted by French and Holden [1971], but it appears to be of great value.

III. ANALYSIS OF ARRIVAL DATA

Yue [1971] has analyzed data consisting of the arrivals of messages at a teleprocessing information retrieval system on several days. The system supports remote terminals scattered across the United States. Here we supplement Yue's analysis with particular reference to spectral analysis of the data. The analysis illustrates what is often a fairly subtle interplay between formal and informal (graphical) statistical analyses.

The plot of the cumulative times to arrivals for messages received on Friday, October 16, 1970 is shown in figure 3. For a constant rate $\lambda(t) = \lambda$ this plot, within sampling fluctuations, should be a straight line. Some falling off in the mornings and evenings should be expected, but is not apparent from figure 3. This is partly due to the fact that the data start at 10 pm and all but one time zone in the U.S. is submitting messages by that time.

A more formal analysis using SASE IV is given in table 1. The quantities computed in INTER can be used to compare the <u>marginal distribution</u> of intervals with the exponential distribution, although this is not a good test for a Poisson process. Note that for an exponential distribution, the coefficient of variation is 1, the coefficient of skewness 2 and the coefficient of kurtosis 6. These are to be compared with the estimated values in table 1 from subroutine INTER.

The tests for a Poisson process shown in table 1 from subroutines EXPO and DURB indicate rejection by some of the test performed in EXPO, but not by those performed in DURB. This is consistent with departures from a homogeneous Poisson process ($\lambda(t)=\lambda$) consisting of fairly slow rates of change in $\lambda(t)$ (Cox and Lewis [1966, ch. 6]). Similarly, the tests for independence of intervals performed in SPEC do not indicate any strong serial dependence between intervals, although these results must be interpreted carefully in light of the possible time trend in the data.

To examine the trend more carefully, the periodogram $I_{t_0}(\omega_p)$ given by Eq. (6) was computed in subroutine BART of SASE IV for frequencies satisfying the relationship $t_0\omega_p = 2\pi p$. Actually, the program computes $I_{t_0}(\omega_p)$ divided by $n/t_0 = \hat{\lambda}$, the estimate of the rate λ in a nonhomogeneous Poisson process. These normalized spectral values are approximately chi-square random variables with two degrees of freedom (Cox and Lewis [1966, ch. 5]), and a smoothed spectrum consisting of the averages of successive non-overlapping groups of periodogram values is plotted in figure 4. The bands for individual values show no deviation from a flat (Poisson) spectrum except at the initial point where the value is 5.46. This could be either due to a long cycle or a monotone trend, but since the latter is not strongly indicated by the U-statistic in table 1, we examine the cyclic hypothesis.

Tests for cyclic trends of <u>known</u> frequency are given in Lewis [1970], but tests for a cycle at any frequency are difficult to implement, partly because of the unlimited frequency range of the spectrum. As an ad hoc procedure, since the data have n degrees of freedom we look for the maximum value of the first $n/2 = 835$ spectral values $I_{t_0}(\omega_p)$, i.e. $p=1,\ldots,835$. The maximum occurs at $p=9$ where $I_{t_0}(\omega_9) = 25.011$. Since this is considerably greater than the .99 quantile of the maximum of 835 chi-square (2 d.f.) random variables we conclude that there is a cycle in the data. The period T_p of the cycle is obtained from the equation $t_0\omega_p = t_0 2\pi/T_p = 2\pi P$; i.e., $T_p \simeq t_0/p$. Since $t_0 = 9.083$ hours, $T_0 = 1.01$ hours.

A cycle of one hour would be consistent with the fact that locations in different time-zones start and stop transmitting messages at times separated by one hour, and also have lunch hours with one-hour phase shifts. The cycle, as might be expected, is not distinct and is more like a modulated hour effect. This can be seen in the first twenty-four values of $I(\omega_p)$ given in table 2. Examination of the spectra on other days confirms the existence of a peak with side-lobes.

The spectrum is a powerful tool for examining data for cyclic effects, although, as with all spectra, smoothing of the spectrum and interpretation of the spectrum at very low frequencies is difficult. The variance-time curve computed in subroutine VART is very useful in this respect, its asymptotic slope being proportional to the derivative of the spectrum at zero. The cycle in the October 16, 1970 data shows up clearly in the variance-time curve plotted by the SASE IV program.

DISCUSSION

Application of the SASE IV program to analyzing event processes in computer systems has been discussed and illustrated. Parts of the program can also be used to analyze time series such as successive service times and successive response times. For further details, see Yue [1971].

REFERENCES

Bartlett, M.S. (1963): The spectral analysis of point processes, J.R. Statist. Soc. B25, 264-96.

Brillinger, D.R. (1972): The spectral analysis of stationary interval functions, Proc. 6th Berkeley Symposium (to appear).

Brown, M. (1972): Statistical analysis of non-homogeneous Poisson processes, Proc. Conf. on Stochastic Point Processes, ed. P.A.W. Lewis, New York, Wiley (to appear).

Cooley, J.W., Lewis, P.A.W., and Welch, P.D. (1970): The application of the fast Fourier transform algorithm to the estimation of spectra and cross-spectra, J. Sound Vib. 12, 339-52.

Cox, D.R. (1972): The statistical analysis of dependencies in point processes, Proc. Conf. on Stochastic Point Processes, ed. P.A.W. Lewis, New York, Wiley (to appear).

Cox, D.R. and Lewis, P.A.W. (1966): The statistical analysis of series of events, London, Methuen; New York, Barnes and Noble.

Cox, D.R. and Lewis, P.A.W. (1972): Multivariate point processes, Proc. 6th Berkeley Symposium (to appear).

French, A.S. and Holden, A.V. (1971): Alias-free sampling of neuronal spike trains, Kybernetik 8, 165-171.

Hannon, E.J. (1970): Multiple time series, New York, Wiley.

Lewis, P.A.W. (1966): A computer program for the statistical analysis of series of events, IBM Syst. J. 5, 202-25.

Lewis, P.A.W. (1970): Remarks on the theory, computation and application of the spectral analysis of series of events, J. Sound Vib. 12, 353-75.

Lewis, P.A.W. (1972): Recent results in the statistical analysis of univariate point processes, Proc. Conf. on Stochastic Point Processes, ed. P.A.W. Lewis, New York, Wiley (to appear).

Lewis, P.A.W., Katcher, A.M. and Weiss, A.H. (1969): SASE IV - an improved program for the statistical analysis of series of events, IBM Research Report RC-2365.

Perkel, D.H., Genstein, G.L. and Moore, G.P. (1967): Neuronal spike trains and stochastic point processes - II. Simultaneous spike trains, Biophysical J. 7, 419-40.

Yue, P.C. (1971): Time-series techniques for measurement analysis. Presented at the System Performance Evaluation Workshop, Argonne National Laboratory, Illinois, October 1971. IBM Research Report.

Lewis, P.A.W. (1964): A branching Poisson process model for the analysis of computer failure patterns, J.R. Statist. Soc. B 26.

Table 1. Analysis of message arrival times (Friday data).

Statistic	'SASE' Subroutine	1600 Messages	Usage	
Mean	INTER	19.2911 secs	first four moments	
St. Dev.		19.9354 secs		
Coef. Var.		1.0334 secs		
Coef. Skew.		2.1896 secs		
Coef. Kurt.		9.6943 secs		
U-Statistic	TREND	-2.055 (.04)	monotone trend	
			5%	1%
(i) K_s^+	EXPO	*1.589 (<.01)	1.224	1.518
K_s^-		0.639 (>.05)	1.224	1.518
K_s		1.589 (>.01)	1.358	1.628
W_n^2		*4.157 (<.01)	2.492	3.857
(ii) Moran		*4.81 (<.01)	$\pm$1.28	$\pm$2.05
(iii) $\tilde{K}_s^+$	DURB	0.697 (>.05)	Testing for Poisson Hypoth.	
$\tilde{K}_s^-$		0.151 (>.05)		
$\tilde{K}_s$		0.697 (>.05)		
$\tilde{W}_n^2$		0.869 (>.05)		
$\hat{K}_s^+$	SPEC	0.0573 (>.05)	Testing for Serial Independence	
$\hat{K}_s^-$		1.0989 (>.05)		
$\hat{K}_s$		1.0989 (>.05)		
$\hat{W}_n$		3.2325 (>.01)		
$\hat{U}$		+1.996 (>.046)		

* indicates rejection at level <1%.

TABLE 2

Table 2. Periodogram values $I(\omega_p)$ for October 16, 1970.

p	$I(\omega_p)$	p	$I(\omega_p)$	p	$I(\omega_p)$
1	9.33	9	25.01	17	1.01
2	9.43	10	3.53	18	1.93
3	10.85	11	5.87	19	1.23
4	0.80	12	9.15	20	3.81
5	9.97	13	2.93	21	0.41
6	9.80	14	0.15	22	0.77
7	2.96	15	1.38	23	1.82
8	8.57	16	0.71	24	4.05

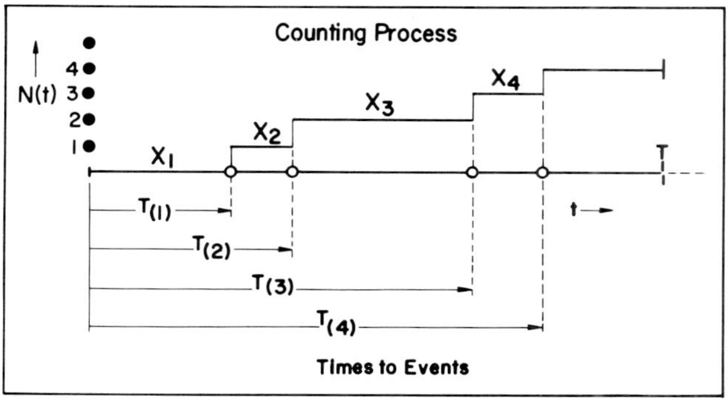

Fig. 1. Equivalent representations of series of events by the counting process N(t) and the times-to-events $\{t_i\}$.

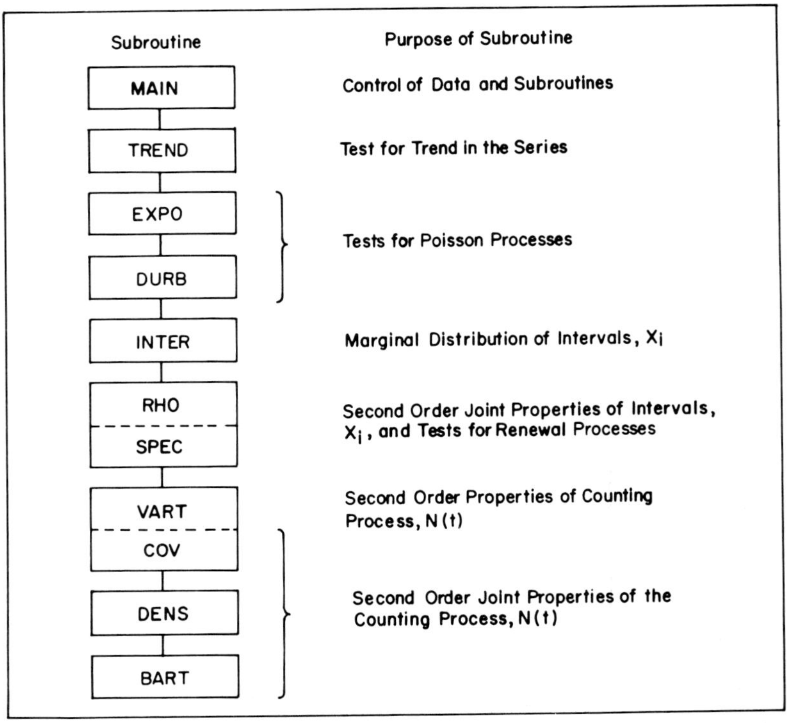

Fig. 2. SASE subroutines.

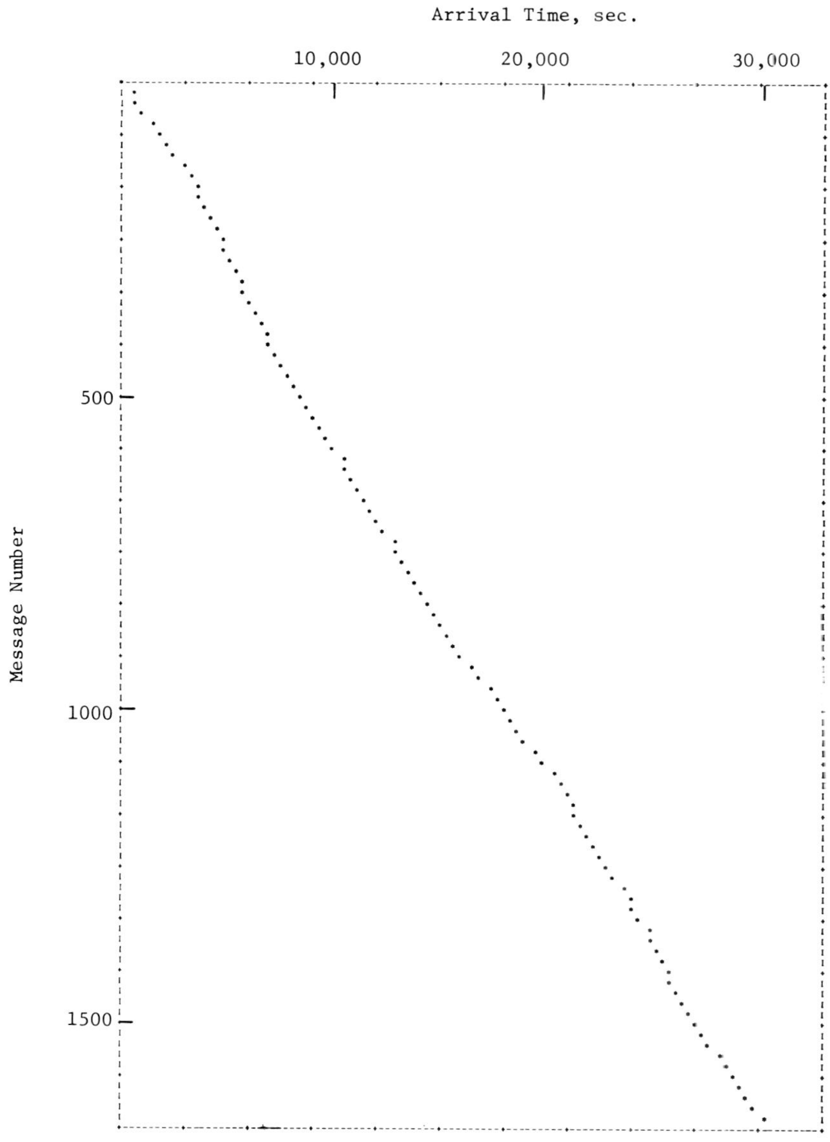

Fig. 3. Cumulative times-to-arrivals for messages received at a teleprocessing information retrieval system. Total number of arrivals, n = 1670.

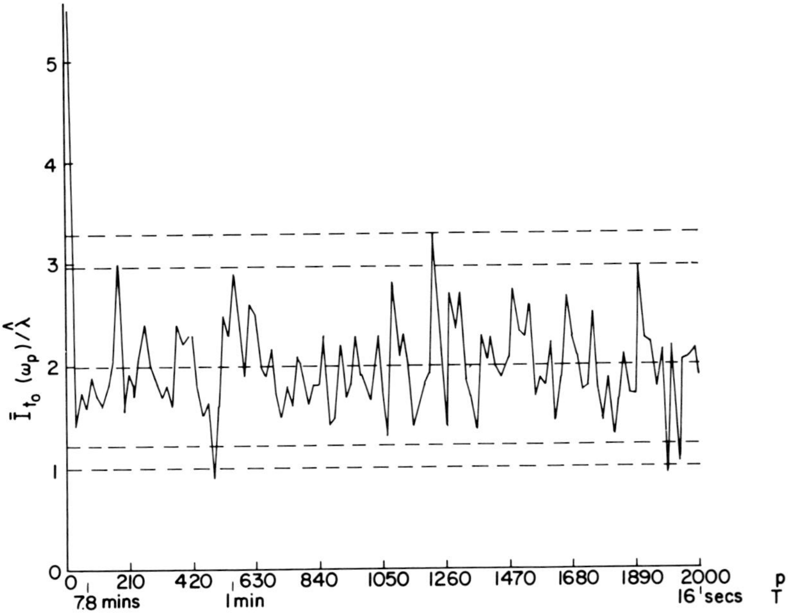

Fig. 4. Smoothed spectrum of counts normalized by dividing by estimated rate n/t_0 sec^{-1}. (20-point uniform weight smoothing. Dashed lines are 0.95 and 0.99 confidence bands. Observed over t_0 = 9.083 hours on October 16, 1970 with n = 1670.)

DISCUSSION OF SESSION IV

1. by Ulf Grenander

Programs (application programs, compilers and interpreters, operating systems) are deterministic but generated by human decisions of more or less stochastic structure. Input data may also be viewed as the result of stochastic processes. I would like to propose the following questions.

1. Is it possible to characterize the request structure on the input side in precise probabilistic terms, perhaps in the language of demand analysis in econometrics? Do the usual stochastic process models suffice for this purpose or do we need radically new stochastic models?

2. Is it possible to describe the logic of programs on a global scale employing what one could call statistical topology? It should account for block structure, looping, data reference systems, etc.

3. Can one use such global descriptions together with demand models to arrive at, at least, order-of-magnitude statements measuring the performance of a computer system? This would be of interest in particular for multiprogrammed systems.

4. Non-trivial computing jobs involve a large number of steps to be executed. This points to the possible existence of asymptotic results for program behavior, similar in spirit to the classical limit theorems in probability theory. They should give, at least, rules of thumb for the expected behavior of programs and systems.

A theory of this kind does not exist today but one can see the beginning of it, some preliminary results having been presented at this conference. It seems likely that the next few years will witness the creation of such a theory.

2. by Emanuel Parzen

I am very pleased to be able to be present at this conference on the statistical evaluation of computers. I come as a neophyte to this area who is convinced that it will be extremely important in the 1970s.

Professor Grenander listed queueing theory, experimental design and regression methods as among the areas of statistics which need to be applied in the statistical evaluation of computers. I would like to add the still underdeveloped area of statistical inference on stochastic processes as an area which will receive much stimulus from our concern to evaluate computers.

I would like to raise a question whether the design of a computer need be decided once and for all when it is built; or can it be parametrized, with suitable parameter values being adaptively chosen so as to optimize the workload which it is handling?

Another aspect of rendering a given computer more efficient is statistical analysis of programming style and characteristics of each job submitted; this could be used both to allocate the jobs and to advise the programmer of how he might alter his programming style to achieve a more efficient mesh of man and machine.

I would like to compliment Professor Peter Lewis on his paper (with P.C. Yue) presented here. It includes an insightful summary of recent developments. Statistical inference on point processes can be regarded as including statistical inference on queues as a special case, and I believe it will be very important for computer evaluation.

Another link between computer science and statistics that I believe ought to be mentioned is that between numerical analysis and statistics (and especially time series analysis). On the one hand, time series analysis (which originated as a theory of analysis of data) is now being applied to develop algorithms to solve purely deterministic problems, such as solution of integral equations. On the other hand, we are learning to formulate time series parameter estimation problems so that their solutions are spline functions.

I hope that this conference heralds the beginning of more intensive cooperation between computer science and statistics. I have mentioned various problems which seem of common interest to the two disciplines and which lie on the current research frontiers of both disciplines. The kinds of data being generated by a computer, which must be analyzed to evaluate and optimize them properly, are typical of the kinds of large systems and large data files which both statistics and computer science are being called upon to collect, handle and analyze.

Let me take this opportunity to indicate what our general aim should be as we evaluate computers; it is to determine how the storage, manipulative and computational capacity of computers can be optimally involved in the research patterns of scientists and scholars and the decision-making patterns of managers and people.

SESSION V

REGRESSION ANALYSIS

CHAIRMAN: CUTHBERT DANIEL

AN APPLICATION OF STATISTICAL METHODOLOGY IN THE STUDY OF COMPUTER SYSTEM PERFORMANCE

A.C. Yeh

ABSTRACT. In planning for future data processing system performance requirements, as well as in the process of evaluating current system performance, there is a need to identify the proper variables of the environment, which will be the prime determinants of the system's performance, and to quantify the relationship. This paper describes a method by which such identification and quantification may be achieved. A real-life example in an installation environment using actual data is given and the resultant empirical model thus derived is described for the particular environment. The results have enhanced our understanding of system performance in terms of CPU utilization with promising potential for applications in system design, performance optimization, performance prediction, and characterization of installations. Some background leading to the particular problem as well as the future directions of the experiment are discussed in the paper.

1. INTRODUCTION

1.1 Historical Background

In planning for future data processing system performance requirements as well as in the process of evaluating current system performance, there is a need to identify the proper variables of the environment which will be the prime determinants of the system's behavior and to quantify the relationship. A survey of the state of the art in performance evaluation and prediction shows that there is no lack of both hardware and software measurement tools and resultant massive data; but as to what the proper variables are to measure and, in particular, how they relate quantitatively to performance, much work is needed. It was suggested by a number of people during the last couple of years, based upon the original definition by Dr. G. Amdahl, that the parameter

E/B* may be one of these key variables. It was observed, for instance, across a number of measured systems, that E/B has a noticeable effect upon CPU utilization of the particular system. In particular, there seems to be a somewhat linear relation between the two. Upon further analysis of more data, however, the correlation between the two appears to be lacking in certain systems and at certain times. It is with this recognition that there is perhaps need for some additional environment or application description parameters that I began to put together an experiment that might yield such parameters.

1.2 Objectives

In view of the above historical background, the primary objective of the experiment which I report in this paper was: to determine whether or not we are able to formalize a mathematical or empirical relationship between CPU utilization and E/B. If not, what additional variables are needed? Furthermore, the experiment was to be exploratory in nature; i.e., to assess the applicability and usefulness of an empirical method in the study of performance evaluation and prediction.

1.3 Data Base

The primary data base for our experiment was a subset of hardware measurement data (except for one variable, all data were collected by a pair of hardware monitor devices so as to minimize any interference with the normal operation of the system) on an IBM/360 model 85 system (System A; see Appendix I for a description of the system) covering a period of approximately two weeks of time. The measurements were conducted during normal operation for three shifts daily (weekends, normal maintenance time excluded) and recording of data by the monitors was initiated automatically at certain time intervals. This subset of data which consists of first shift only is tabulated in table I. In addition, a secondary source consisting of hardware measurement data from system B (see Appendix IV) covering approximately 3 days (first shift only) was used for derivation of a new equation and for validation.

2. METHODOLOGY

A linear relationship of the form

*E/B = number of instructions executed/number of bytes transferred. Amdahl's original definition has the form: number of bits transferred/number of instructions executed.

$$y = a_0 + \sum_{i=1}^{n} a_i x_i \qquad (1)$$

was sought, where y is the response (dependent) variable (CPU utilization of a given system in our case), the x_i's are the factors (independent variables) and the a_i's the coefficients to be determined. The following criteria are applied wherever possible in the process of our selection of variables and final formation of the model equation:

(i) System considerations
- system configuration independence
- known influence upon system's behavior
- ease of measurement in obtaining the data

(ii) Statistical
- miscellaneous statistical tests and criteria

(iii) Invariance over time

The following is a brief account of the procedure. Nine variables* were selected out of a total of forth-two measured and/or derived variables to form the basic set of variables from which the equation would be determined. (The variables and their definitions are listed in table II.) These variables were selected on the basis of their generally known influence upon system performance.

2.1 Correlation Matrices and Profiles

The simple correlation coefficient r defined below

$$r_{12} = \frac{\sum_{j=1}^{N} (x_{1j}-\bar{x}_1)(x_{2j}-\bar{x}_2)}{\sqrt{[\sum_{j=1}^{N} (x_{1j}-\bar{x}_1)^2][\sum_{j=1}^{N} (x_{2j}-\bar{x}_2)^2]}} \qquad (2)$$

is a measure of the linear interdependence between two variables. A correlation matrix consisting of such coefficients was constructed for the nine variables for each shift of each day during the measurement period. (Table III gives one such matrix and Appendix II contains some of the other matrices used in our study.) The index j where j=1,2,...,N in Eq. (2) above is the number of data points collected at certain time intervals for each of our data sets. From

*Channel/channel overlap variable was later added for our regression analysis.

these matrices a correlation profile consisting of correlation "weights" is constructed for each shift. The "weights" were assigned between two variables according to rules in table IV and their magnitudes indicate how consistent they are as well as the degree of correlation between the variables across different days. This was done to minimize the effect of variations in system workload from day to day. (It is known, on the other hand, that the workload changes considerably from shift to shift.) (Table V lists one of these profiles; others are contained in Appendix III.)

Finally, a summary list, taking into consideration also shift variations, was constructed from these profiles (table VI). Rank 1 in the table indicates the best-correlated pairs and so on in descending order. It showed that both the CPU/CH overlap variable and storage fetch variable were among the best-correlated with our response variable, CPU utilization. However, since these two variables were themselves highly correlated ($r>.8$), only one of them ought to be chosen as a candidate for our independent variable. The reason for choosing CPU/CH overlap in our case was based upon the fact that people have a much better feel for CPU-I/O overlap of their system than for number of storage fetches. Thus, applying our criteria of best correlation, statistical independence, and system considerations repeatedly, CPU/CH overlap and E/B became the two best variables for further analysis.

2.2 Regression

The final analysis of data was performed by including in the regression the two variables selected above and two additional variables, namely the number of initiators active and channel/channel overlap. Neither variable was supported by sufficient and/or available correlation data, but was included due to system considerations, i.e. for their generally known influence on system performance. In our regression analysis, variables were chosen one at a time starting from the one giving the greatest reduction in the sum of squares of the residuals* and adding on variables giving the

*The sum of squares of residuals is defined as

$$S = \sum_{i=1}^{N} r_i^2 \tag{3}$$

where residuals $r_i = y_i - \hat{y}_i$, y_i = observed y, and $\hat{y}_i$ = fitted or predicted y (by the regression equation (1)).

next greatest reduction and so on; this is commonly known as the stepwise forward regression technique. The procedure used is the one in STATPAC of the APL library.

Initially one set of data (M85-1-0 in table I) containing 30 data points was used for our computation (mainly due to the availability of data on number of initiators). The result showed variable E/B to be the most influential variable, with CPU/CH overlap ranked next and the other two variables almost negligible (see table VII). R^2, the multiple correlation coefficient defined as

$$R^2 = \frac{\sum_{i=1}^{n} (\hat{Y}_i - \bar{Y})^2}{\sum_{i=1}^{n} (Y_i - \bar{Y})^2} \tag{4}$$

where $\hat{Y}_i$ = the predicted value of Y, Y_i = the observed value of Y, and $\bar{Y}$ = the mean of Y, is a measure of the percentage predictable by the regression equation. Recognizing the limitations of the stepwise selection of variables, we tried eight combinations of variables (alghouth there are 2^4 possible combinations for 4 variables) to see if indeed the two variables (E/B and CPU/CH overlap) gave the best results (see table VIII). The table confirmed the stepwise selection of the two variables as the best candidates for our regression equation. Although the resulting equation (with 2 independent variables only) when used to predict CPU utilizations from five other data sets (M85-1-1 through M85-1-5 in table I) was found to be very satisfactory (over 90% of the points fall within ±10% of observed Y and over 70% within ±5%), the estimated regression coefficients were found to be varying (the relative degree of influence by the two variables also varied) when different data sets (from different days) were used to derive our regression equations, probably because of a change of workload from day to day. To improve the stability of the coefficients, approximately 5 days' worth of data (M85-1-0 through M85-1-5), which consists of 174 data points, were pooled together to comprise our new data base. The resulting regression equation and its accompanying significant statistics are summarized in the next section.

3. RESULTS, DISCUSSION, VALIDATION AND PHYSICAL INTERPRETATION OF THE MODEL

3.1 Results

General equation:

$$Y = a_0 + a_1x_1 + a_2x_2 \tag{5}$$

Resulting equation:

$$Y = 3.30 + 1.24x_1 + .84x_2 \tag{6}$$

where Y = CPU utilization (percent), x_1 = E/B and x_2 = CPU/CH overlap (percent).

Some interesting and significant statistics and plots are shown in tables IX and X and in figure 1. Another form of the same equation may be obtained by means of the following transformation of the regression equation:

$$Y = \bar{Y} + a_1(x_1-\bar{x}_1) + a_2(x_2-\bar{x}_2) \tag{7}$$

where $a_0 = \bar{Y} - a_1\bar{x}_1 - a_2\bar{x}_2$, $\bar{x}_1$ = mean of x_1 and $\bar{x}_2$ = mean of x_2. Thus our equation becomes

$$Y = 53.1 + 1.24(x_1-9.2) + .84(x_2-45.7) \tag{8}$$

Figure 1 (plot of residuals versus predicted Y of the original data) shows that 94% of the values predicted fall within ±10% and 83% within ±5% of observed Y.

3.2 Outliers

An examination of figure 1 (the residual plot) indicated two possible so-called "outliers" (here defined as $R = 3 * \sigma$ where σ is the estimated standard deviation (root mean square) of the residuals): namely, observations 88 and 170 (see also table X). When these two outliers were eliminated from our data base and a new equation fitted, the change in coefficients was, however, small. (The new coefficients are a_0': 3.15, a_1': 1.30, a_2': .83, and R^2: .957.) It is possible, nevertheless, to associate these two points with unusual system activities at that time. It was found, for example, that at observation 88 all selector channel utilizations were very low, and at observation 170 not only were the selector channel utilizations low but also there was a very low MIP and an unusually high multiplex channel utilization.

3.3 Validation

Two aspects of validation of the goodness of our model may be considered here:

(i) How well the regression equation fits data collected from days other than those from which the equation is derived.

(ii) Whether the same variable relationship holds true for different systems.

For (i) above, 134 data points were used (M85-2-1 through M85-2-5 in table I) and the result of applying our prediction equation is plotted in figure 2. It shows that 90% of all points fall within ±10% and 80% fall within ±5% of observed Y. For (ii) above, stepwise regression was applied to 90 data points on an entirely different system (see Appendix IV) using the same variables. The result is tabulated below and in table XI:

$$Y = 4.36 + 3.0x_1 + .93x_2 \tag{9}$$

The alternate form is

$$Y = 66.2 + 3.0(x_1-3.5) + .93(x_2-55.3) \tag{10}$$

where Y, x_1 and x_2 are as previously defined. Figure 3 is a plot of residuals versus Y. It shows that 97% of data points fall within ±10% and 90% within ±5% of observed Y. This proved that the same two variables are at least as good in accounting for the CPU utilization component as in the case of our previous system. Although lack of available data prevented us from starting with 9 variables as in the previous case, the regression result indicated to us that these two variables would have been the best ones had the same procedure been applied.

It is noticed that since CPU/CH overlap becomes the dominant factor (judging from R^2 contribution in the tables) in our model, the question arises whether it alone is sufficient to account for CPU utilization. Both the actual residual plots and multiple correlation coefficients (see table XII) indicate that the addition of the E/B term significantly enhanced the predictions of the model in both systems.

3.4 Physical Interpretation

The empirical equations (Eqs. (6), (8), (9) and (10)) that we obtained may be given the following very simple physical approximation. The fact that the equation is composed of these two variables can be viewed thus: in any computer system, the total time that a certain workload will take to complete can be divided roughly into two portions, as shown in figure 4. Clearly, what determines CPU utilization of the system is a function of CPU time spent and how much overlap is achieved, given the two components of CPU time and I/O time. Our variable in the model, E/B, which is a characteristic of the workload, measures how CPU- or I/O-bound our system is and our second variable, CPU/CH overlap, is a direct measure of the overlap achieved. In a single-thread environment, where CPU-I/O overlap is almost nonexistent, E/B then becomes the prime determinant of CPU utilization. In a multi-thread environment, the coefficient of the variable CPU/CH overlap may be viewed as a reflection of the degree of I/O-boundness for a given system; i.e., the more I/O-bound the system, the greater the coefficient (approaching 1, perhaps) as the first term of our equation becomes negligible.

In Eqs. (8) and (10), it is interesting to note that when x_1 (E/B) and x_2 (CPU/CH overlap) fall below their mean values, the two terms give negative contributions to Y (CPU utilization). This seems to say that one cannot expect the performance of the particular system under consideration to be above a certain level (performance level here is measured by the mean of CPU utilization) if the parameters that characterize the workload fall below their mean values. How much of this performance would be affected is determined by the coefficients of the equation.

Figures 5 through 7 show that for the same net change in E/B and CPU/CH overlap, system B is more sensitive (resulting in a larger change in CPU utilization) than system A because of larger coefficients. This is a quantitative measurement relating the particular environment with its performance.

4. POSSIBLE APPLICATIONS OF THE MODEL AND GENERAL CONCLUSIONS

4.1 Possible Applications

(i) Enhance our understanding of the determinants of CPU utilization of a system. E/B is one of the prime determinants of CPU utilization, but it is not possible to formalize an accurate relationship all by itself across different environments and systems. In predicting CPU utilization, the CPU-I/O overlap factor must be taken into consideration. The equation confirms and quantifies our intuitive notion of the influence of CPU-I/O overlap and our observation of E/B effects upon CPU utilization in certain environments.

(ii) Prediction. We are able to derive or interpolate within the range of our model the values of the response variable given the values of the other two that appear in the equation. Since the relative effects of the two variables are quantified in our model, the trade-off can be made between altering E/B or CPU/CH overlap characteristics in order to maximize CPU utilization. (This utility is, however, contingent upon the stability of the regression coefficients when one or more of the variables is changed.) For example, we can increase CPU/CH overlap by running more initiators, relocation of data sets or I/O buffering; or we can change E/B by making certain data sets resident in core. How much change in each of these variables is necessary to effect a given change in CPU utilization can be determined from the equation. At the design level, ranges of these values may be derived from the model as guidelines. In Appendix V it is shown that a similar model may be derived for CPU/CH overlap (as the response variable) using CPU utilization and E/B as the independent variables. This may be more desirable at times since it takes more to measure CPU/CH overlap. It is also conceivable that after deriving X number of regression equations for X systems, we may be able to derive an equation for the (X+1)st system by a certain relation of the coefficients between hardware, workload and the regression equations.

(iii) Characterization of an environment or application may be achieved by means of the coefficients of the model. It may be possible to calibrate workloads and environments by comparing the coefficients in their representative equations. This would be a step forward in the quantitative understanding of the relation between workload and performance. At present, however, more data (see section 5 below) are needed to make a general assessment of this utility.

(iv) Location of "outliers" by means of the model points to abnormal or unusual system conditions (barring the possibility of gross measurement error), as mentioned in section

3. This may lead to better understanding of our system and potential areas deserving further investigation.

4.2 Some General Conclusions Resulting from the Experiment

(i) It is possible to derive a highly accurate CPU utilization model in terms of two system variables only by regression analysis for different systems. Evidences seem to indicate that this is applicable for all systems with the possible exception of systems with multiple CPUs where we run into definition problems with our variables.

(ii) The empirical method outlined in this paper has been successfully applied in one form or another in various disciplines. The result of this experiment gave further evidence that it also has applicability in computer performance evaluation.

5. FUTURE DIRECTIONS

5.1 Improvement of the Model

(i) One major drawback of the model in its present form is the inherent system dependency of one of the independent variables, CPU/CH overlap, as well as its inherent association with our response variable. (E/B, on the other hand, is a characteristic of the workload which varies only slightly from system to system.) Therefore, if CPU/CH overlap can be broken down into components which are more directly measurable (without also having to measure CPU time) and which are more independent of the system configuration, it would be a great step forward for its utility as a prediction tool at the design level. I might add that the methodology employed in this study may well be used for the task of identifying such variables. It would be interesting to replace CPU/CH overlap with supervisor time, MIP, and/or storage fetch, as our study indicated that these variables are also correlated with CPU utilization.

(ii) Calibration of the regression coefficients. How the coefficients in the model vary relative to changes in environment and/or applications needs to be determined. This may be accomplished through a series of measurements and through applying regression analysis to a given system with varying jobstreams characterizing various applications. The jobstream may be a synthetic one whose parameters (E/B and CPU-I/O overlap) can be varied by means of programming. Also, equations need to be derived for different systems.

5.2

It was due to the availability of data and a theoretical as well as historical interest that CPU utilization was chosen as our response variable in the experiment. Needless to say, there are other more interesting and perhaps much needed variables that deserve our attention. For example, what are the variables and how much do they contribute to the response time in a time-sharing environment? (Probable candidates for independent variables: number of active tasks; CPU utilization.) It was reported recently that similar methodology was used successfully in the prediction of data set movements in data staging study. These are but two examples of possible uses under very different operation mode and environment. Therefore, experiments in the future should be directed towards embracing more variables of interest (perhaps by means of both software and hardware measurement techniques), and the resultant data should be analyzed by the general methodology outlined in this paper. This will enable us to establish a basic set of key variables in various environments, to quantify their effects, and to provide insight and a realistic data base for the formulation and validation of analytical and/or simulation models.

Much work needs to be done before generalized conclusions can be drawn as to the specific applications of the above results, aside from their theoretical significance. The author hopes, however, that this paper may serve the purpose of pointing out the feasibility and promising potential of an empirical approach, the usefulness of which has hardly been explored up to now in the realm of evaluation of computer systems performance.

Acknowledgments. The author is particularly indebted to Dr. R.F. Tsao and Mr. Herman P. Friedman for their generous assistance and many valuable suggestions. The author also wishes to thank Mr. Ints Delgalvis for his help and valuable comments.

REFERENCES

1. Daniel, C., and Wood, F.S. (1971): Fitting equations to data, Wiley-Interscience.
2. Draper, N.R., and Smith, H. (1967): Applied regression analysis, John Wiley & Sons, Inc.

Table I: Data base from system A. In addition, a subset of hardware measurement data from system B (see Appendix IV) covering approximately 3 days (1st shift only) was used for validation purposes.

Date	Data Designation	# of Data Points	Measurement Interval Length
3/8	M85-1-0	30	120 secs
3/9	M85-1-1	30	120 secs
3/1	M85-1-2	29	1000 secs
3/2	M85-1-3	30	1000 secs
2/25	M85-1-4	30	100 secs
2/23	M85-1-5	25	1000 secs
2/18	M85-2-1	18	1000 secs
2/19	M85-2-2	24	1000 secs
2/24	M85-2-3	29	1000 secs
2/25	M85-2-4	33	100 secs
3/1	M85-2-5	30	1000 secs

Table II: Variables selected for our experiment.

Variable	Description	Unit
CPU Utilization	Total CPU Time/Total Elapsed Time	%
Supervisor	CPU Time in Supervisor State / Total Elapsed Time	%
MIP	(Actual) Millions of instructions executed per seconds	-
Buffer Hit Ratio	# of Buffer Fetches / # of Total Fetches	Fraction
E/B	# of Instructions Executed / # of Bytes Transferred	-
Average Record Size	Average Record Length on a DASD, Assuming Average Rotational Delay	Bytes
CPU/CH Overlap	Time in CPU State Overlapped with Any of the Selector Channel Busy/Total Elapsed Time	%
Storage Fetch	# of Fetches to Main Storage	-
INITIATOR	# of Initiators Active	-

Table III: Correlation matrix.

Date: 3/8

Time Interval:120 sec

Shift: 1st

# of Data Points: 30	CPU	Supv	MIP	Buffer Hit Ratio	E/B	Av. Record Size	CPU/CH Overlap	Stor Fetch	# of Init
CPU	1	.001	.59	.46	.87	.22	.57	.49	-.35
Supv		1	-.49	-.70	-.37	-.56	.59	.84	.72
MIP			1	.80	.75	.66	.07	-.19	-.66
Buffer Hit Ratio				1	.61	.65	-.13	-.47	-.76
E/B					1	.39	.17	.12	-.63
Av. Record Size						1	-.26	-.42	-.51
CPU/CH Overlap							1	.79	.32
Stor Fetch								1	NA
# of Initiators									1

Table IV: Rules for assigning correlation profile weights.

Weights	Rules
+3	All correlation coefficients (from correlation matrices) are above .5
+2	50-50 above and below .5
+1	Mostly or all below .5
0	No apparent correlation

NOTE: 1) Negative weights are assigned the same way
2) Those rules are simply guidelines. There are ambiguous situations where rules are difficult to apply.

Table V: Correlation profile.

Date: 3/8, 3/9 Time Interval:120 sec Shift: 1st # of Data Points:	CPU	Supv	MIP	Buffer Hit Ratio	E/B	Av. Record Size	CPU/CH Overlap	Stor Fetch	# of Init
CPU	*								
Supv	1	*							
MIP	3	0	*						
Buffer Hit Ratio	2+	-1	3	*					
E/B	3	0	3	3	*				
Av. Record Size	0	-2	1	1	1	*			
CPU/CH Overlap	3	2	1	1	2	-1	*		
Stor Fetch	3	3	1	0	2	-1	3	*	
# of Initiators	0	3	-1	-1	-2	-2	1	1+.	*

Table VI: Summary of correlation weights from correlation profiles.

Rank	Variable Pair		Rules
1	CPU/CH Overlap	- CPU Utilization	All + 3
	Stor Fetch	- CPU Utilization	All + 3
	CPU/CH Overlap	- Stor Fetch	All + 3
2	Buffer Hit Ratio	- E/B	2(+3), 3(+2)
	Buffer Hit Ratio	- MIP	3(+3), 2(+2)
3	CPU/CH Overlap	- Supervisor	1(+3), 3(+2), 1(+1)
	Stor Fetch	- Supervisor	1(+3), 3(+2), 1(+1)
	E/B	- MIP	5(+2)
	CPU Utilization	- Supervisor	4(+2), 1(0)
	CPU Utilization	- Buffer Hit Ratio	4(+2), 1(+1)
	CPU Utilization	- MIP	1(+3), 2(+2), 2(+1)
4	CPU Utilization	- E/B	1(+2), 4(+1)
	CPU/CH Overlap	- MIP	1(+3), 1(+2), 3(+1)
	Stor Fetch	- MIP	2(+2), 3(+1)
	CPU/CH Overlap	- Buffer Hit Ratio	2(+2), 3(+1)

NOTE: Pairs of variables are arranged in descending order according to their correlation weights.

Table VII: Contribution of each variable toward variation in the dependent variable as a result of stepwise regression.

Variable	Contribution Towards R^2
E/B	.79
CPU/CH Overlap	.16
CH/CH Overlap	.006
# of Initiators	.0005
Total	.97

Table VIII: Regression combinations.

Variable (I)*	R^2 (Multiple Correlation Coefficient)
1	.76
2	.33
4	.30
5	.23
1, 2	.94
2, 4	.89
1, 2, 4	.96
1, 2, 5	.95

*Variable (1) is E/B
Variable (2) is CPU/CH overlap
Variable (4) is # of initiators
Variable (5) is CH/CH overlap

Table IX: Summary of significant statistics.

Variable (I)	Coeff.	Standard Error (Coeff.)	T-Value	R^2 (Contribution)	Min X(I)	Max X(I)	Range X(I)	Confidence Interval (at 95%)
0	3.30							3.3 ± 4.29
1	1.24	.062	19.8	.125	1.7	40.6	38.9	1.24 ± .26
2	.84	.019	43.8	.819	1	98	97	.84 ± .08

Min Y = 4 Max. Y = 100 Range Y = 96

# of Observations	174
# of Independent Variables	2
Residual Degree of Freedom	171
F-Value	1483.4
Residual Root Mean Square	5.31
Residual Mean Square	28.21
Residual Sum of Square	4824.9
Total Sum of Square	88537.7
Multiple Correlation Coefficients (R^2)	.945

Table X: Ordered by residual.

OBSV. NO.	*OBSERVED Y*	*PREDICTED Y*	*ORDERED RESIDUAL*
88	12	37.9172	25.9172
148	100	112.5624	12.5624
86	35	46.1556	11.1556
89	7	17.3676	10.3676
126	65	75.2448	10.2448
87	66	74.706	8.706
101	42	50.4384	8.4384
133	95	103.2508	8.2508
150	29	37.0756	8.0756
134	74	81.8164	7.8164
137	78	85.2752	7.2752
158	11	17.528	6.528
157	20	25.308	5.308
138	100	105.188	5.188
32	22	27.1716	5.1716
128	97	102.1652	5.1652
24	46	51.1068	5.1068
7	53	57.9788	4.9788
125	49	53.788	4.788
56	99	103.6404	4.6404
116	10	14.636	4.636
51	75	79.4876	4.4876
136	88	92.4232	4.4232
84	86	90.4108	4.4108
142	66	70.396	4.396
8	29	33.3796	4.3796
124	22	26.3684	4.3684
121	9	13.3372	4.3372
6	62	66.2492	4.2492
166	70	74.1872	4.1872
120	10	14.174	4.174
72	86	89.6608	3.6608
61	35	38.5748	3.5748
37	39	42.536	3.536
123	16	19.3724	3.3724
4	35	38.1096	3.1096
40	24	26.9972	2.9972
129	100	102.872	2.872
90	31	33.86	2.86
38	36	38.8536	2.8536
171	40	42.7964	2.7964
160	38	40.574	2.574
143	64	66.5368	2.5368
140	62	64.4508	2.4508
23	44	46.3144	2.3144
139	92	94.304	2.304
115	4	6.2852	2.2852
167	43	45.2792	2.2792
53	51	53.2484	2.2484
93	49	51.24	2.24
104	20	22.2152	2.2152
135	89	91.1404	2.1404
151	100	102.1308	2.1308
48	35	37.0092	2.0092
141	58	59.8696	1.8696
81	55	56.8564	1.8564
9	47	48.844	1.844
52	44	45.5456	1.5456
49	38	39.3648	1.3648
22	43	44.3212	1.3212

Table X (continued)

OBSV. NO.	*OBSERVED Y*	*PREDICTED Y*	*ORDERED RESIDUAL*
117	70	71.23	1.23
43	14	15.2216	1.2216
2	35	36.2188	1.2188
82	48	49.1724	1.1724
46	46	47.1636	1.1636
50	44	45.1612	1.1612
98	54	55.112	1.112
83	49	50.0868	1.0868
28	34	35.0688	1.0688
59	32	32.9452	0.9452
33	41	41.9408	0.9408
91	41	41.916	0.916
5	58	58.8528	0.8528
74	42	42.852	0.852
100	59	59.8472	0.8472
130	61	61.8468	0.8468
27	21	21.7152	0.7152
55	100	100.652	0.652
103	17	17.646	0.646
36	47	47.61	0.61
60	47	47.5292	0.5292
62	53	53.4968	0.4968
39	33	33.4196	0.4196
114	42	42.3724	0.3724
42	28	28.3516	0.3516
30	37	37.3376	0.3376
44	32	32.2296	0.2296
66	49	49.222	0.222
47	40	40.2172	0.2172
65	61	61.1652	0.1652
57	56	56.1248	0.1248
156	35	35.1184	0.1184
63	64	64.0484	0.0484
127	54	54.0392	0.0392
163	67	66.946	¯0.054
111	74	73.8832	¯0.1168
10	24	23.7736	¯0.2264
168	45	44.7492	¯0.2508
75	55	54.7144	¯0.2856
70	79	78.7124	¯0.2876
45	37	36.69	¯0.31
161	46	45.6664	¯0.3336
108	36	35.6512	¯0.3488
92	58	57.6344	¯0.3656
172	44	43.6084	¯0.3916
147	63	62.532	¯0.468
97	59	58.5244	¯0.4756
58	42	41.5096	¯0.4904
154	22	21.4708	¯0.5292
41	31	30.388	¯0.612
144	60	59.3328	¯0.6672
1	45	44.29	¯0.71
71	54	53.2888	¯0.7112
159	15	14.2672	¯0.7328
162	67	66.258	¯0.742
169	54	53.2576	¯0.7424
26	40	39.1756	¯0.8244
25	48	47.1728	¯0.8272
21	72	71.1272	¯0.8728
96	50	49.1008	¯0.8992
113	49	48.0968	¯0.9032
35	54	53.0964	¯0.9036

Table X (continued)

OBSV. NO.	*OBSERVED Y*	*PREDICTED Y*	*ORDERED RESIDUAL*
80	47	46.0792	¯0.9208
113	57	56.0784	¯0.9216
109	36	34.96	¯1.04
153	41	39.9168	¯1.0832
149	78	76.8	¯1.2
152	55	53.7752	¯1.2248
64	52	50.7688	¯1.2312
106	92	90.7236	¯1.2764
165	46	44.6124	¯1.3876
34	51	49.6032	¯1.3968
174	44	42.5636	¯1.4364
94	52	50.5424	¯1.4576
112	60	58.512	¯1.488
107	52	50.2512	¯1.7488
146	84	82.2272	¯1.7728
79	53	50.9856	¯2.0144
20	51	48.9708	¯2.0292
31	34	31.9448	¯2.0552
95	58	55.7996	¯2.2004
29	60	57.7372	¯2.2628
155	29	26.7188	¯2.2812
68	58	55.4804	¯2.5196
78	55	52.2252	¯2.7748
54	100	97.1768	¯2.8232
76	95	92.1248	¯2.8752
67	37	33.1536	¯3.8464
69	97	92.7356	¯4.2644
132	96	91.7228	¯4.2772
131	66	61.658	¯4.342
145	75	70.538	¯4.462
12	77	72.038	¯4.962
14	45	39.8248	¯5.1752
119	57	51.7792	¯5.2208
105	50	44.7652	¯5.2348
110	71	64.9928	¯6.0072
13	76	69.7972	¯6.2028
164	57	50.5976	¯6.4024
17	93	86.0688	¯6.9312
73	65	57.718	¯7.282
173	72	64.5968	¯7.4032
102	71	62.5064	¯8.4936
18	49	39.7408	¯9.2592
11	58	48.1628	¯9.8372
77	74	64.0756	¯9.9244
85	58	47.8884	¯10.1116
19	48	36.2288	¯11.7712
99	71	58.572	¯12.428
15	95	81.534	¯13.466
122	49	35.3576	¯13.6424
3	72	57.2224	¯14.7776
16	99	82.6208	¯16.3792
170	40	19.0504	¯20.9496

Table XI: Summary of significant statistics (system B).

Variable (I)	Coeff.	Standard Error (Coeff.)	T-Value	R^2 (Contribution)	Min X(I)	Max X(I)	Range X(I)	Confidence Interval (at 95%)
0	4.36							4.36 ± 6.46
1	3.00	.173	17.3	.142	1.84	26.26	24.42	3.00 ± .74
2	.93	.021	44.4	.816	1	89	88	.93 ± .09

Min Y = 5 | Max Y = 97 | Range Y = 92

# of observations	90
# of Independent Variables	2
Residual Degree of Freedom	87
F-Value	1004.2
Residual Root Mean Square	4.01
Residual Mean Square	16.05
Residual Sum of Square	1396.7
Total Sum of Square	33640.9
Multiple Correlation Coefficients	.958

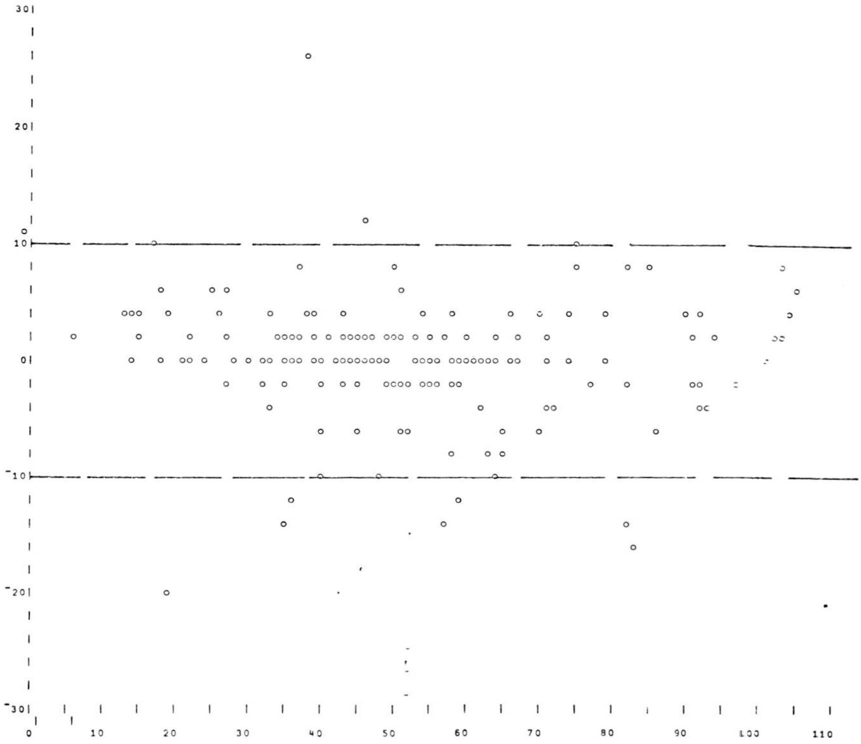

Fig. 1. Residuals versus predicted Y (original data, system A).

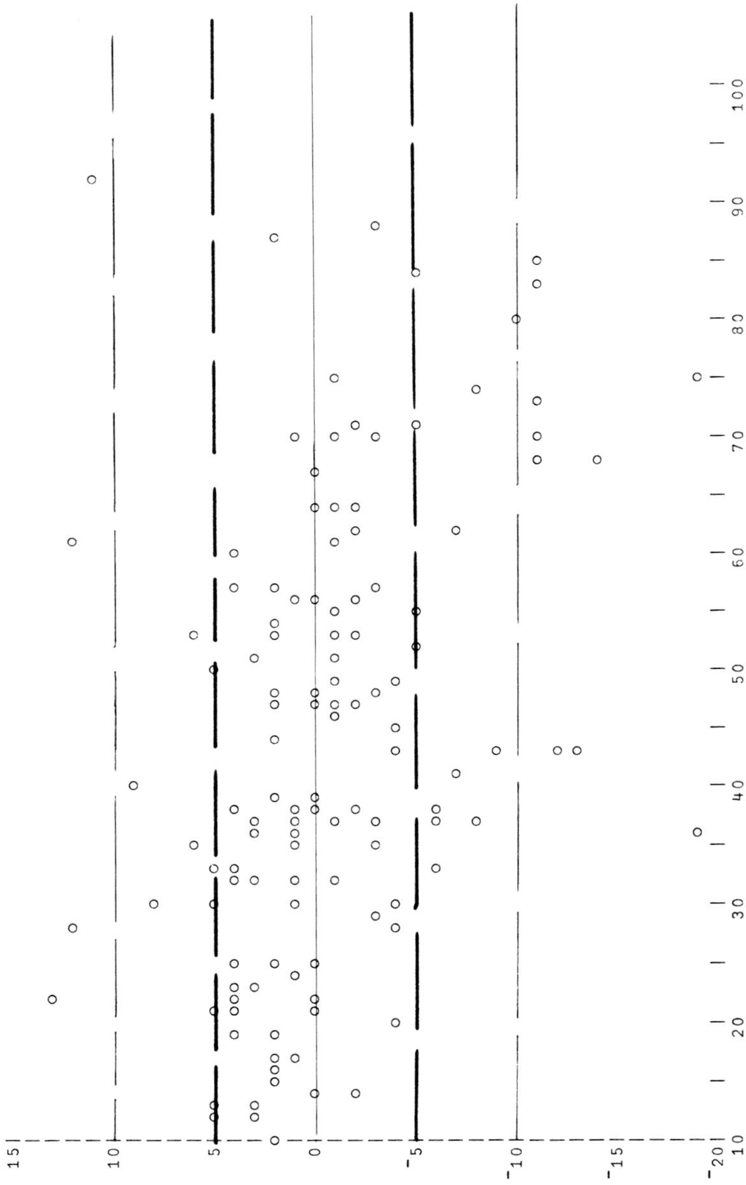

Fig. 2. Residuals versus predicted Y (validation data, system A).

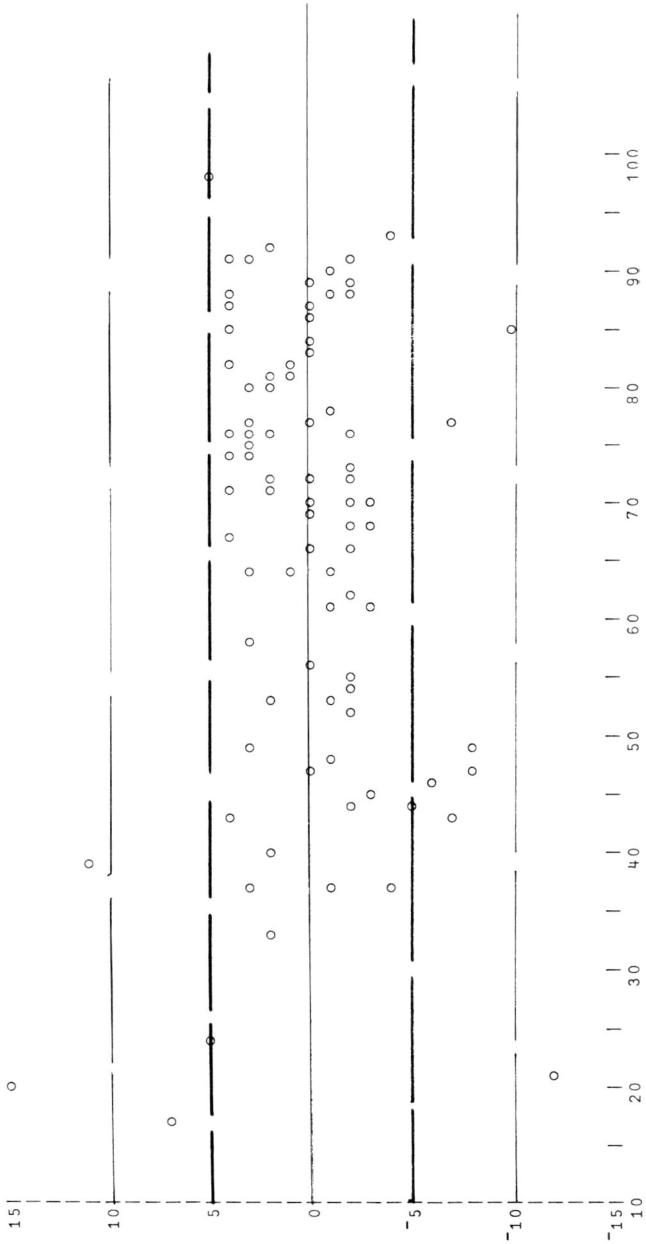

Fig. 3. Residuals versus predicted Y (original data, system B).

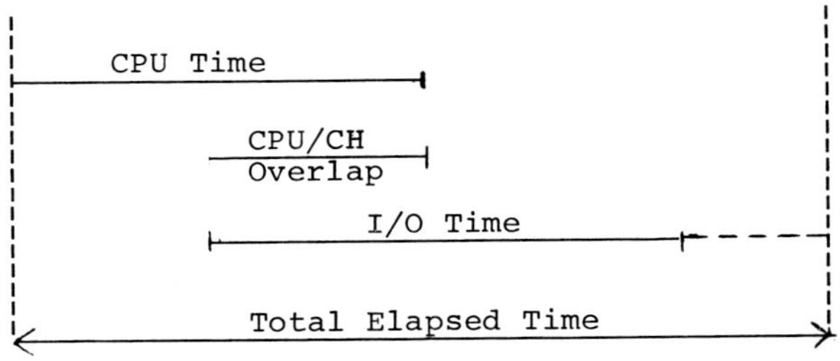

Fig. 4.

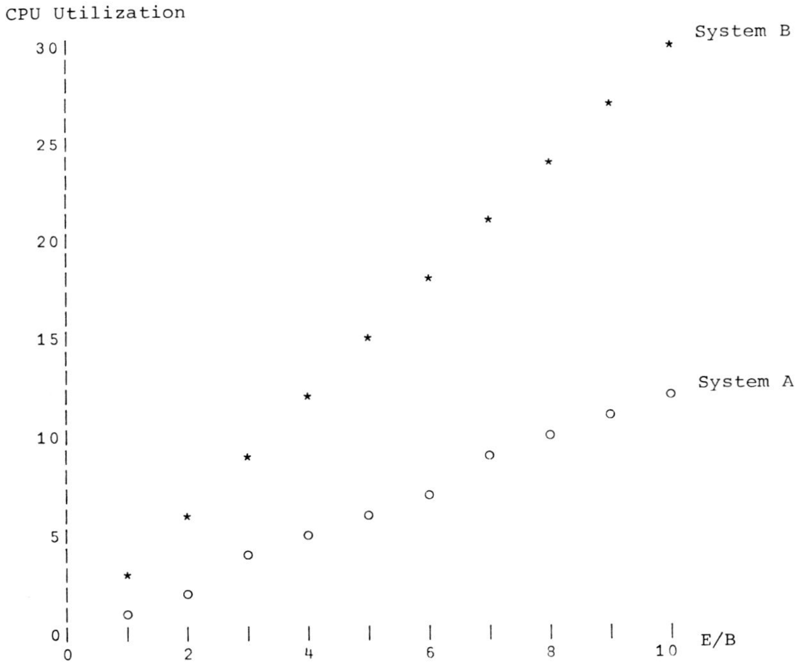

Fig. 5. Change in CPU utilization versus net change in E/B.

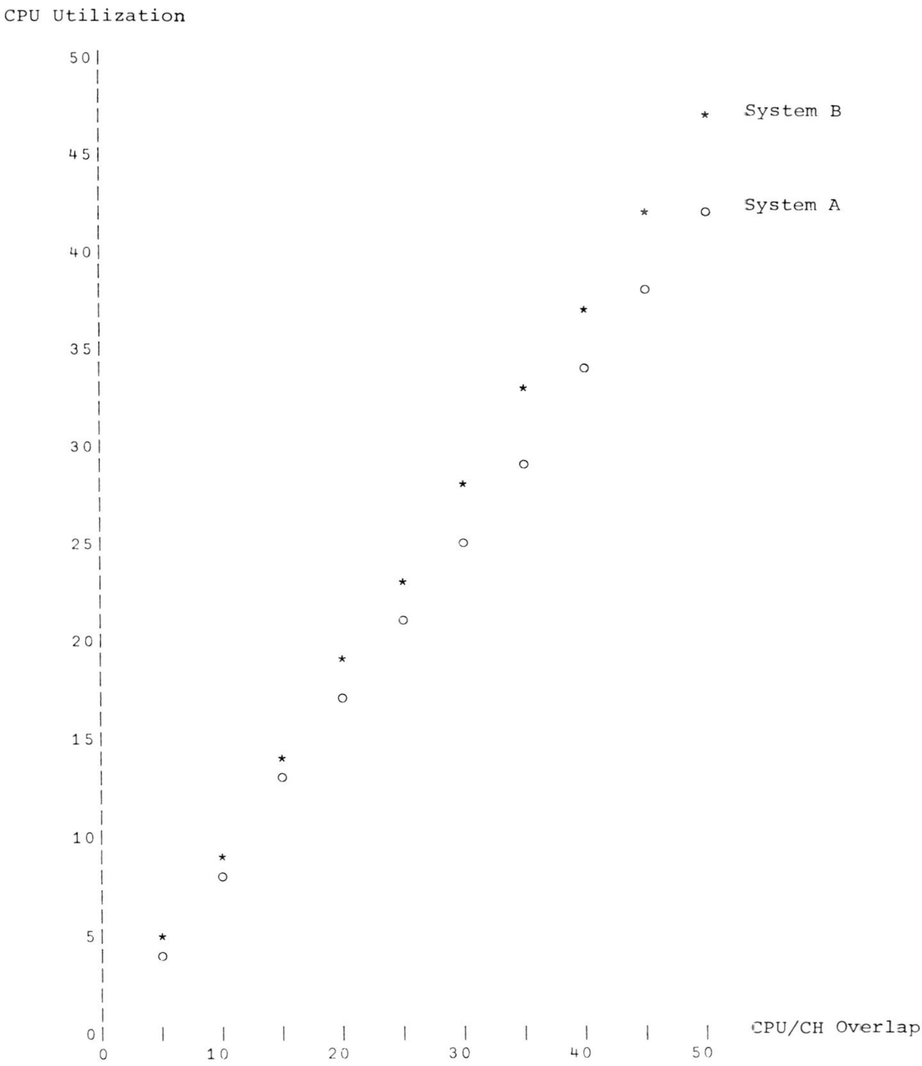

Fig. 6. Change in CPU utilization versus net change in CPU-I/O overlap.

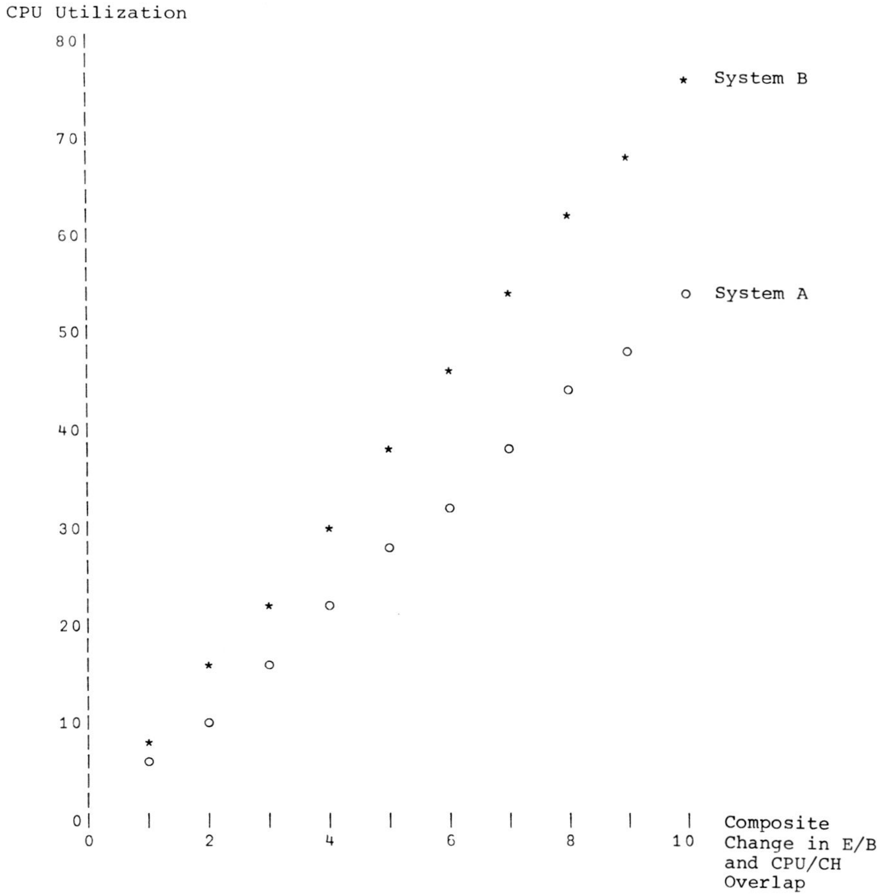

Fig. 7. Change in CPU utilization versus net change in both E/B and CPU-I/O overlap.

APPENDIX I: DESCRIPTION OF SYSTEM A

System A is a general computing center complex. The basic system configuration is summarized below.

Hardware:

- 360/M85 CPU with 16K cache and HSM
- 2M main memory
- 6 selector channels (in practice only 5 are actually utilized)
  - 7 2314 facilities
  - 2 2301 drums
  - 24 tape drives
- 1 multiplex channel
  - 43 APL lines, 300-400 terminals
  - 21 CRBE lines
  - RJE
  - 9 printers
  - 3 RDR/Punch

Software:

- OS/MVT release 18.6
- LASP
- APL
- CRBE
- BATCH region

Workload Description (selected parameters and first shift only):

	Range Min	Range Max	Average
Number of jobs processed*	322	386	355
Average number of initiators active*	3.5	4.6	4
Average core used*	390K	570K	470K
Average APL users	15	20	18
CPU utilization	43	57	50
CPU/CH overlap	37	46	40
E/B	6	9	7

*Batch processing only.

APPENDIX II

Table I. Correlation matrix (3/9/71).

Date: 3/9/71 Time Interval: 120 sec Shift: 1st # of Data Points: 30	CPU	Supv	MIP	Buffer Hit Ratio	E/B	Av. Record Size	CPU/CH Overlap	Stor Fetch	# of Init
CPU	1	.45	.75	.86	.80	-.12	.99	.80	.19
Supv		1	.17	.14	.22	-.25	.47	.86	.53
MIP			1	.83	.77	-.25	.74	.51	.11
Buffer Hit Ratio				1	.71	-.15	.84	.49	.08
E/B					1	.08	.75	.57	-.13
Av. Record Size						1	-.13	-.25	-.48
CPU/CH Overlap							1	.81	.23
Stor Fetch								1	.42
# of Initiators									1

Table II. Correlation matrix (2/18/71).

Date: 2/18
Time Interval: 1000 sec
Shift: 1st

# of Data Points: 18	CPU	Supv	MIP	Buffer Hit Ratio	E/B	Av. Record Size	CPU/CH Overlap	Stor Fetch	# of Init
CPU	1	.15	.59	.64	.35	.12	.85	.81	NA
Supv		1	-.22	-.13	-.60	-.16	.56	.33	NA
MIP			1	.85	.73	-.39	.39	.16	NA
Buffer Hit Ratio				1	.66	-.31	.41	.12	NA
E/B					1	-.28	-.09	-.02	NA
Av. Record Size						1	.11	.35	NA
CPU/CH Overlap							1	.78	NA
Stor Fetch								1	NA
# of Initiators									1

Table III. Correlation matrix (2/19/71).

Date: 2/19
Time Interval: 1000 sec
Shift: 1st

# of Data Points: 24	CPU	Supv	MIP	Buffer Hit Ratio	E/B	Av. Record Size	CPU/CH Overlap	Stor Fetch	# of Init
CPU	1	.46	.15	.29	.19	.32	.95	.88	NA
Supv		1	.13	.26	.21	.39	.39	.36	NA
MIP			1	.93	.86	-.10	-.07	-.25	NA
Buffer Hit Ratio				1	.83	-.03	.08	-.16	NA
E/B					1	-.10	-.09	-.18	NA
Av. Record Size						1	.29	.36	NA
CPU/CH Overlap							1	.93	NA
Stor Fetch								1	NA
# of Initiators									1

Table IV. Correlation matrix (2/23/71).

Date: 2/23
Time Interval: 1000 sec
Shift: 1 st

# of Data Points: 25	CPU	Supv	MIP	Buffer Hit Ratio	E/B	Av. Record Size	CPU/CH Overlap	Stor Fetch	# of Init
CPU	1	.68	.48	.39	.55	-.49	.95	.85	NA
Supv		1	-.22	-.13	-.60	-.16	.56	.65	NA
MIP			1	.03	.40	-.27	.64	.65	NA
Buffer Hit Ratio				1	.76	-.02	.21	-.07	NA
E/B					1	-.16	.43	.20	NA
Av. Record Size						1	-.57	-.58	NA
CPU/CH Overlap							1	.93	NA
Stor Fetch								1	NA
# of Initiators									1

Table V. Correlation matrix (2/24/71).

Date: 2/24 Time Interval: 1000 sec Shift: 1st # of Data Points: 29	CPU	Supv	MIP	Buffer Hit Ratio	E/B	Av. Record Size	CPU/CH Overlap	Stor Fetch	# of Init
CPU	1	.83	.40	-.06	.20	-.43	.97	.91	NA
Supv		1	-.02	-.41	-.08	-.54	.88	.95	NA
MIP			1	.50	.45	.31	.31	.18	NA
Buffer Hit Ratio				1	.71	.45	-.19	-.39	NA
E/B					1	.24	.12	-.04	NA
Av. Record Size						1	-.47	-.49	NA
CPU/CH Overlap							1	.94	NA
Stor Fetch								1	NA
# of Initiators									1

APPENDIX III

Table I. Correlation profile (first shift).

Date: 2/18, 2/19, 2/23, 2/24 Time Interval: 1000 sec Shift: 1st # of Data Points:	CPU	Supv	MIP	Buffer Hit Ratio	E/B	Av. Record Size	CPU/CH Overlap	Stor Fetch	# of Init
CPU	*								
Supv	2	*							
MIP	1+	-1	*						
Buffer Hit Ratio	1	0	2+	*					
E/B	1+	0	2	3	*				
Av. Record Size	0	-1+	-1	-1	-1	*			
CPU/CH Overlap	3	2+	1	1	0	0	*		
Stor Fetch	3	2	1	-1	0	0	3	*	
# of Initiators	NA	NA	NA	NA	NA	NA	NA	NA	*

Table II. Correlation profile (second shift).

Date: 2/18, 2/19, 2/23, 2/24
Time Interval:1000 sec
Shift: 2nd

# of Data Points:	CPU	Supv	MIP	Buffer Hit Ratio	E/B	Av. Record Size	CPU/CH Overlap	Stor Fetch	# of Init
CPU	*								
Supv	2	*							
MIP	1	0	*						
Buffer Hit Ratio	2	0	2	*					
E/B	1+	-1	2	3	*				
Av. Record Size	0	-1	0	-1	-1	*			
CPU/CH Overlap	3	2	1	1	0	1	*		
Stor Fetch	3	2+	1	-1	-1	1	3	*	
# of Initiators	NA	NA	NA	NA	NA	NA	NA	NA	*

Table III. Correlation profile (APL shift).

Date: 3/1, 3/2, 3/3
Time Interval: 1000 sec
Shift: APL*

# of Data Points:	CPU	Supv	MIP	Buffer Hit Ratio	E/B	Av. Record Size	CPU/CH Overlap	Stor Fetch	# of Init
CPU	*								
Supv	0	*							
MIP	2	-1	*						
Buffer Hit Ratio	2	-1	3	*					
E/B	1	-1	2	2	*				
Av. Record Size	-1	-1	0	0	0	*			
CPU/CH Overlap	3	1	1	1	0	-1	*		
Stor Fetch	3	1	1	0	-1	-1	3	*	
# of Initiators	NA	NA	NA	NA	NA	NA	NA	NA.	*

* APL Shift starts 8:00 a.m., ends 7:30 p.m., i.e., during the time APL is running

APPENDIX IV: DESCRIPTION OF SYSTEM B

System B is a dedicated on-line large data base inquiry system. The basic system configuration is summarized below.

360/M65 CPU
1M main memory
6 selector channel (channel 6 is utilized very little)
- 11 2314 facilities
- 1 2301 drum
- 13 tape drives

Multiplexor channel
- 50 1050 telephone lines
- 1 2740 telephone line
- 4 printers
- 1 reader/punch

APPENDIX V: CPU/CH OVERLAP MODEL

The following is a summary of regression applied to data (M85-1-0 through M85-1-5 in table I) with CPU/CH overlap as our response variable.

$$y = -.51 - 1.28x_1 + 1.09x_2 \qquad (11)$$

where y = CPU/CH overlap (percent), x_1 = E/B and x_2 = CPU utilization (percent). The alternate equation is:

$$y = 45.7 - 1.28(x_1 - 9.2) + 1.09(x_2 - 53.1) \qquad (12)$$

Figure **I** is a plot of residuals versus predicted y of the original data (M85-1-0 through M85-1-5). It shows that 91% fall within ±10% and 72% within ±5% of observed values of y.

Validation: (134 data points, M85-2-1 through M85-2-5): 92% of predicted values fall within ±10% and 75% within ±5% of observed values of y (see figure II).

APPENDIX V (continued)

Table I. Summary of significant statistics (system A, CPU/CH overlap model).

Variable(I)	Coeff	Standard Error	T-Value	R^2	Min X(I)	Max X(I)	Rang X(I)
0	- .51						
1	-1.28	.084	-15.2	.104	1.7	40.6	38.9
2	1.09	.025	43.8	.819	4	100	96

Min Y=1 Max Y=98 Range Y=97

# of Observation	174
# of Independent Var.	2
Residual d.f.	171
F-Value	1035.2
Residual Root Mean Sq.	6.07
Residual Mean Sq.	36.84
Residual Sum of Sq.	6301.2
Total Sum of Sq.	82601.3
Multiple Correlation Coeff.	.923

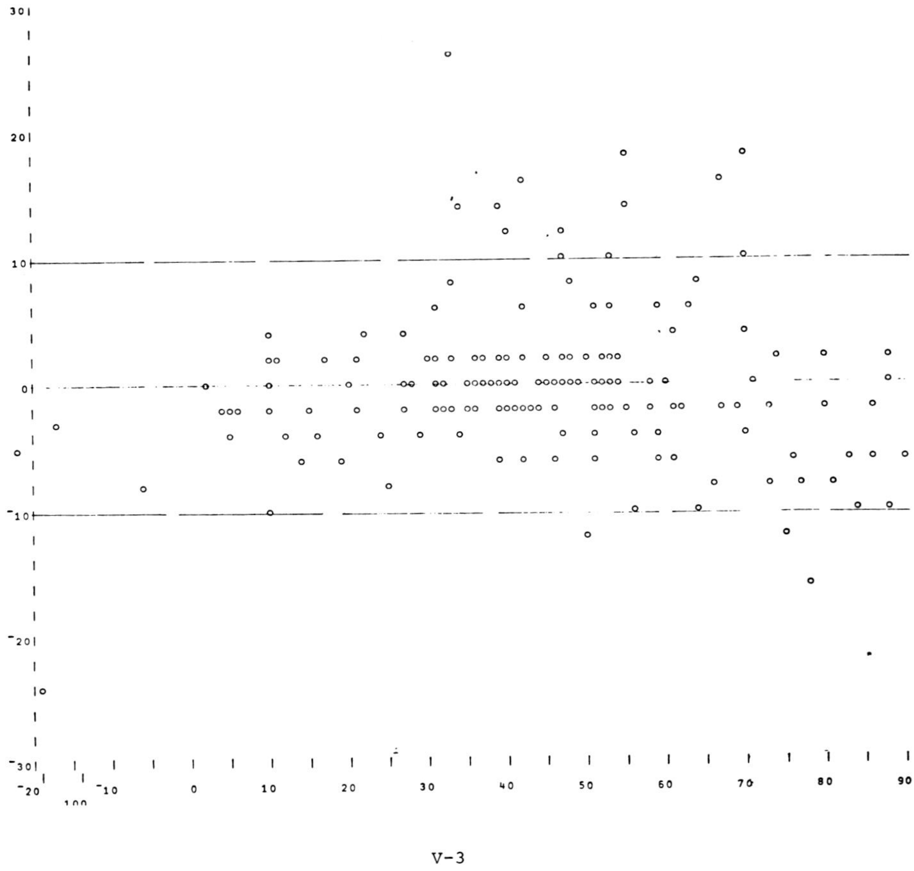

Fig. 1. Residuals versus predicted Y (original data, system A).

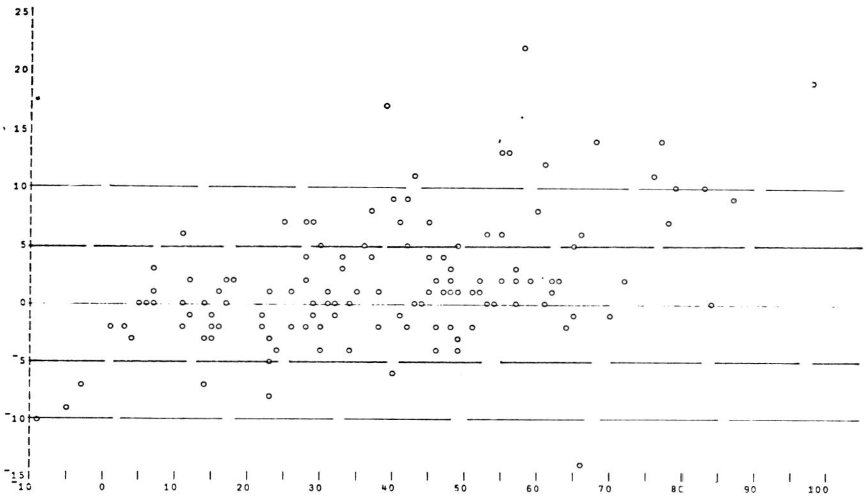

Fig. 2. Residuals versus predicted Y (validation data, system A).

ON THE STRUCTURE OF CP-67 OVERHEAD

Y. Bard and K.V. Suryanarayana

INTRODUCTION

An important factor in the evaluation of system performance is the overhead incurred by the system in performing its various functions. In the case of a time-sharing computer system, the overhead is measured in terms of CPU time consumed by the operating system. One is then interested in attributing the total observed overhead to the various specific functions performed by the system. In previous reports [1,2] we suggested that if overhead time and counts of various events are monitored over a large number of sampling periods, then the overhead per event can be estimated by regressing the observed overhead on the event counts. This method was applied to data records for a CP-67 system [3], and it was shown [2] that the overhead estimates varied in a reasonable manner from one version of CP-67 to another. It was also shown that the regression model could predict error with a mean square error of 85-90% given the counts of five or so basic event types, provided that a few outlying observations were ignored (about 3% of the total). While these results were encouraging, we felt that they could be improved upon. First, we hoped to get an even better fit to the data; second, we observed that if the data available were broken up into several subsets, and if the regression model was applied to each subset, then the estimates of overhead per event of each type varied from one subset to the other far more drastically than expectations based on the standard regression model. Our aims, then, were to improve the overall fit, and at the same time to obtain more stable estimates of the unit overhead times.

THE VARIABLES

A primary deficiency in the original regression model was the omission of events which were not counted but which were known to contribute to overhead. The omission of significant variables from the regression limits the quality of the fit that can be obtained. Furthermore, since the uncounted events are correlated with the counted ones, a bias is introduced into the estimated coefficients. Since this bias varies from sample to sample, it could account for the instability of the estimates. For these reasons, additional event counters were programmed into CP-67. We now list those measured variables which were included in the analysis described below.

1. Dependent variable: CP time: y = msec of CPU time in supervisor mode per second of real time.

2. Primary event variables (all are given as number of events per second of real time):
VSIO: x_1 = successful start IO instructions issued by users to virtual selector channel devices. Generally disk and tape IOs.
VMIO: x_2 = start IO instructions issued by users to valid virtual multiplexer channel devices. Generally terminal and peripheral IOs.
DIAGIO: x_3 = a special kind of VSIO which is handled more efficiently by CP-67.
SPOOL: x_4 = SIO instructions issued by CP-67 to spooling devices.
PAGES: x_5 = pages read into core and swapped out of core.

These primary variables correspond more or less to the variables used in the previous studies. They are also the ones in terms of which a user might describe the demands his program makes on the system.

3. Secondary event variables (all are given as number of events per second of real time):
UNSIO: x_6 = unsuccessful Start IO instructions issued by users to virtual selector channel devices.
TIO: x_7 = test IO instructions issued by users.
PGMINT: x_8 = user program interrupts.
IOINT: x_9 = user IO interrupts.
DISP: x_{10} = number of calls to dispatch.
PAGEIO: x_{11} = start IO instructions issued by CP-67 to paging devices. Since several pages may be read or swapped with a single SIO instruction, we have $x_5 \geq x_{11}$.
PAGEIN: x_{12} = pages read into core. Clearly $x_5 \geq x_{12}$.
HIO: x_{13} = halt IO instructions issued by users.
TCH: x_{14} = test channel instructions issued by users.

The secondary variables are a mixed bag. x_6 and x_7 are user demands, but of a type which is often overlooked. x_8 and x_9 are events resulting indirectly from user demands. x_{10} and x_{11} represent the manner in which CP-67 organizes its work internally. Finally, x_{12} through x_{14} are included here for the sake of completeness, but will not be used in the regressions, the last two because of their insignificant contribution and the first because of its high correlation with x_5.

THE DATA

The data used in this study were collected as described previously [2] on the CP-67 installation at the IBM Cambridge Scientific Center. At the time of this study, the system was running under an experimental version of the control program (x_3 and x_{11} did not exist on previous versions). Data were sampled at intervals varying nominally between 60 and 1000 seconds: the more users on the system, the shorter the sampling period. The formula used for choosing a sampling period was:

$$T = \max \{60, 1000 \times 5^{-\max[0,(N-10)/10]}\}$$

where T is the nominal sampling period in seconds and N is the number of logged-on users. The actual sampling period always exceeds T because of the time taken to run the sampling program.

A preliminary analysis was performed on 992 observations taken on June 22 - June 24, 1971. This established primarily the fact that the number of signed-on users had no systematic effect on the CP regression coefficients. The main results, reported here, were obtained on the following data sets:

I. 3629 observations, July 12 - 28, 1971
II. 4158 observations, July 29 - August 19, 1971.

Data set I was subdivided into six subsets entitled I-1 to I-6, each containing 605 consecutive observations (604 in the last). Similarly, II was divided into II-1 to II-7, each containing 594 consecutive observations.

REGRESSION ON PRIMARY VARIABLES

In order to obtain results comparable to those previously reported [2], regressions were run using only the five primary variables. That is, we assumed a functional relation of the form

$$y = \Theta_0 + \sum_{i=1}^{5} \Theta_i x_i$$

and estimate the Θ_i by means of a forward selection stepwise regression method. Selection of variables is halted when further reduction in the sum of squares of residuals is less than 0.1% of the sum of squares around the mean. The process used was the following:

(1) Fit the entire data set (I or II).
(2) Remove observations whose residuals exceed 2.5σ.
(3) Refit the entire data set.
(4) Fit each subsidiary data set, using only the observations retained after step (2).

Clearly, Θ_i (i=1,2,...,5) represents our estimate of the marginal amount of CPU time (in milliseconds) required by CP-67 to handle an event of type i. The results are summarized in tables 1 and 2. In addition to the values of all the coefficients for the full and subsidiary data sets, we also give the following statistics which were computed by regarding the estimated primary coefficients (Θ_1 to Θ_5) of the subsidiary data sets as 6 (for I) or 7 (for II) samples from a population of estimates:

(1) The average value of each coefficient.
(2) The standard deviation of each coefficient.
(3) The standard deviation expressed as a percentage of the average.

In computing these quantities, any missing coefficients were replaced by the average of the nonmissing values. We also provide estimates of what we would expect the standard deviations to be if the usual regression model were valid. These are obtained by multiplying the estimated standard error of the coefficients for the entire data sets by the square root of the number of subsidiary data sets. The magnitude by which the actual standard deviations exceed the expected ones is a clear indication of the invalidity of the usual regression assumptions in this case.

REGRESSION ON ALL VARIABLES

Using the same procedure as above, estimates of the coefficients ϕ_i were obtained in the following equation:

$$y = \phi_0 + \sum_{i=1}^{10} \phi_i x_i$$

The variables x_{11} and x_{12} had to be omitted because of their very high correlation with x_5, and x_{13}, x_{14} were never selected by the stepwise regression procedure. Results are displayed in tables 3 and 4. The additional statistics were computed only for the coefficients of the primary variables. In comparing these results with those of tables 1 and 2, we are led to the following observations:

(1) A considerable improvement in R^2, i.e. in the quality of the fit, has been achieved.
(2) The coefficients of the primary variables have generally decreased. This is because some CP time, previously falling to the primary variables through their correlation with the secondary variables, has now been attributed to the latter.
(3) There is, on the whole, a decrease in the standard deviations of the primary coefficients. However, there is no decrease on a percentage basis.

We thus seem to have made progress on one front - a better fit. However, the variability of the coefficients remains as bad as ever. Furthermore, to predict the amount of overhead required for a given work load, we now need to know the values of some variables which are internal to CP-67 and which themselves would have to be predicted in some way or another.

PRINCIPAL COMPONENT ANALYSIS

In order to determine what, if any, dependencies existed among the "independent" variables, we performed a principal component analysis on the twelve variables $x_1, x_2, \ldots, x_{12}$. The correlation matrices of these variables were formed for data sets I and II, and the eigenvalues and vectors of these matrices were computed. Any eigenvector whose eigenvalue is less than one represents a possible linear relationship among the variables, and this linear relationship is the more exactly adhered to, the smaller the eigenvalue. Specifically, let λ be some eigenvalue and let $u_1, u_2, \ldots, u_{12}$ be the associated normalized ($\Sigma u_i^2 = 1$) eigenvector. Then λ is the variance of the quantity $\Sigma u_i x_i / \sigma_i$, where σ_i is the standard deviation of the observed values of x_i. Hence, if

λ is very small, it follows that the following equation is very nearly satisfied:

$$\sum_i \alpha_i x_i = c$$

where $\alpha_i = u_i/\sigma_i$ and c is a constant (specifically, $c=\Sigma\alpha_i\bar{x}_i$, $\bar{x}_i$ being the average value of x_i).

By finding those u_i which are not nearly zero we isolate the variables which really enter the relation. The relations thus uncovered often correspond to relations that we expect to find from our prior knowledge about the system. We can then isolate one variable in the relation as a dependent variable, and we can solve the above equation to express this variable in terms of the others.

The eigenvectors corresponding to the small eigenvalues for data sets I and II are displayed in tables 5 and 6. The following relations appear:

(1,2) The two smallest eigenvalues yield relations between x_5, x_{11} and x_{12}, i.e. the three variables related to paging. Indeed, the correlation between x_5 and x_{12} is 99.9% in data set I and 99.8% in data set II. This reflects the fact that has been observed in all measurements taken on CP-67 to date, namely that the ratio between pages read in and pages swapped out remains very constant under virtually all operating conditions. Because of this, x_{12} is an almost completely redundant variable, and will not be considered further. The second relation among the paging variables arises from the fact that the majority of page requests (x_5) give rise to page IO's (x_{11}) on a one-to-one basis. Only if page requests for different sectors on the same drum arrive during a channel-busy period are they then chained together to a single Start IO instruction. For the configuration used in the course of this study (three drums on two channels), the expected rate of page IO's can be shown to be given approximately by the following expression:

$$E(x_{11}) = \frac{x_5}{x_5 t + 3\exp(-x_5 t/3)} + \frac{2x_5}{2x_5 t + 3\exp(-2x_5 t/3)}$$

where t = 0.0172 sec is the drum revolution time. This formula does indeed fit the observed data well (R^2 = 99.5% and 98.7% for data sets I and II respectively).

(3) The number of IO interrupts is shown to depend

strongly on the number of VSIO and VMIO and to a lesser extent on some of the other variables. By performing regressions of IO interrupts on VSIO and VMIO one obtains the following estimates:

$$E(x_9) = 8.35 + 0.853x_1 + 0.485x_2 \qquad \text{(data set I)}$$

$$E(x_9) = 1.63 + 0.911x_1 + 0.845x_2 \qquad \text{(data set II)}$$

with R^2 = 58 and 91%.

(4) The number of entries into dispatch depends on several other variables. Again, by regression we found:

$$E(x_{10}) = 77.6 + 5.76x_1 + 3.57x_3 + 2.37x_5 + 1.66x_8 \qquad \text{(I)}$$

$$E(x_{10}) = 78.9 + 5.95x_1 + 6.62x_3 + 1.93x_5 + 2.10x_8 \qquad \text{(II)}$$

with R^2 = 50 and 59%. In all cases, no outliers were rejected.

Note that we have used the principal component analysis only as a general guide for the detection of dependencies. The actual relations used were chosen on the basis of knowledge of the system and direct study of the correlation matrix, as well as on the results of the principal component analysis.

REGRESSION ON MODIFIED VARIABLES

We have verified above that some of the secondary variables depend, at least partly, on the primary and on other secondary variables. This incomplete dependence is certainly one of the causes for the variation in the primary coefficient estimates. Since we are interested in the total average contribution of an occurrence of a primary event to CP-67 overhead, it would appear logical to include within the primary coefficients the expected contribution of any secondary events which accompany the primary events. The remaining CP-67 overhead then can be attributed to the remaining secondary variables, and to any departures of the secondary events from their expected number. For instance, we expect each VSIO to be accompanied by approximately one IO interrupt and six entries into dispatch, and the contributions of these to overhead should be accounted to the VSIO. If, however, on some specific occasion there were two IO interrupts and only four dispatches, these departures from the norm would help explain why the observed overhead differed from the expected value. We accomplish our objectives by replacing x_9, x_{10} and x_{11} with their residuals relative to their expected values. Since these residuals are uncorrelated with any variables used in the subsidiary regressions, the coefficients computed for the latter variables will be stabilized. Our overhead regression now takes the form:

$$y = \psi_0 + \sum_{i=1}^{8} \psi_i x_i + \sum_{i=9}^{11} \psi_i z_i$$

where $z_i = x_i - E(x_i)$ (actually, in forming z_9 and z_{10} the constant terms were ignored). The results of these regressions are displayed in tables 7 and 8.

Perusal of these results shows that a considerable reduction in the variability of the estimates of the primary event coefficients has been achieved in replacing the dependent secondary variables by their residuals. This is further demonstrated in table 9, where we give the generalized variance (determinant of variance-covariance matrix of the estimates) of the five primary coefficients for the three sets of regressions.

CONCLUSIONS

By adding new measurements to those previously available, we improved the precision with which CP-67 overhead can be predicted. By replacing some of the new variables with their residuals relative to their conditional expectations, more stable estimates for the coefficients of the primary variables were obtained. Even these coefficients, however, show more variation than would be expected from the usual regression assumptions. In particular, we cannot uphold the assumption that the time spent by CP-67 in servicing a primary event plus the expected number of accompanying events is constant. More sophisticated models, which take into account the random character of these service times, are required if a completely satisfactory statistical model is desired.

Acknowledgement. The authors are indebted to L. Wheeler of the Cambridge Scientific Center for implementing the new measurements in CP-67.

Disclaimer. The measurements reported here were taken on an experimental version of CP-67 at one particular installation. No representation is made that the results will apply to any released version of the system.

REFERENCES

1. Y. Bard, B. Margolin, T. Peterson and M. Schatzoff: CP-67 measurements and analysis I: regression studies. IBM Cambridge Scientific Center Report, G320-2061 (1970).
2. Y. Bard: Performance criteria and measurements for a time-sharing system, IBM Syst. J. 10, 193-216 (1971).
3. CP-67/CMS version 3 system description manual, Form No. GH20-0802-1, IBM, White Plains, New York (1970).

Table 1. Primary-variables regression coefficients, in msec CPU time per event. Data set I.

Data Set	No. of observations after removal of outliers	R^2 (%)	Θ_0	Θ_1 VSIO	Θ_2 VMIO	Θ_3 DIAGIO	Θ_4 SPOOL	Θ_5 PAGES
I	3525	84.5	37.66	5.48	1.78	1.80	3.07	1.49
I-1	599	91.0	19.58	6.02	1.53	2.40	3.76	1.53
I-2	590	83.6	32.25	5.62	1.93	1.80	3.68	1.14
I-3	601	87.6	31.52	5.28	2.10	2.60	1.56	1.56
I-4	591	89.5	41.00	5.96	2.41	2.92	1.41	0.89
I-5	570	86.5	38.89	4.73	1.78	1.37	3.97	1.65
I-6	574	61.1	92.26	4.79	1.59	0.68	2.91	1.32
Average				5.40	1.89	1.96	2.88	1.35
s.d.				0.56	0.33	0.84	1.14	0.29
s.d. (% of average)				10	18	43	40	22
expected s.d.				0.15	0.08	0.34	0.75	0.07

Table 2. Primary-variables regression coefficients, in msec CPU time per event. Data set II.

Data Set	No. of observations after removal of outliers	R^2 (%)	Θ_0	Θ_1 VSIO	Θ_2 VMIO	Θ_3 DIAGIO	Θ_4 SPOOL	Θ_5 PAGES
II	4055	80.6	51.54	5.64	1.74	2.36	3.17	1.27
II-1	581	80.0	53.77	5.26	1.43	0.76	4.14	1.59
II-2	586	75.0	39.39	5.87	2.09	1.25	2.92	1.30
II-3	558	79.7	79.13	6.01	1.88	2.25	1.97	0.94
II-4	583	80.5	45.08	5.17	2.00	1.74	1.11	1.39
II-5	592	79.8	51.35	6.19	1.91	1.81	1.87	1.03
II-6	571	82.4	60.82	4.92	1.86	1.23	1.61	1.64
II-7	584	86.2	51.48	6.25	1.44	3.17	6.97	1.17
Average				5.67	1.80	1.74	2.94	1.29
s.d.				0.53	0.26	0.79	2.04	0.27
s.d. (% of average)				10	15	45	69	21
expected s.d.				0.18	0.10	0.38	0.96	0.07

Table 3. All-variables regression coefficients, in msec CPU time per event. Data set I.

Data Set	No. of observation after removal of outliers	R^2 (%)	ϕ_0	ϕ_1 VSIO	ϕ_2 VMIO	ϕ_3 DIAGIO	ϕ_4 SPOOL	ϕ_5 PAGES	ϕ_6 UNSIO	ϕ_7 TIO	ϕ_8 PGMINT	ϕ_9 IOINT	ϕ_{10} DISP
I	3528	95.7	14.46	3.35	1.30	1.24	1.69	1.49	0.38	0.48	*	0.40	0.23
I-1	594	97.1	5.44	3.04	1.39	1.33	1.95	1.53	0.64	0.23	-0.18	0.10	0.33
I-2	573	95.8	7.03	2.57	0.59	0.42	2.96	1.14	0.49	0.71	-0.12	0.64	0.34
I-3	596	95.1	14.54	2.70	0.80	1.48	1.78	1.56	0.36	0.51	10.29	0.84	0.24
I-4	577	96.9	10.39	2.70	0.49	1.86	1.04	0.89	0.47	0.76	*	1.47	0.22
I-5	592	97.4	13.42	2.68	0.85	1.19	2.69	1.65	0.35	0.45	-0.10	0.65	0.22
I-6	596	93.6	30.22	1.60	*	0.84	1.98	1.32	0.42	1.02	-0.24	1.54	0.24
Average				2.55	0.82	1.19	2.07	1.10					
s.d.				0.49	0.31	0.50	0.68	0.17					
s.d. (% of average)				19	38	42	33	16					
expected s.d.				0.12	0.05	0.19	0.43	0.03					

*Not selected by stepwise regression procedure.

Table 4. All-variables regression coefficients, in msec CPU time per event. Data set II.

Data Set	No. of observations after removal of outliers	R^2 (%)	ϕ_0	ϕ_1 VSIO	ϕ_2 VMIO	ϕ_3 DIAGIO	ϕ_4 SPOOL	ϕ_5 PAGES	ϕ_6 UNSIO	ϕ_7 TIO	ϕ_8 PGMINT	ϕ_9 IOINT	ϕ_{10} DISP
II	4042	94.7	16.92	2.88	0.72	1.48	1.79	1.07	0.45	0.77	*	0.88	0.24
II-1	578	96.6	17.68	2.01	*	0.75	3.05	1.26	0.37	0.99	-0.03	1.45	0.24
II-2	577	94.8	11.85	2.91	0.60	1.22	2.30	1.03	0.50	0.77	*	0.96	0.24
II-3	555	93.4	33.14	2.46	0.29	1.42	1.09	0.82	0.56	0.74	-0.07	1.30	0.27
II-4	580	95.9	16.76	1.33	-0.30	0.83	0.62	1.15	0.40	1.78	*	2.01	0.23
II-5	583	93.3	14.26	2.30	*	1.00	1.64	1.10	0.52	0.93	-0.84	1.51	0.24
II-6	589	96.2	18.95	1.78	-0.35	0.80	1.62	1.26	0.35	1.20	*	1.90	0.24
II-7	580	96.4	18.44	3.23	0.89	1.74	3.45	1.02	0.58	0.46	*	0.62	0.26
Avg.				2.29	0.23	1.11	1.97	1.09					
s.d.				0.65	0.46	0.37	1.02	0.15					
s.d. (% of avg.)				29	196	33	52	14					
expected s.d.				0.13	0.09	0.22	0.56	0.04					

*Not selected by stepwise regression procedure.

Table 5. Principal component analysis, data set I.

Variable	Average (events per second)	s.d. (events per second)	Eigenvalue of Correlation matrix: 0.00109	0.0155	0.178	0.302
			Coefficient in normalized eigenvector*			
x_1 VSIO	21.9	12.0	---	---	0.510	0.354
x_2 VMIO	18.5	27.3	---	---	0.401	0.128
x_3 DIAGIO	6.15	5.76	---	---	---	---
x_4 SPOOL	2.68	2.73	---	---	0.137	---
x_5 PAGES	46.5	29.2	-<u>0.757</u>	0.299	---	---
x_6 UNSIO	17.3	90.3	---	---	---	---
x_7 TIO	10.5	25.7	---	---	---	---
x_8 PGMINT	6.85	53.0	---	---	---	0.492
x_9 IOINT	36.0	21.6	---	---	-<u>0.663</u>	---
x_{10} DISP	347	199	---	---	0.194	-<u>0.765</u>
x_{11} PAGEIO	40.6	23.2	0.113	-<u>0.815</u>	---	---
x_{12} PAGEIN	27.9	17.8	0.643	0.496	---	---
x_{13} HIO	0.06	0.09	---	---	---	---
x_{14} TCH	4.94	9.29	---	---	-0.258	---

*Only coefficients in excess of 0.1 are displayed. Largest coefficient in each eigenvector is underlined.

Table 6. Principal component analysis, data set II.

Variable	Average (events per second)	s.d. (events per second)	Eigenvalue of correlation matrix			
			0.0011	0.0146	0.126	0.539
			Coefficient in normalized eigenvector*			
x_1VSIO	20.3	11.8	---	---	0.454	0.257
x_2VMIO	17.2	23.8	---	---	0.508	---
x_3DIAGIO	5.69	5.79	---	---	---	---
x_4SPOOL	2.47	2.50	---	---	---	---
x_5PAGES	49.2	32.7	-0.763	0.286	---	---
x_6UNSIO	14.2	79.9	---	---	---	---
x_7TIO	9.11	19.1	---	---	0.234	---
x_8PGMINT	5.04	44.9	---	---	---	0.531
x_9IOINT	34.1	21.7	---	---	-0.685	0.193
x_{10}DISP	363	243	---	---	---	-0.753
x_{11}PAGEIO	42.5	26.4	0.131	-0.809	---	---
x_{12}PAGEIN	29.7	20.0	0.633	0.513	---	---
x_{13}HIO	0.06	0.10	---	---	---	---
x_{14}TCH	5.04	8.46	---	---	---	---

*Only coefficients in excess of 0.1 are displayed. Largest coefficient in each eigenvector is underlined.

Table 7. Modified-variables regression coefficients, in msec CPU time per event. Data set I.

Data Set	No. of observations after removal of outliers	R^2 (%)	ψ_0	ψ_1 VSIO	ψ_2 VMIO	ψ_3 DIAGIO	ψ_4 SPOOL	ψ_5 PAGES	ψ_6 UNSIO	ψ_7 TIO	ψ_8 PGMINT	ψ_9 IOINT RESIDS.	ψ_{10} DISP RESIDS.	ψ_{11} PACEIO RESIDS
I	3523	95.8	15.80	5.03	1.50	2.06	1.74	1.63	0.38	0.47	0.37	0.40	0.24	1.08
I-1	594	97.0	9.04	5.10	1.44	2.57	1.92	1.71	0.62	0.22	0.35	0.11	0.33	1.20
I-2	573	95.8	7.03	5.07	0.90	1.63	2.96	1.65	0.49	0.71	0.44	0.63	0.34	*
I-3	594	95.2	16.54	4.86	1.29	2.35	1.73	1.67	0.36	0.47	11.63	0.78	0.25	1.60
I-4	575	97.0	10.53	5.22	1.21	2.56	1.15	1.47	0.47	0.74	0.39	1.46	0.22	*
I-5	591	97.4	16.38	4.61	1.19	1.94	2.81	1.71	0.35	0.45	0.31	0.70	0.22	1.68
I-6	596	93.6	32.91	4.32	0.70	1.75	2.04	1.70	0.41	1.07	0.16	1.65	0.24	*
Average				4.86	1.12	2.13	2.10	1.65						
s.d.				0.34	0.27	0.41	0.68	0.09						
s.d. (% of avg.)				7	24	19	32	6						
expected s.d.				0.11	0.05	0.24	0.53	0.05						

*Not selected by stepwise regression procedure.

Table 8. Modified-variables regression coefficients, in msec CPU time per event. Data set II.

Data Set	No. of observations after removal of outliers	R^2 (%)	ψ_0	ψ_1 VSIO	ψ_2 VMIO	ψ_3 DIAGIO	ψ_4 SPOOL	ψ_5 PAGES	ψ_6 UNSIO	ψ_7 TIO	ψ_8 PGMINT	ψ_9 IOINT RESIDS.	ψ_{10} DISP RESIDS.	ψ_{11} PAGEIO RESIDS.
II	4042	94.7	16.89	5.12	1.48	3.08	1.75	1.54	0.45	0.77	0.50	0.90	0.24	*
II-1	578	96.6	17.61	4.73	1.22	2.31	3.06	1.72	0.37	1.01	0.47	1.52	0.24	1.15
II-2	577	94.9	11.88	5.20	1.41	2.79	2.31	1.49	0.50	0.77	0.48	0.95	0.24	1.41
II-3	555	93.4	33.14	5.23	1.39	3.19	1.09	1.33	0.56	0.73	0.49	1.30	0.27	1.22
II-4	580	95.9	16.69	4.55	1.40	2.37	0.61	1.60	0.40	1.79	0.61	2.01	0.23	-0.58
II-5	582	93.3	14.01	5.14	1.29	2.60	1.70	1.55	0.52	0.91	-0.23	1.49	0.24	-1.33
II-6	590	96.1	19.18	4.92	1.32	2.36	1.31	1.71	0.35	1.22	0.56	1.97	0.24	0.66
II-7	580	96.4	18.46	5.34	1.42	3.45	3.45	1.52	0.58	0.46	0.54	0.62	0.26	0.68
Average				5.02	1.35	2.72	1.93	1.56						
s.d.				0.29	0.08	0.45	1.05	0.13						
s.d. (% of avg.)				6	6	16	54	9						
expected s.d.				0.10	0.06	0.21	0.56	0.04						

*Not selected by stepwise regression procedure.

Table 9. Generalized variance of primary event coefficient estimates, $(\text{msec/event})^{10}$.

Regression	Data set I	Data set II
Primary Variables	14000	120000
All Variables	2000	2100
Modified Variables	83	350

THE USE OF PURE AND MODIFIED REGRESSION TECHNIQUES FOR DEVELOPING SYSTEMS PERFORMANCE ALGORITHMS

M.P. Racite

ABSTRACT. This paper describes the use of various curve-fitting techniques for development of systems performance algorithms. Nonlinear simple and multiple linear and non-linear regression methods are described. For the nonlinear two-variable case, where the resulting normal equations are also nonlinear, a novel search and modified method of least squares approach is illustrated. These techniques are examined and compared with other existing techniques from the standpoint of CPU time (and hence cost) requirements for algorithmic development. Several algorithms (or models) generated from real systems data will be described and emphasis will be placed on development, practicality of use and criteria for goodness of fit. In addition, the use of terminal-accessible databank programs which greatly minimize the labor of data handling and curve fitting in problems of this type is described.

The techniques of systems performance evaluation are many and varied. They range all the way from judgment and intuition on the one hand to ultra-sophisticated methodologies on the other. This paper will describe the use of statistical regression techniques to develop systems performance algorithms in this area of systems evaluation.

In a later section of this paper several algorithms (or models) will be presented; however, at this point let us define a systems performance algorithm, at least as far as it is understood in this area of systems evaluation. An algorithm is a statistical cranking device for estimating or predicting some system performance parameter on the basis of inputting it to one or more other known parameters. The models described in this paper were developed from measured systems data and the techniques of linear and nonlinear regression were used to determine appropriate fitted functions.

Before describing the actual models I would first like to discuss some of our tools and procedures used for model development. At the conclusion of the data collection phase of an experiment aimed at development of specific performance algorithms, the measured data are read into a computer databank accessible by a terminal. Also accessible by the same terminal is a package of conversational-mode curve-fitting programs which operate on the data and assist the user in developing the desired models.

Let's consider first the use of these tools and procedures in model development for the nonlinear two-variable case. The collected data are first examined from the standpoint of their patterns and an ordinary x-y plot of them is made on the terminal. This is done for the purpose of selecting the group of functions most likely to result in a good fit to the observed data. The functions available for curve fitting in this terminal-accessible curve-fitting program package consist of 27 exponential types and the polynomial set from degree 1 through 9. The functions are broken up into six different groups, where each group was formed on the basis of the data patterns most likely to be well fit by the functions in a particular group. A list of all functions and the function groupings used in these programs can be found in table 1, Appendix A.

When using the function groups available at the terminal, the functions within each of the six groups are defined and information as to what specific group should be used for a particular data pattern is given. For example, if the data plot suggests a function which is everywhere increasing and concave downward, group 4 should be used to fit the data. As another example, if the data pattern suggests a function which is everywhere increasing and passes from concave upward through a point of inflection to concave downward, group 5 should be used. These function groups were defined on the basis of both analytical and empirical considerations. By evaluating first and second derivatives, functions were defined which could fit certain data patterns. The functions were then fitted to various sets of data patterns in an effort to find which functions consistently gave good results to particular patterns. Thus the resulting function groupings.

After the user has chosen the appropriate function group to fit the observed data, the curve-fitting programs commence. For the functions within any group that are polynomials the straightforward method of least squares is applied, since the resulting normal equations are linear and easily solved for the polynomial function coefficients. The remaining functions in the groups are for the most part exponential in type and when the method of least squares is applied the resulting normal equations are nonlinear and therefore not solvable in the ordinary linear sense.

For functions of this type, some have tried the technique of first linearizing the function through use of logarithms and then applying least squares to the resulting normal equations. Our experience in attempting to use this method was very unsatisfactory because the use of logarithms tended to give greater weight to the smaller values of the dependent variable y. This bias caused poor-fitting functions from an ESS (error sum of squares) standpoint. In addition, we tried the further refinement of weighting the logarithms in proportion to the value of y, where the weighted least squares condition is:

$$\sum_{i=1}^{n} y_i \, d_i^2 = \text{minimum}$$

where $d_i = y_i - \hat{y}_i$ and y_i and $\hat{y}_i$ are, respectively, the observed and model-predicted values of the dependent variable y. Weighting in this manner did improve results, but even with this refinement the technique was still unacceptable.

The technique employed in the terminal program package for functions of this type is a modified method of least squares. The technique was first developed and used by R.V. Frei and W.F. Mikhail of IBM and later refinements were added by personnel in this systems evaluation group. To describe how it works an example will be used.

Consider the function

$$\hat{y} = \alpha + \beta e^{\gamma x} \tag{1}$$

which has three parameters α, β and γ which must be evaluated when fitting this function to observed data. As mentioned previously, the normal equations that result from applying least squares are nonlinear and not solvable except through trial-and-error iterative procedures. In using our technique we in effect eliminate these nonlinear normal equations, as further described.

If we assume that γ has some domain of values bounded both negatively and positively, say ± 10, we can use a combination search technique for finding γ and then solve for its counterpart parameters α and β through use of the method of least squares. This is mathematically valid sinee by searching for γ, γ is "known" and Eq. (1) reduces to a simple linear form to which least squares can be readily applied to obtain the related (for this γ) values of α and β. With this idea in mind we can look at Eq. (1) in this form:

$$\hat{y} = \alpha + \beta c \tag{2}$$

where

$$c = e^{\gamma x} \tag{3}$$

The bounds on γ are found by inputting to (3) the vector (group) of observed values on x, the independent variable, and a starting value very close to 0 for γ. The resulting vector or group of values for c is examined to see if any value in c is approaching a magnitude such that the computer would encounter a domain error (a number larger or smaller than the machine can handle arithmetically). If there are no values of c that fall into this category the value of γ is made equal to 1 and the vector on x is again inputted to c and the resulting vector on c is again examined for domain error encroachment. This procedure is continued until the upper limit or bound on γ is reached. The same technique is then applied to find the negative or lower bound on γ. For most data, resulting limits on γ are usually well within ± 10.

The next phase of the program is to calculate an ESS for each value of γ inclusive within its established bounds. It should be noted that at this time all values for γ are integers with the exception of the value close to but not equal to 0. The value of 0 is not used because some of the functions in the program package contain a γ/x term. The method of least squares is used to calculate the ESS for each value of γ. This is now possible since, as stated previously, γ is "known" and (2) is simple linear. This vector of ESS values is now examined and the smallest value in the group is identified with the value of γ that brought it about.

To explain more easily how the programmed technique works from here, let us extend our example and assume that the value of γ that brought about the smallest ESS was γ = 2. The program would then set γ to values slightly smaller and larger than 2, say 1.999 and 2.001, and calculate the pair of resulting ESS. These two ESS values are compared and, depending upon which one is smaller, the direction in which to search further for the optimum value of γ (that value for γ which results in a smaller ESS than any other value for γ) is determined. Suppose the ESS for γ = 2.001 was larger than the ESS for γ = 1.999. This would mean that the optimum value of γ would lie between 1 and 2, as opposed to 2 and 3 if the larger value of γ had resulted in the smaller ESS.

At this time the region between 1 and 2 is broken up into the equi-spaced values 1.0,1.1,...,1.9,2.0 and γ is in turn assigned each one of these values. For these assigned values of γ the resulting ESS group of values is examined to determine, again, which direction to move γ on a finer basis to find the optimum (smallest) ESS. In continuing with our example, suppose the value of γ = 1.3 yielded the smallest ESS. Suppose further that values for γ slightly smaller and larger than γ = 1.3 yielded values of ESS that showed that the optimum for γ lies to the right of 1.3. This would mean that the next region to continue the search would be between 1.3 and 1.4. This region would then be broken up into the equi-spaced values of 1.30,1.31,...,1.39,1.40 and the procedure described previously would be continued to find the optimum value of γ to whatever degree of accuracy is desired (usually 3 decimal places). For this final and optimal value of γ, the associated values for α and β, the ESS and the function ID number to which they apply are printed on the terminal and also stored in the program for further considerations. After all the functions in the selected group have been similarly evaluated the "best-fitting" function must be selected.

Before discussing the aspect of "best" function selection it should be noted that the described technique implies that a single absolute minimum value of ESS exists for some value of γ within the original bounds established. This has not, to my knowledge, been proven; however, for the exponential functions in this package it has been empirically observed that the sequence of the ESS values has some monotonic properties for γ within the range of applicability. For example, if the ESS is observed to decrease (when γ is calculated as described) until it reaches a turning point after which it increases, it has not been observed to decrease again. In addition, this technique, which we have applied to a plethora of systems performance data (and other applications), has consistently yielded considerably better fits than the logarithmic linearization procedure when applied to the same data. Also, some comparisons were made between this technique and a derivative-free method based upon the Levenberg, Marquart and Gauss algorithms. In all cases tried, both methods yielded identical results, with the exception that the last-named technique did not always converge to a solution. It should also be mentioned that the derivative-free method requires considerably more CPU time (and hence is costlier to use) than the technique described in this paper. From a terminal standpoint the response time as well as the excessive cost make the derivative-free method prohibitive in this area of systems evaluation. Thus our justification of the technique of "modified method of least squares".

Now let's get back to the selection of the "best" function to use for the model. The model with the smallest ESS may not necessarily be the "best" or most desirable one to use. It should be noted that while the ESS is a measure of goodness of fit (the smaller the better) in the least-squares sense, it is only a measure of fit within the confines of the inputted data. Many of the functions in this package can take on undesirable and unexpected shapes outside of these confines. To safeguard against this possibility the program package has plotting and forecasting features which enable the user to input to a selected fitted function the largest and/or smallest values of x (the independent variable) the model could ever encounter in a realistic situation. These features then provide the user with a plot and values for y (the dependent variable) over the whole range of the inputted extreme values of x.

In addition to the plot and forecast features, which afford protection from the function shape problem defined, many of the functions in the program package have been basically analyzed with respect to what possible shapes they can assume over the domain on x for the open interval $(0<x<\infty)$. This information is available to the user and renders further assistance in function selection. Two examples of basic shape analysis follow.

Consider Function 1 of the package, which is of the form:

$$y = \alpha + \beta e^{\gamma x} \quad (4)$$

where

$$dy/dx = \gamma\beta e^{\gamma x} \quad (5)$$

and

$$d^2y/dx^2 = \gamma^2\beta e^{\gamma x} \quad (6)$$

It can be seen by examination of (5) that the shape of (4) is dependent upon the sign of $\gamma\beta$ as to whether it is everywhere increasing or decreasing, and dependent upon the sign of β as seen in (6) as to whether (4) is everywhere concave upward or downward.

For the second example, consider this function from the package:

$$y = \alpha + \beta e^{\gamma x}(1 - e^{\gamma x}) \quad (7)$$

where

$$dy/dx = \beta\gamma e^{\gamma x} - 2\beta\gamma e^{2\gamma x} \quad (8)$$

and

$$d^2y/dx^2 = \beta\gamma^2 e^{\gamma x} - 4\beta\gamma^2 e^{2\gamma x} \quad (9)$$

By setting (8) to 0 we see that (7) has a maximum or minimum at $e^{\gamma x} = 1/2$. Further, by evaluating (9) at $e^{\gamma x} = 1/2$ we obtain:

$$d^2y/dx^2 = \frac{1}{2}\beta\gamma^2 - \beta\gamma^2 = - \beta\gamma^2/2$$

which tells us that a minimum exists at $e^{\gamma x} = 1/2$ when β is negative and a maximum exists at this point when β is positive.

Let us assume that we used this function (7) to fit the data plotted in figure 1. For this data and function (7) β would be forced into being positive because a minimum is impossible at $e^{\gamma x} = 1/2$. Since β must be positive, then (9) is negative when evaluated at $e^{\gamma x} = 1/2$, thereby causing the function (7) to have a single maximum at $e^{\gamma x} = 1/2$ and no minimum anywhere in the domain on x. Further, it can be seen that when (9) is set to 0 (when β is +) we get $e^{\gamma x}$ =

1/4, which means that a point of inflection exists at $e^{\gamma x}$ = 1/4 and that (7) is concave upward in the interval ($0<e^{\gamma x}<1/4$) since (9) is positive throughout this domain. Also, (7) is concave downward in the interval ($1/4<e^{\gamma x}<\infty$) since (9) is negative in this interval when β is positive. In addition, (7) takes on the value   at $e^{\gamma x} = 1$ and intersects the x-axis when $e^{\gamma x}(1-e^{\gamma x}) = -\alpha/\beta$.

For the case described (where β would be +) the shape of (7) is shown in figure 2.

For the data involved in this fit (data shown in figure 1), function (7) could yield a good fit from an ESS standpoint; however, it is obvious that the model could be a disaster if it were selected and used to predict values for y in the region $e^{\gamma x} > 1/2$. Again, though, if the observed data were as pictured in figure 3, this function could serve as an excellent prediction model.

A similar approach is taken with most of the other functions in the package with respect to shape analysis. It is obvious that information of this sort minimizes the probability of selecting a poor or even a bad model and enhances the likelihood of choosing a good one. This is especially true when a prediction model is being sought for forecasting well beyond the available data.

This concludes the section in this paper on tools and techniques for curve fitting for the two-variable case. The next section will deal with multiple regression methods where three or more variables are involved. The material in the next section will deal with multiple regression for the nonlinear case since the use of the technique for the linear case is rather routine; however, in the final section of this paper, which presents several models, a model of this type is included.

About nonlinear multiple regression not much has been observed in the literature as to its use or applicability. This is no doubt due in part to the increased complexity of functions for this case when compared with those functions for the simpler two-variable case. In addition, the problem is further made difficult with regard to what functional forms one should use in attempting to fit data involving more than two variables.

G.E. Beckman and H. Hoogendoorn of IBM Components Division chose the five functions shown in table 2 of Appendix

A somewhat arbitrarily and made trial fits of these functions to data (for the 3-or-more-variable case) that had previously resulted in poor or unacceptable fits when multiple linear regression had been applied. In some cases the fits arrived at using these nonlinear functions were remarkably good. Mr. Beckman is currently involved in evaluating new functions to supplement further the existing group. Our own use of these nonlinear functions to date, while not extensive, has been quite successful and very promising. We have succeeded in several cases in using these functions to obtain good to excellent fit where the multiple linear technique was totally unacceptable. In the next section of this paper one of these nonlinear multiple regression models is discussed.

THE MODELS

In developing a model or algorithm, many factors may be involved. In some cases much data may be available while in others there may be very little. Some applications may require an extremely accurate model in order to be in any way useful, while still yet some other applications can be profitably satisfied with less well-fitting models. This latter case is often one in which very little data are available and it is often necessary to supplement the minimal data with well-thought-out theoretical considerations and/or known boundary values or limitations. The models which follow will be related to the above as well as other criteria with respect to their development.

Model 1: CPU to I/O time ratio versus percent non-overlap between CPU and I/O

Overlap relationships between CPU and I/O are of major importance in the understanding of computer operation. The algorithm used to determine this is defined above as Model 1. By inputting to this model the CPU to I/O time ratio the estimate of the unoverlapped I/O time with CPU time can be established. This algorithm is used in several sub-systems models which are programs used to determine various systems performance criteria.

The model (actually several models were developed from systems measurements, since a separate one is required for each of the several different workload activity type categories built into the computer programs referred to above) was developed on the basis of many measurements over a wide range of systems. This model, along with the data upon which it was developed, is shown in functional and plotted forms in Appendix B. The measurements were made with a hardware monitor and for each job step within the jobstream the CPU time, I/O time and I/O time overlapped with CPU were recorded. The resulting data were fed into the terminal-accessible curve-fitting program package referred to in the previous section of this paper. Several of the functions in the package showed reasonably good fit and the model shown in Appendix B was best from the standpoint that it had the least of all ESS for all functions tried. In addition, this model also showed a pattern of good fit, outside the domain of the available data, that was consistent with known theoretical boundaries of this domain. These known theoretical boundaries were 0% non-overlap when the CPU to I/O time ratio was high, say approximately 10, and 100% non-overlap when this ratio was 0.

<u>Model 2</u>: Numbers of initiators versus channel-to-channel overlap

Model 2 is actually a set of three models for the 2-, 3- and 4-channel system cases. This model is also used as an algorithm in the programs referred to in the description of Model 1. These programs use this model to determine, for the system "given" for comparison with a base system, the proportion of time all channels on the system are overlapped or are concurrently active. This proportion of time (expressed as a percent) is with respect to the specific channel having the smallest individual time. The development of this model is an example of one where very few data were available and had to be supplemented with theoretically sound extrapolations.

The measurements that were available were for the two-channel case with 1, 2, 3 and 4 initiators and the three-channel case with 3 initiators. In comparing the two known points for the 3-initiator case with 2 and 3 channels it was seen that the improvement in channel-to-channel overlap time was 10%. On the basis of other reliable considerations it was determined that for the 1-initiator case the curves for the three conditions (2, 3 and 4 channels) should converge at approximately 8% overlap. In addition, these other considerations showed that for the 2-initator case, improvements of approximately 8% and 4% could be expected in going from 2 to 3 and 3 to 4 channels, respectively. Also, for the 3-initiator case an improvement of approximately 5% should result in going from 3 to 4 channels. Figure 4 illustrates the known and extrapolated data points. The known data points are shown as k and the extrapolated ones as e. Where a known and extrapolated point coincide the symbol used is c.

Taking the data for the 2-channel case as shown in this figure, a function was fitted. Then, using the same function (it was felt that the set of curves should be fitted with the same function), fits were obtained for the 3- and 4-channel cases. The model(s) can be seen in Appendix B along with extrapolated and measured data points.

Of course, it should be realized that the validity of this model is suspect on the basis of limited data and the use of such extended extrapolations. Models of this sort must be used with great care until they can be validated and/or calibrated with more data. At the present time, an experiment aimed at validating and/or calibrating this model is in progress and uses of this model to date have shown it to be accurate within the bounds of its use.

Model 3: Un-overlapped seek time versus number of channels and number of initiators

Model 3 is a three-dimensional multiple nonlinear regression model. The dependent variable is disk seek time (for a particular device) that is un-overlapped with any CPU or I/O activity. The independent variables are number of channels on the system and number of initiators used by the operating system. The measurements were made for a specific workload activity type using a jobstream selected for this purpose. A hardware monitor was used to record the data which resulted in this model. As in the previous model, some of the data points for this model were calculated using theoretical considerations. In addition, this model is also currently being evaluated with new measurements so as to calibrate and/or validate it for further use in the systems evaluation area. The model, data points and a table of residuals can be seen in Appendix B. At the present time the use of this model for predicting un-overlapped seek time for initiators in excess of 4 for the 2-, 3- and 4-channel cases yields values that seem reasonable and realistic.

Model 4: Net channel busy time versus supervisor CPU time and EXCP count

Model 4 is a three-dimensional multiple linear regression model. It was developed as a result of a need to make large numbers of systems measurements in the field over a wide geographical area without the cumbersome, difficult and expensive task of shipping large hardware monitors for this purpose. Available at the time was an IBM-developed software monitor which could be run under the operating system. Among the many measurements possible for the software monitor to measure were supervisor CPU time and EXCP (execute channel program) count; however, no facility existed within it to measure net channel busy time (any I/O activity on any channel). Among the measurements being sought in the field was net channel busy time and the question arose: could a model be developed which could predict this parameter using as input the related variables which the software monitor could measure? If the answer to this question was yes, it would eliminate the need of a hardware monitor for the desired measurements in the field since the software monitor and the model would suffice for this purpose.

An experiment was set up and measurements were made using a hardware monitor to measure net channel busy time concurrently with the software monitor measuring supervisor CPU time and EXCP count (many other measurements were made concurrently but were not related to this specific model, and discussion of them will be omitted). Using selected jobstreams, 119 measurements were made in which factors such as run time, numbers of initiators and numbers of readers and writers were varied over the gamut of permutations of small, intermediate and large values.

Prior to running the experiment it was determined that a model which could predict net channel busy time to within an accuracy of ±10% would be very satisfactory for the application intended. In developing this model the supervisor CPU time and the EXCP count data measured by the software monitor were combined with the net channel busy time data recorded by the hardware monitor for each of the 119 runs.

The data were fed into a multiple linear regression program and the resulting model, along with an ANOVA (analysis of variance) table, is shown in Appendix B. As can be seen, this is a good model when considered from the viewpoint of its intended use and accuracy requirements.

<u>Model 5</u>: System parameter Y versus system parameter X

This model is an example of one derived from a large mass of data in which the data pattern is very reliably known and specific. The only reason for fitting a function to it in this case is to have an algorithm which can be incorporated into a program with a single line of code. The actual data for this model are not defined since they are not relevant to the reason for inclusion of the model in this paper.

Without an algorithm (expressed as a function), coding a plot into a program for use in the program is tedious and can require a page or more of instructions. The usual method of coding a plot consists of breaking the curve up into several straight lines. When a value of x is inputted the program must first find the appropriate straight-line equation to use and then give a straight-line value for y in that portion of the coded plot. In addition to being time-consuming, this method can be quite inaccurate depending upon the shape of the plot and the number of straight-line segments used to approximate it. In many cases where the plot is smoothly continuous a function can be fitted to it with practically no ESS. Many of the algorithms in our subsystems models are of this type, and an example of a model of this type is shown as Model 5 in Appendix B.

Acknowledgements. I would like to gratefully acknowledge the assistance and contributions of the following people with respect to the techniques and procedures described in this paper. Mr. G.E. Beckman and Dr. H. Hoogendoorn for the data bank and nonlinear multiple regression programs, Mr. R.V. Frei and Mr. W.F. Mikhail for the development of the curve-fitting technique described in the paper, and Mr. R.A. Swanson for his valuable insight into system parameter interrelationships and assistance in development of the models in this paper, as well as many others in this area of systems evaluation.

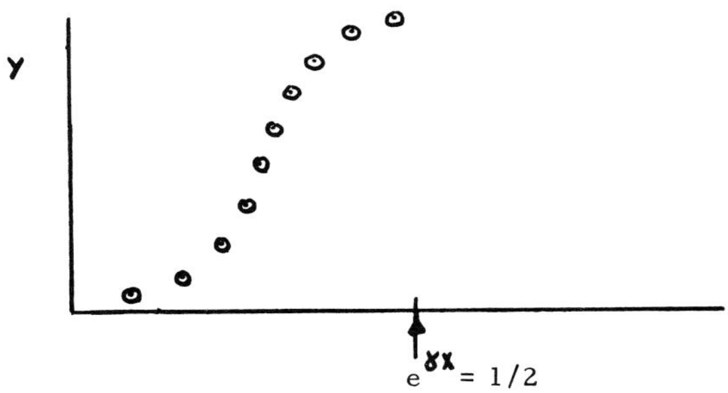
Y
$e^{\gamma x} = 1/2$

Fig. 1.

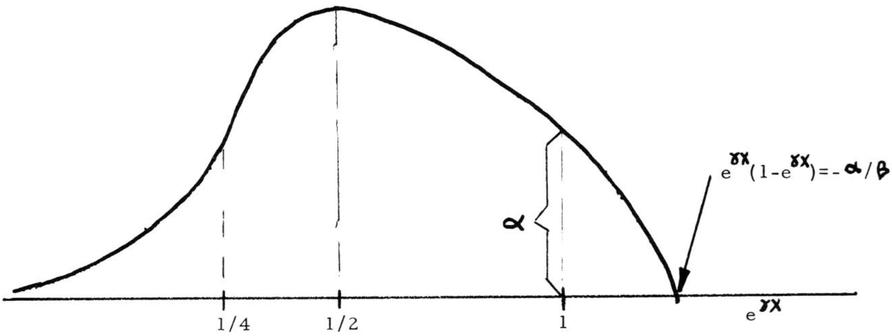
$e^{\gamma x}(1-e^{\gamma x}) = -\alpha/\beta$
α
1/4
1/2
1
$e^{\gamma x}$

Fig. 2.

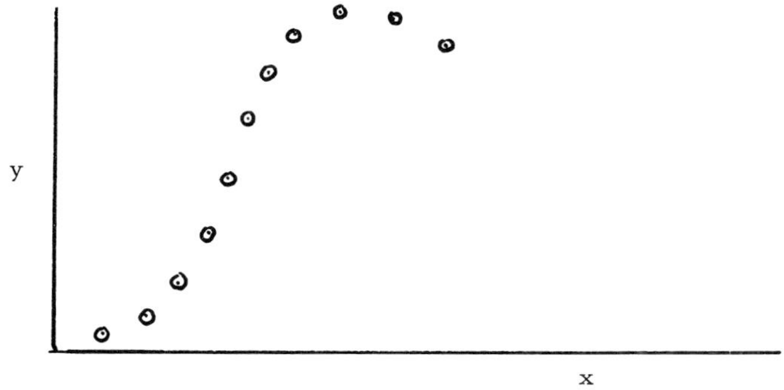
y
x

Fig. 3.

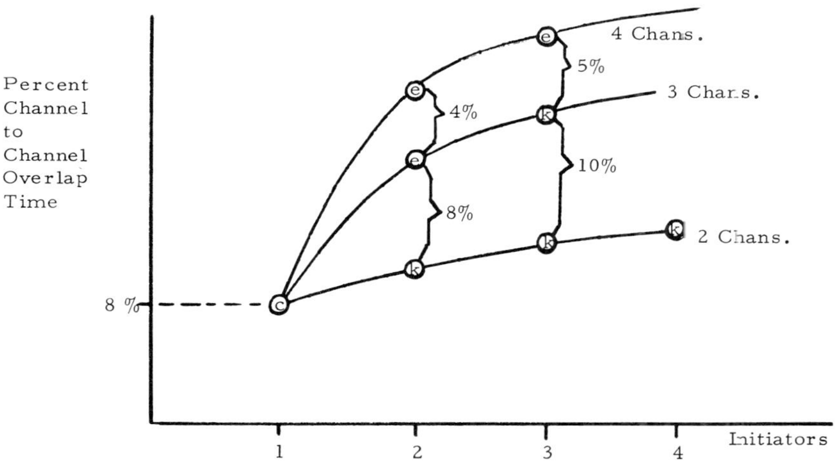
Percent Channel to Channel Overlap Time
8 %
4 Chans.
3 Chans.
2 Chans.
5%
4%
10%
8%
e
k
c
1
2
3
4
Initiators

Fig. 4.

APPENDIX A

Table 1A. Functions in terminal curve-fitting program package. Notation for the functions is standard algebraic with the following exceptions: = is expressed as ←, e^{cx} is expressed as *c×x, $\sqrt{x}$ is expressed as x*0.5, × stands for multiplication.

```
NO.   FUNCTION

1     Y ← A + B×(*C×X)
2     Y ← A + B×(*C×X*0.5)
3     Y ← A + B×((*C×X)×(1-*C×X))
4     Y ← A + B×(X×(1-*C×X))
5     Y ← A + B×((X*0.5)×(1-*C×X))
6     Y ← A + B×((*C÷X)×(1-*C×X))
7     Y ← A + B×((*C×X*0.5)×(1-*C×X))
8     Y ← A + B×((*C÷X*0.5)×(1-*C×X))
9     Y ← A + B×(*C÷X)
10    Y ← A + B×(*C÷X*0.5)
11    Y ← A + B×((*C÷X)×(1-*C÷X))
12    Y ← A + B×(X×(1-*C÷X))
13    Y ← A + B×(1-(1÷X)*C)
14    Y ← A + B×(1-((*C×X)×(1-*C×X)))
15    Y ← A + B×((*C÷X*0.5)×(1-*C÷X))
16    Y ← A + B×(1-((*C÷X)×(1-*C÷X)))
17    Y ← A + B×(1-((*C÷X)×(1-*C×X)))
18    Y ← A + B×(*C×(X-1÷X))
19    Y ← A + B×((X*0.5)×(1÷X)*C)
20    Y ← A + B×((1÷X)*C)
21    Y ← A + B×((1÷X)*C÷X*0.5)
22    Y ← A + B×((1÷X)*C×X*0.5)
23    Y ← A + B×(X×(1-*C÷X*0.5))
24    Y ← A + B×(X×(1-*C×X*0.5))
25    Y ← A + B×((X*0.5)×(1-*C×X*0.5))
26    Y ← A + B×((1-(1-*C×X))*X)
27    Y ← A + B×((1-(1-*C×X*0.5))*X*0.5)

101   Y ← A + B×X
102   Y ← A + (B×X) + (C×X*2)
103   Y ← A + (B×X) + (C×X*2) + (D×X*3)
ALSO INCLUDED IN THIS PROGRAM PACKAGE BUT NOT IN ANY OF THE
GROUPS ABOVE ARE THE POLYNOMIALS FROM DEGREE 4 THRU 9.  THESE
POLYNOMIALS HAVE FUNCTION NUMBERS 104 THRU 109 AND MAY BE USED
AFTER A LITTLE MORE FAMILIARITY WITH THESE PROGRAMS.
```

Table 1B. Function groupings.

```
GROUP 1 CONSISTS OF FUNCTIONS 1, 7, 3, 22 ,14, 18, 2, 25, 5, 24, 101, 102 AND 103
GROUP 1 SHOULD BE USED IF THE DATA PATTERN SUGGESTS A FUNCTION WHICH
IS EVERYWHERE INCREASING AND CONCAVE UPWARD.

GROUP 2 CONSISTS OF FUNCTIONS 27, 1, 10, 5, 25, 2, 19, 12, 13, 20, 101, 102 AND 103
GROUP 2 SHOULD BE USED IF THE DATA PATTERN SUGGESTS A FUNCTION WHICH
IS EVERYWHERE DECREASING AND CONCAVE UPWARD

GROUP 3 CONSISTS OF FUNCTIONS 2, 7, 3, 14, 9, 1, 25, 27, 5, 24, 101, 102 AND 103
GROUP 3 SHOULD BE USED IF THE DATA PATTERN SUGGESTS A FUNCTION WHICH
IS EVERYWHERE DECREASING AND CONCAVE DOWNWARD

GROUP 4 CONSISTS OF FUNCTIONS 1, 27, 12, 22, 10, 8, 6, 17, 9, 2, 101, 102 AND 103
GROUP 4 SHOULD BE USED IF THE DATA PATTERN SUGGESTS A FUNCTION WHICH
IS EVERYWHERE INCREASING AND CONCAVE DOWNWARD:

GROUP 5 CONSISTS OF FUNCTIONS 26, 11, 16, 21, 15, 17, 6, 8, 10, 22, 101, 102 AND 103
GROUP 5 SHOULD BE USED IF THE DATA PATTERN SUGGESTS A FUNCTION WHICH
IS EVERYWHERE INCREASING AND PASSES FROM CONCAVE UPWARD TO DOWNWARD.

GROUP 6 CONSISTS OF FUNCTIONS 26, 11, 15, 16, 17, 9, 6, 8, 10, 4, 101, 102 AND 103
GROUP 6 SHOULD BE USED IF THE DATA PATTERN SUGGESTS A FUNCTION WHICH
IS EVERYWHERE DECREASING AND PASSES FROM CONCAVE DOWNWARD TO UPWARD.
```

Table 2. Functions used for multiple nonlinear regression.

(1) $y = B_0 e^{B_1 x_1 + B_2 x_2 + \cdots + B_n x_n}$

(2) $y = B_0\, x_1^{B_1}\, x_2^{B_2} \cdots x_n^{B_n}$

(3) $y = B_0 + B_1/x_1 + B_2/x_2 + \cdots + B_n/x_n$

(4) $y = \dfrac{1}{B_0 + B_1 x_1 + B_2 x_2 + \cdots + B_n x_n}$

(5) $y = \dfrac{x_1 + x_2 + \cdots + x_n}{B_0 + B_1 x_1 + B_2 x_2 + \cdots + B_n x_n}$

APPENDIX B

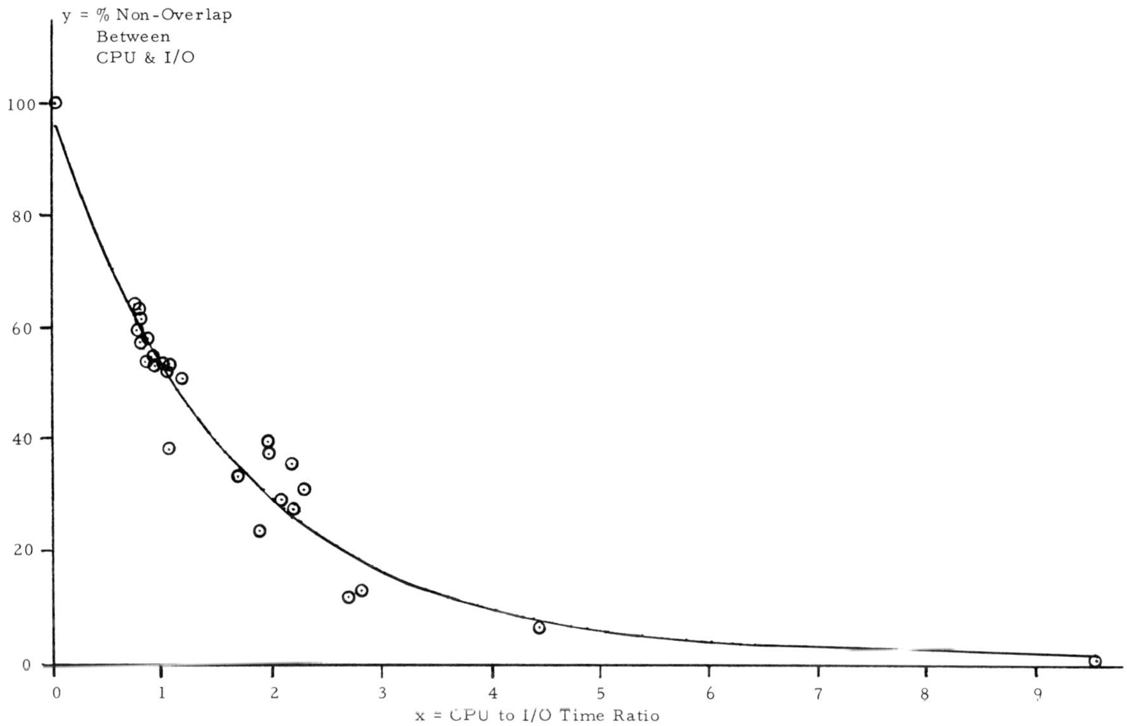

Model 1: CPU to I/O time ratio versus percent non-overlap between CPU and I/O. $y = a + be^{gx}$; $a = 1.2367$, $b = 95.141$, $g - -0.61551$. ESS = 723.

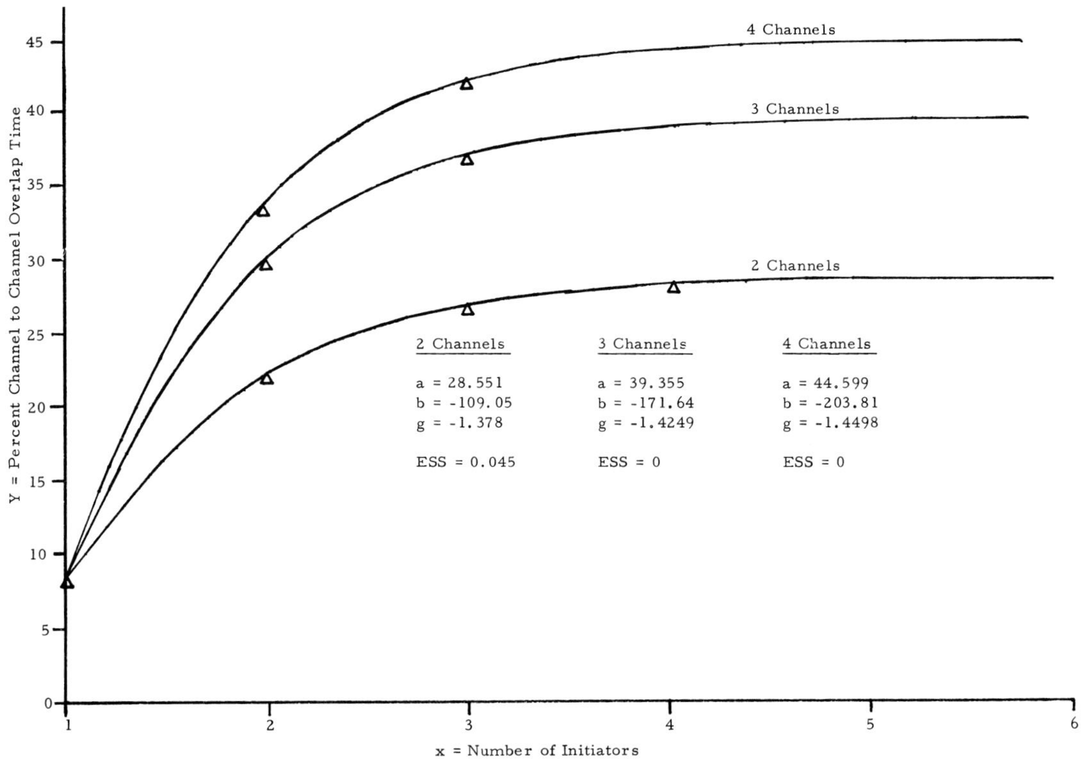

Model 2: Number of initiators versus percent channel to channel overlap time. $y = a + be^{gx}(1 - e^{gx})$.

ESS = 0.02102

Observed Data			y pred. using model	Residuals
y	X_1	X_2	$\hat{y}$	$y - \hat{y}$
1.250	1	1	1.357	-0.107
1.000	1	2	0.968	0.032
0.890	1	3	0.884	0.006
0.864	1	4	0.847	0.017
1.386	2	1	1.398	-0.012
1.000	2	2	0.989	0.011
0.891	2	3	0.901	-0.010
0.863	2	4	0.863	0.000
1.475	3	1	1.412	0.063
1.000	3	2	0.996	0.004
0.892	3	3	0.907	-0.015
0.863	3	4	0.868	-0.005
1.477	4	1	1.419	0.058
1.000	4	2	1.000	0.000
0.892	4	3	0.909	-0.018
0.864	4	4	0.871	-0.007

Model 3: Un-overlapped seek time versus number of channels and number of initiators. y = unoverlapped seek time, X_1 = number of channels, X_2 = number of initiators. Model 3: $\hat{y} = (X_1+X_2)/(B_0+B_1X_1+B_2X_2)$; $B_0 = 1.285339689$, $B_1 = 0.04336081615$, $B_2 = -0.5915839112$. ESS = 0.02102.

Independ. Variable Name	Independ. Variable I.D. No.	Mean	Standard Deviation	Simple Correlation x vs. y	Regression Coefficient	Std. Error of Reg. Coeff.	Computed t value
Sup. CPU T.	13	433.595	486.413	0.984	0.02159	0.00029	72.123
EXCP Count	9	6.916	13.350	0.676	0.00015219	0.01090	13.953

Depend. Variable Name	Depend. Variable I.D. No.	Depend. Variable Mean	Depend. Variable Std. Dev.
Net Chan. Busy T.	8	11.091	11.849

Model 4: Pred. Value Variable(8) = 0.02159(Variable(13)) + 0.00015219(Variable(9)) + 0.67641

Multiple Corr. Coeff. = 0.994
Std. Error of Estimate = 1.3

ANOVA (Analysis of Variance for the Regression)				
Source of Variation	Degrees of Freedom	Sum of Squares	Mean Square	F Value
Attributable to Regression	2	16369.59	8184.79	4837.09
Deviation from Regression	116	196.28	1.69	
Total	118	16565.87		

Model 4: Net channel busy time versus supervisor CPU time and EXCP count.

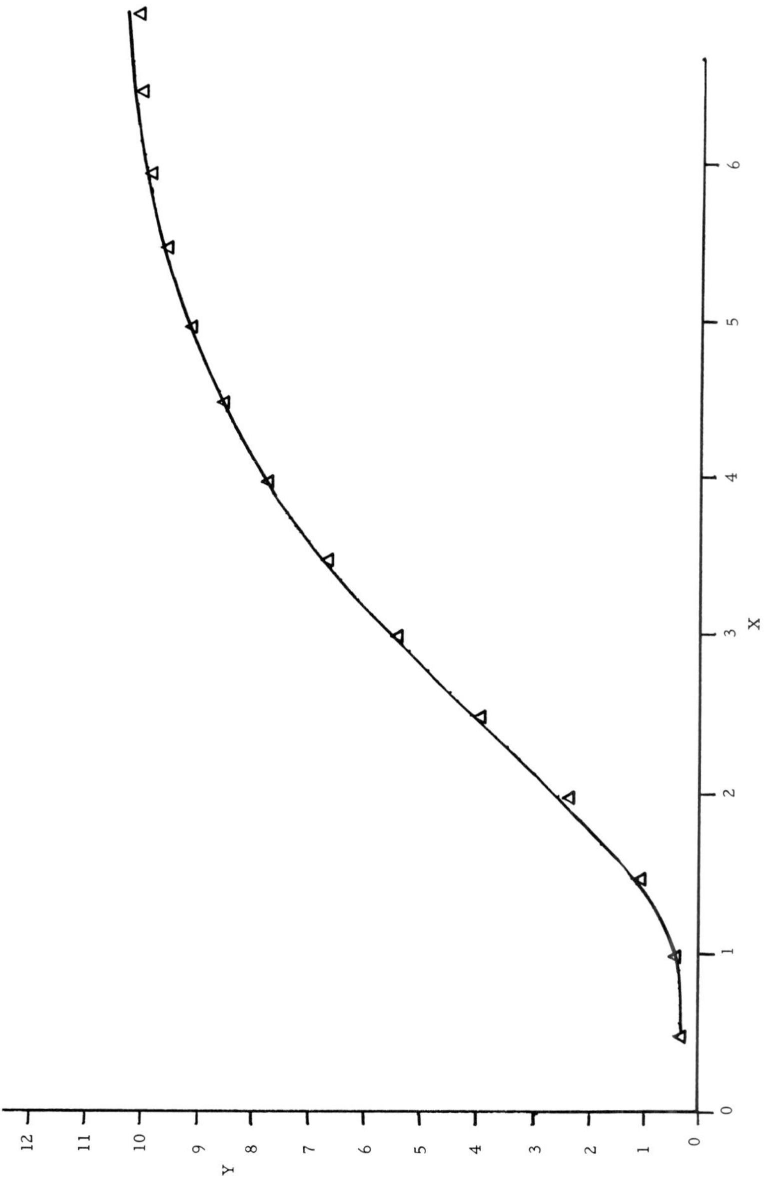

Model 5: Y versus X. $y = a + be^{g/x}(1 - e^{g/x})$; $a = 0.25404$, $b = 10.451$, $g = -5.6659$. ESS = 0.104. Function I.D. = 11.

DISCUSSION OF SESSION V

1. by H.P. Friedman

The paper by Bard and Suryanarayana comes closest to achieving the objectives of this conference. In this paper we see an innovational use of existing data analysis techniques in attempting to provide answers to computer performance and evaluation problems. The authors are well aware that assumptions are wrong - however, they do push the existing techniques fairly far and are able to achieve some useful approximate answers, with some indication of a measure of variability. My main objection to this paper is that the utility of their "equation" does not come through in this paper. One has to read the early paper (their reference [2]) together with this one to find the "need" for such an equation.

They do mention an important use of the equation in the present paper, i.e. to predict the amount of overhead required for a given workload. However, as pointed out by the authors, further work is required to relate program behavior to the individual events occurring in the system. The innovation is not in this use, however, but in using the event counter (hardware measurements) and regression methods to "define" a set of derived measurements in the set of estimated regression coefficients.

Possibly these "derived measurements" could be embedded in a more inclusive model that describes their relation to program behavior (workload) so that the model may be useful for prediction in the future.

The paper by Yeh is a good first attempt of a systems engineer applying standard techniques in the investigation of a "general problem". To me, the general utility of the result remains in doubt. I feel the author has to push further in indicating the utility of his derived regression equation.

This paper is important in showing that regression methods competently handled can produce interesting equations that can focus on computer systems problems. There is a similarity between this paper and the Bard paper, namely that until one can relate workload or program behavior on a given system to the E/B and channel/channel overlap there is no utility in the equation for prediction. Once you have to go in and measure these two variables you might as well measure the one you are trying to predict.

The paper by Racite exhibits competence in demonstrating a tool for doing nonlinear regression. It really does not deal with a problem. In the spirit of this conference, I'll wait for the next paper when the authors or anyone else impacts a computer problem with these methods.

In summary, I feel the Bard paper has shown that by using existing methods in an innovative fashion one can achieve useful, even though approximate, results and provide direction toward the development of new techniques. The Yeh paper has shown that existing procedures applied in a fairly straightforward manner by a competent systems engineer can begin to focus on gaps in our understanding of the behavior or complex computer systems. The paper by Racite demonstrates once again that computers have provided us with many more techniques than we have been able to exploit fruitfully and with understanding.

2. by Peter Wegner

Cuthbert Daniel, in the discussion of an earlier paper at this conference, characterized the statistical study of computing systems as the use of a wellknown set of verbs arising in statistics with a new set of nouns deriving from computer science. I would like to question whether we are in fact combining the right verbs and nouns in using regression analysis to model system behavior.

In developing macro models of system behavior we run the danger of treating as a black box a system which may be more naturally explained by a qualitative analysis of its components than by external behavior which ignores the internal structure of the system. It has become fashionable to assert that computer systems are too complex to analyze structurally. However, I personally feel that we have not yet exhausted the potential of qualitative analysis of system structure, and find it much more satisfying to explain system behavior in terms of system structure than in terms of observed but unexplained quantitative relations.

Computer science is different in character from empirical disciplines such as agriculture or physics. Agriculture and physics are concerned with the study of natural phenomena, while computer science is concerned with the study of man-made phenomena. A computer system generally has a far larger number of independently variable components than the systems studied in agriculture or physics.

Techniques such as linear regression analysis can model only very limited kinds of quantitative variability among variables of a system. They are applicable only when there is a reasonable assurance that system parameters which are not being varied are at their correct values. Relations among system parameters such as the E/B ratio and channel-CPU overlap are not generally helpful in system design because the important relations among system parameters are qualitative structural relations among lower-level system components.

Statistical studies of computing subsystems are useful when we are justified in making fairly rigid assumptions about the structure of the subsystem, as in the case of scheduling and queueing models. They are useful in performing certain forms of local optimization when structural characteristics of the system have been determined by qualitative analysis. However, such methods may be counter-productive and counter-intuitive when they are used to model quantitative relations among system components that are more naturally explained by explicitly considering the system structure than by treating the system as a black box.

The above methodological considerations apply also to other structurally complex areas of investigation such as artificial intelligence. In artificial intelligence there are certain restricted situations where "intelligence" can be improved by the quantitative adjustment of parameters. For example, the "skill" of a checkers-playing program developed by Arthur Samuels was improved by a learning program which encoded "experience" by means of quantitative weights. However, chess has turned out to be a more complex game in which qualitative components of strategy are more important than quantitative parameters.

One of the chief characteristics which distinguishes computing systems from other kinds of systems is their great structural complexity. In modelling such complex systems it is important to preserve a proper balance between quantitative statistical techniques and qualitative techniques of structural analysis.

SESSION VI

SYSTEMS MANAGEMENT

CHAIRMAN: P. WEGNER

ANALYSIS OF FREE-STORAGE ALGORITHMS

B.H. Margolin, R.P. Parmelee and M. Schatzoff

ABSTRACT. Dynamic management of free storage in a time-sharing operating system was studied empirically by the techniques of monitoring, emulation, and on-line experimentation. A new algorithm, based on observed usage patterns of different block sizes, was implemented and evaluated. On-line experiments demonstrated that supervisor time spent in free-storage management was reduced by seven or eight to one.

Central to successful operation of a computer system based heavily on reentrant (or pure) procedures is reliable and efficient dynamic management of free storage. Such systems must allocate, use, and release one or more blocks of free storage for each task or system operation, e.g., each I/O task or each request for supervisor services. For any list-processing system such as AED or LISP, efficient management of free storage is a fundamental problem. The consequence of errors in allocation and release is usually total collapse of the system; that of mismanagement is usually processor inefficiency or under-utilization of the free-storage pool. The processor inefficiency resulting from a poor or ill-chosen management algorithm is usually tolerated, even though it is high relative to that of other system functions; under-utilization of free storage is less tolerable, as "lock-up" can be encountered. This circumstance must be averted by task deferral or "garbage collection" procedures, both of which are costly to system performance.

To date, discussions of free-storage management have treated either: (1) a specific algorithm, which is developed and supported post-hoc, e.g. on the basis of a simulation study, or (2) a general algorithm presented without much information about its parameters. Thus Ross [1], in describing what is perhaps the most widely applicable and general approach to free-storage management, indicates the nature of the tools and techniques for very detailed programmer control of the management algorithm; the question of how to determine which algorithm to use is not considered. An excellent review and discussion of free-storage management is given by Knuth [2]. A recent paper by Campbell proposes a strategy based on the solution of the optimal-stopping problem on a Markov chain of fixed length, and shows that this strategy is superior to the first-fit method under certain conditions [3].

The context in which the present free-storage management research was conducted is CP-67, a virtual machine control program, which provides for each logged-in user the environment of a System/360 computer [4]. Broadly speaking, our use of CP-67 can be viewed as a large general-purpose time-sharing system. It was in operation 24 hours a day, with 20 to 40 users logged-in during prime shift throughout the observation period. The user load is mainly interactive programming, e.g. editing, compiling, loading, and executing; however, batch computing does play an important part in the load on the system.

The general goal of the project was to explore techniques of system analysis, modeling and redesign. The specific goal was not only to produce a free-storage management algorithm that would improve upon the existing system, but also to predict certain aspects of the improvement and then to validate and further quantify the improvement resulting from the new algorithm. The approach adopted was entirely empirical. The initial phase of the effort consisted of the monitoring, collecting and reducing of trace data pertaining to the allocation and release of blocks of free storage. The second phase consisted of off-line development of, and experimentation with, various algorithms. The algorithm ultimately developed [5] depends heavily on the observed statistical properties of the data collected in phase one. Final validation and quantification of the improvement was effected through an on-line experiment with the two algorithms competing over an eight-day period. This methodological approach is general; free-storage management in the particular system is merely one of a class of problems to which it can be applied.

BACKGROUND

Free Storage

CP-67 has three separate pools of free storage:

The page space: 4K-byte blocks of the system's address space made available via the system's relocation hardware on a dynamic basis.

SVC linkage space: 24-byte blocks of storage, used by the system for call linkages between its internal routines and modules.

Free storage: a pool of address space made up of separate, though possibly contiguous, 4K-byte blocks. It is from this pool that the control program obtains storage for describing and controlling all system tasks. This pool can be extended (but not retracted) at the expense of page space.

Storage Management

Programs acquire storage blocks via the subroutine FREE (and return them via FRET) in integer multiples of double-words (8 bytes). Each such block is treated distinctly; if n double-words are acquired, n double-words are returned in whole, not in part. Further, each such block represents a specific request. Not only does a requirement for exactly this block size exist, but when the requirement has passed, the block is returned. Thus there is neither splitting nor recirculating of the blocks by the requesting routine.

In its initial implementation, the FREE/FRET algorithm was very similar to one described by Knuth [2]. In brief, each block of free storage is threaded on a chain ordered by increasing addresses. A request is satisfied by the first exact fit found; if none is found, the first larger block is "split". On return, the block is rechained and "melded" with adjacent free blocks (if any). An unpleasant property of this procedure is that the depth of search as a result of a FREE (or FRET) is a random variable whose expectation and variance may increase to unacceptable levels during periods of heavy usage. Depending upon the load, the system managed a space of from 48K to 100K bytes during the period of observation, or from 7 to 14% of the total memory available to both free storage and page space. The range of requested sizes varied from 1 to 193 double-words; the time between request and release from milliseconds to hours. The FREE/FRET routine took between 2 and 45% of total control program (CP) time during 15-minute periods, with an average of 14.6%. (CP time is that time during which the CP is in supervisor state.)

Time versus Storage Space

The essence of the storage management problem is the desire to minimize simultaneously storage space wastage and control program overhead time, which are frequently diametric. One might hope to construct a crude relationship that would equate x wasted double-words with y units of wasted time (given the number of active users and the size of storage), but no work has been done on this.

Since it is unlikely that one can reduce both space and time ineffficiencies simultaneously, and since no function equating space to time existed, the project goal set was to develop an algorithm to substantially reduce time inefficiency yet maintain roughly the same space requirements as the original algorithm.

DATA GATHERING

Motivation

The first step taken in the study was to monitor the real demands on the FREE/FRET routine. These data were crucial for two reasons. First, the data might possess striking statistical properties that would suggest a desirable algorithm; they did. Second, initial comparisons of any new algorithm with the original algorithm were to be made off-line to prevent system degradation. One standard approach to this type of comparison is to simulate the process. However, this suffers from the need to specify statistical models for the demands. Other researchers who have taken this approach have assumed relatively simple models for the demands, such as independence of successive requests, well-behaved distributions of request sizes and request durations, or independence of request size and request duration. It was important to determine if these simplifying assumptions were appropriate for our free-storage problem; they were not.

An alternative to simulation is to capture free-storage trace data from the running system. The sequence of varying requests and their subsequent releases can then be used as input to "drive" a program that duplicates the original algorithm or a specified variation of it. We use the term "emulation" to refer to this alternative to simulation.

During emulation, measurements were made to compare algorithms. The off-line emulation suggested that the algorithm ultimately developed would greatly improve upon the speed of the original algorithm. This was later validated by an on-line eight-day experiment.

Implementation

Since storage management is centralized in the FREE/FRET routine, it was a simple matter to trace each transaction made. A minor change was made in CP-67 which enabled the system operator to cause the capture on tape of the following data describing each call to FREE/FRET:

- type of call (FREE or FRET)
- time of day
- location of call
- size of block handled
- location of block handled

In addition, each occurrence of an extension of free storage, i.e. the acquisition of 4K bytes from page space, was recorded. The data capture program used two 4096-byte buffers and, in the interest of minimizing system slow-down, made no attempt to detect or prevent buffer overrun. Rather, the records were numbered serially, and during subsequent data reduction a check was made for lost records.

System slow-down, although discernible, was within acceptable limits. A full reel of data (4800 buffers recording 960,000 calls to FREE/FRET) was typically obtained in about 2 hours, depending upon the level of activity. The only difficulty in acquiring a full trace of FREE/FRET calls was that those made during system start-up, before the trace program could be activated, were not recorded. It was assumed that after the first thirty minutes of emulated system operation this loss was negligible.

Data Reduction

The raw data provided a great deal of relevant information for designing an efficient free-storage management algorithm. Essentially, the data describe a two-dimensional representation of the demands upon free storage in terms of storage space and storage time. The utility of such data derives in large measure from the marked demand patterns they depict and the strong reproducibility of these patterns from day to day.

Size Distribution

Perhaps the simplest summary of the data is provided by a tabulation of the number of requests for each block size, as shown in table 1. These data were gathered on a particular day in the approximately 2.5-hour period from 9 am to

11:30 am. The request sizes are ordered by decreasing frequency of requests, and reveal that a small number of sizes account for a large proportion of all requests. Thus, nearly 90% of all free storage requests in the 2.5-hour interval were accounted for by the half-dozen sizes 4, 5, 29, 1, 8 and 10. Addition of the next ten most frequently requested block sizes would bring this up to 99%. Table 2, which contains similar data for each of four different days, shows that there is a high degree of reproducibility. Although these data have been sampled from a single installation, the nature of the system tasks giving rise to frequently requested block sizes suggests that similar phenomena occur in other CP-67 installations. For example, blocks of sizes 4 and 5 are associated with user-initiated I/O, whereas blocks of size 29 are associated with paging I/O. Thus an efficient algorithm for CP-67 installations in general might be one that would service the above frequently requested sizes very efficiently in time and space without sacrificing reasonably efficient handling of the less frequently requested sizes.

Demand Duration

In addition to the relative frequencies of requests by block size, it is important also to know the storage time requirements of requests, i.e. how long various block sizes are used before they are returned. This information is presented in table 3, which contains the frequency distributions of the durations that each block size is retained for data taken on day 4.

Most free-storage requests are short. For example, 94% of all blocks requested are returned in less than 5 seconds. For many sizes, the requests are almost exclusively short. For example, sizes included in the ranges 1-2, 6-9, 11-15, 18-29, 32, 34-48, 50-63, 66-98, 103, and 109-129 are primarily short-term requests, as evidenced in table 3. Only one block size gives rise to predominantly long occupancies - 65. The remaining block sizes, such as 3, 4, 5, 10, 16, 17, 31 and 146, which account for about two-thirds of all calls, are mixtures of long- and short-duration requests.

The interest in identifying long-duration block sizes derives from the potential profit in recognizing and servicing such requests in a separate area of free storage. This would reduce the fragmentation in main free storage and improve chances of returning empty pages to the control program during slack periods. This strategy was not pursued in the work reported here.

Distribution of Blocks Extant

A third aspect of free-storage utilization is the distribution of the number of each block size extant. More explicitly, think of incrementing a counter whenever a particular size is requested and decrementing it whenever that size is returned. Then the distribution of the number of blocks extant for that size is simply obtained by a frequency count of the number of times the counter was equal to each of the integers 1,2,... . For example, figure 1 shows the cumulative distributions of the numbers extant for sizes 1, 2, 3, 4, 5, 17 and 29. Such distributions are useful for estimating the probabilities that the number of blocks extant for given sizes will exceed some specified values. It is interesting to note that some sizes with relatively high frequencies tend to have only small numbers of requests extant at any time and, conversely, some sizes with relatively low frequencies tend to have moderate numbers of requests extant at any time. Examples of these phenomena are sizes 29 and 17, respectively.

Blocks Extant by Size and Time

While data such as that presented in figure 1 depict the variability of the number of blocks extant for individual sizes, they do not provide any information concerning the joint variation of the number of blocks extant for different sizes. Such information may be obtained by time-sampling the counters of the numbers extant for the various sizes. Table 4 shows a small portion of such data, recorded at ten-second intervals. Because of the transient nature of the system, in addition to information describing conditions at the end of the ten-second intervals, it is desirable to know something about activity during the intervals. One useful indication of such activity is provided by table 5, which shows the maximum number of blocks extant for each of the 14 most frequently requested sizes during successive ten-second intervals. The data in table 5 were generated from the same raw data used to generate the data in table 4. The analysis of data such as in table 5 was central to determining efficient free-storage management techniques.

The frequency distributions obtained during the data-gathering phase indicated that simplifying statistical assumptions, such as exponentially distributed duration times for the storage requests or independence of duration and size of the request, would not be tenable. This aspect of the problem made simulation an unattractive investigatory tool; there would be no simple way to characterize and then alter the demands on free storage and still be sure of maintaining a realistic study. Instead, as indicated previously, we chose to monitor the real demands upon storage and use the recorded demands as input to a program that emulated and evaluated two broad classes of algorithms. (We continue to use the term "emulation" to refer to the use of real data from a running system as input to a program that duplicates a subset of the running system or specified variations of it.)

Algorithm Variations

The first class included the original FREE/FRET algorithm and variations of it, as follows:

- Specification of a constant to be used as a rounding-up factor in satisfying FREE calls, as opposed to exactly fitting the request (the constant equal to zero). This differs somewhat from rounding up storage requests to, say, the nearest multiple of a given quantum of storage, so as to reduce the number of different sizes requested. (We will comment on this latter idea later.)
- Elimination of melding or permission to meld across page boundaries on extended pages.
- Specification of the number of pages to be allocated to FREE storage at system start-up time.

Pre-allocation

The second class of algorithms studied by emulation was one in which areas of free storage are pre-allocated for particular sizes. Sizes that are not pre-allocated are handled by a version of the original algorithm, as are pre-allocated sizes that overflow the assigned areas. The advantage of pre-allocation is that the handful of sizes that account for the vast majority of all free-storage requests can be handled quickly without a list search. One need only maintain a separate single-chained list of available blocks (a push-down stack) for each pre-allocated size. A request for a pre-allocated size is satisfied by the top element in the appropriate list. If none is available, it is serviced from the regular free-storage chain servicing nonallocated sizes. When a block from a pre-allocated area is no longer needed, it is returned to the top of the appropriate pre-allocation chain. Thus, by handling the top 5 or 10 sizes in separate chains, 90 to 95% of all requests are serviced in one step each.

One purpose of the emulator was to provide some indication of the trade-offs between time and space for these two classes of algorithms. The space needed by a particular algorithm may be computed exactly for a given stream of input data. The translation from emulation time to real time, however, is quite complicated, since the emulator was written in FORTRAN whereas any actual algorithm would be written in System/360 assembler language. Fortunately, the magnitude of the differences in list search lengths (depths) for the two classes of algorithms obviated the need for precise timing comparisons.

Data Reduction

After each block of 2000 transactions (FREEs plus FRETs), the emulation program provided the following data:

- Frequency distribution, mean, and standard deviation of depth of FREE search by size and for all sizes combined
- Frequency distribution, mean, and standard deviation of depth of FRET search by size and for all sizes combined
- Meld counts, number of unsuccessful melds, number of top-only melds, number of bottom-only melds, number of both top and bottom melds, and number of melds across page boundaries
- Frequency distribution of blocks in use, by size
- Number of double-words in use
- Frequency distribution of available blocks, by size, together with mean and standard deviation
- Number of available blocks
- Percentage utilization of available space

It should be noted that in computing the means and standard deviations of depth of search, both FREEs and FRETs of a size in a pre-allocated area were considered equivalent to a search of depth one. Thus, in terms of depth of search for FREE (or FRET), the optimally efficient algorithm would have an average search depth of one.

For algorithms of the second class, data taken on a typical day were used to provide specification of the pre-allocated sizes and of the number of double-words to be allocated for these sizes. Thus, when five sizes were to be pre-allocated, the top five most frequently requested sizes were determined from the frequency distribution by size (e.g., table 1); then the number of blocks to be allocated by size was determined as a specified percentage point, say the 50% point, of the cumulative distribution of the number extant by size. For example, in figure 1, the 50% points of the distribution of block sizes 3 and 5 are 27 and 57 respectively. (This prespecification of sizes and percentages plus the entirely static nature of the algorithms of this class appeared to be the major drawbacks of the pre-allocation scheme.)

Emulation Results

The results of only one set of emulation experiments are reported. Five different algorithms were involved, one from the first class and four from the second. The algorithm chosen from the first class was the exact replica of the original algorithm of CP-67 and was labeled CONTROL. Algorithms 2 and 3 pre-allocated at the 50% and 95% levels, respectively, for the five most frequently requested sizes. Algorithm 4 pre-allocated at the 95% level for the top 15 sizes, whereas algorithm 5 partitioned these 15 sizes into subgroups by rounding up certain of the request sizes. These pre-allocated sizes and percentiles were determined from one day's tape of data, labeled Day 1. The five algorithms were then run against Day 1 and four other days' tapes, labeled Day 2 through Day 5. The responses measured were the average depth of search for the FREE and the FRET subroutines in consecutive intervals of 200 blocks, plus the number of pages used in the free-storage area. Table 6 depicts the results of the twenty-five emulation experiments. For purposes of comparison, the responses were recorded for blocks 3601-3800 on each tape, since the shortest available tape (terminated by a system malfunction) contained only 3800 blocks. Measurements at this point should be free of any transient start-up effects and should provide a sizeable sample for comparisons.

The results of the experiments indicate that increasing the extent of the pre-allocation for each size results in a marked reduction in depth of search for both FREE and FRET, at an initially small cost in the number of pages used in the free-storage area. For example, results for algorithm 5 indicate that average search depths can be reduced by a factor of twenty to one at the cost of only 2 or 3 additional pages. These observations led to the design of a new algorithm, which has been implemented in CP-67 and is described later in this paper.

DYNAMIC CHAINS AND POOLING STUDIES

Dynamic Chain Growth

The major drawback to the fixed-allocation algorithms discussed is their dependence on static pre-allocation. Each time the system starts up, a chain of the exact same number of blocks is set up for each of the sizes to be pre-allocated. A realistic determination of the number of blocks needed in each chain requires a sizeable amount of data concerning the current state or degree of system usage. These algorithms cannot adjust dynamically to altered profiles of the installation's daily requests for storage nor, for that matter, to another installation's profile.

To eliminate the need for static pre-allocation in the algorithms of the second class, and thereby to make them more flexible, these algorithms were altered so that chains servicing the specially treated request sizes would grow dynamically as the need for each particular size increased. The special sizes would never overflow their chains since the chains would increase in length without limitation to accommodate new requests. For example, sizes 4, 5, 8, 10 and 29 might be selected for special handling; these sizes would then be serviced with five separate growing chains. The remaining sizes would be treated in accordance with the original CP-67 algorithm.

This new hybrid class of algorithms, the third class, has the property that at any time T the chain for a specially treated size is as long as the maximum number of blocks of that size that have been extant up to time T since the most recent system start-up. This feature could ultimately cause a sizeable waste of storage space, which is the main trade-off in removing the static chain-length limitation. If this waste occurs after the day's peak of activity, it poses less of a problem, since time can be afforded for "garbage collection". The garbage collection procedures in the literature tend to be costly, since they require much time-consuming bookkeeping and checking of addresses. This aspect was not investigated; no specific garbage collection techniques other than that in the original CP-67 algorithm are discussed in this paper.

Space Allocation and Subpool Arrangements

The concept of allocation of storage space is easily quantified. Let $f_n(t)$ denote the number of blocks of size n extant at time t, let S represent the set of request sizes to be handled dynamically in their own separate chains, and let t = 0 be the time of the most recent system start-up. The allocation of space at time T, T > 0, by the special chains is

$$A_S(T) = \sum_{n \in S} n \cdot \max_{0 \leq t \leq T} f_n(t) \tag{1}$$

where

$$\max_{0 \leq t \leq T} f_n(t)$$

is the largest number of blocks of size n extant from the last system start-up, t = 0, to the present time, t = T. The waste of space at T due to the special chains is the allocation at T minus the need at T, namely

$$W_S(T) = \sum_{n \in S} n\{[\max_{0 \leq t \leq T} f_n(t)] - f_n(T)\}$$

$$= \sum_{n \in S} n \cdot \max_{0 \leq t \leq T} [f_n(t) - f_n(T)] \tag{2}$$

If one were to try to serve every size with its own chain, the waste of space would increase greatly. Much of this waste would be due to those sizes that rarely have a single block extant. If any of these sizes has appeared since the last start-up, then its chain has at least one block available, frequently unused.

The method chosen to combat the space waste caused by the special chains is to create nonoverlapping subpools, or disjoint subsets, among the request sizes contained in S and slated for special handling.

Each subpool is served by a single chain, where the blocks in the chain are as large as the largest request size in the subpool. Thus, if the requested sizes forming subpool p, p=1,2,...,s, are denoted by $(n_{p1},\ldots,n_{pk_p})$, with

$$n_p = \max_r \cdot n_{pr}, \qquad r=1,2,\ldots,k_p$$

then the chain for the pth subpool consists of blocks of size n_p.

The allocation of space for the pth subpool at time T is:

$$A_p(T) = n_p \cdot \max_{0 \le t \le T} \sum_{r=1}^{k_p} f_{n_{pr}}(t) \tag{3}$$

The total allocation of space by the sizes in the pth subpool, when each is handled separately by its own chain, is

$$\sum_{r=1}^{k_p} n_{pr} \cdot \max_{0 \le t \le T} f_{n_{pr}}(t) \tag{4}$$

No simple relation such as an inequality exists between Eqs. (3) and (4), but a judicious choice of subpools should effect a reduction in space allocated, and hence in space wasted.

The key question is how to determine the subpools so as to reduce the wastage reflected in Eq. (3). Minimizing the wastage for any one body of data is not practical because of the stochastic nature of the problem from day to day, and the desire to avoid unwarranted simplifying assumptions about the pertinent stochastic variables $[f_n(t)]$. Rather, the approach used was again empirical.

One single tape of data was used to study the subpooling question with the intention not of finding a near-optimal subpooling arrangement but rather of finding some simple rule of thumb that might generally be used to construct the subpools.

If there are no *a priori* restrictions on the number of subpools, or on any other aspect of the subpools, then the number of possible subpool arrangements is enormous. For a total of N sizes, the number of distinct subpool arrangements into M subpools can be shown to be:

$$\frac{\sum_{j=1}^{M} (-1)^{M-j}\, j^N \binom{M}{j}}{M!} = \sum_{j=1}^{M} (-1)^{M-j}\, j^N \frac{1}{j!(M-j)!} \tag{5}$$

many of which would be nonsensical in the problem at hand.

The total number of possible subpooling arrangements is Eq. (5) summed over M, namely

$$\sum_{M=1}^{N} \sum_{j=1}^{M} (-1)^{M-j} j^{N} \frac{1}{j!(M-j)!} \tag{6}$$

Thus, for N = 10, the number of distinct subpool arrangements is 115,975. By the time N = 25, the number of arrangements is 4.64×10^{18}. Clearly, in the case under discussion, N is greater than 50 and hence an investigation of all subpools for even one day's tape is impossible.

One restriction that seems reasonable and decreases the number of distinct subpool arrangements considerably is to have subpools consist of consecutive sizes only. This is equivalent to the general procedure of rounding up request sizes. Again, for a total of N sizes, there are now 2^{N-1} distinct subpool arrangements that obey the requirement of "consecutiveness" or consistent "rounding up" of sizes.

The study of all 2^{N-1} arrangements is still an enormous undertaking, and so the empirical investigation was simplified further by checking possible subpooling arrangements among ten consecutive sizes at a time. Thus, for sizes 1,2,3,...,10, the $2^9 = 512$ possible consecutive subpools were constructed. Then the number of double-words that would have been needed by each subpooling arrangement to accomodate the day's tape was computed. This procedure was repeated for sizes 10,11,...,19, for sizes 19,20,...,28, etc.

<u>Computation of Space Allocation</u>

Some comments are in order on how the computation of the total usage of space was carried out for the 512 possibilities for each set of ten consecutive sizes. Each subpool arrangement of ten ordered sizes satisfying the requirement of consecutiveness can be represented by an ordered string of nine zeros and ones. The first number will be a one if the smallest and next smallest sizes in the set of ten sizes are to be pooled together, and a zero otherwise. The <u>kth</u> number will be a one if the <u>kth</u> and <u>k+1st</u> sizes in the set of ten are to be pooled together, and a zero otherwise. Thus, for example, one of the 512 subpool arrangements of the sizes 1,2,...,10 satisfying the requirements of consecutiveness is

$$1,\ 2\text{-}3,\ 4,\ 5,\ 6\text{-}7\text{-}8,\ 9\text{-}10, \tag{7}$$

where sizes connected by a dash are to be pooled together. There are six subpools in this arrangement: sizes 1, 4 and

5 are not pooled with any other size; sizes 2 and 3 are pooled together; sizes 6, 7 and 8 are pooled together; and sizes 9 and 10 are pooled together. In terms of rounding up, size 2 is rounded up to 3; sizes 6 and 7 are rounded up to 8; and size 9 is rounded up to 10. The vector of zeros and ones representing this subpool arrangement of the ten sizes is

0 1 0 0 0 1 1 0 1,

and the allocation of space up to time T is:

$$\begin{aligned}[1 \cdot \max_{0 \leq t \leq T} f_1(t)] &+ \{3 \cdot \max_{0 \leq t \leq T} [f_2(t) + f_3(t)\} \\ &+ [4 \cdot \max_{0 \leq t \leq T} f_4(t)] + [5 \cdot \max_{0 \leq t \leq T} f_5(t)] \\ &+ \{8 \cdot \max_{0 \leq t \leq T} [f_6(t) + f_7(t) + f_8(t)]\} \\ &+ \{10 \cdot \max_{0 \leq t \leq T} [f_9(t) + f_{10}(t)]\} \qquad (8)\end{aligned}$$

For the evaluation of space requirements, it is desirable to order the 512 arrangements to be checked so that in moving from the ith to the i+1st arrangement, there is either a single joining of two subpools at the ith stage into one subpool at the i+1st stage or a single division of one subpool at the ith stage into two subpools at the i+1st stage. This minimizes the amount of computation needed to evaluate the allocation for each of the 512 arrangements. For example, if one had evaluated the allocation for the arrangement in Eq. (7) and wanted the allocation for the arrangement:

1, 2-3, 4, 5, 6-7-8-9-10 (0 1 0 0 0 1 1 1 1),

namely

$$\begin{aligned}[1 \cdot \max f_1(t)] &+ \{3 \cdot \max [f_2(t) + f_3(t)]\} \\ &+ [4 \cdot \max f_4(t)] + [5 \cdot \max f_5(t)] \\ &+ \{10 \cdot \max [f_6(t) + f_7(t) + f_8(t) \\ &+ f_9(t) + f_{10}(t)]\}\end{aligned}$$

then one would have stored from the computation of Eq. (8) all but the last term of Eq. (9), which would require obtaining

$$\max_{0 \le t \le T} [f_6(t) + f_7(t) + f_8(t) + f_9(t) + f_{10}(t)]$$

One way systematically to run through all the possible 512 vectors of zeros and ones, i.e. through all subpool arrangements, so that at each stage either a single 0 is changed to a 1 or vice versa, is known in switching theory as the Gray code [6]. A program was written to follow the Gray code through the 512 possible arrangements, compute the storage space allocation for each, and then reorder the vectors of zeros and ones in terms of increasing space allocation. The output consisted of ten columns and 512 rows for each set of ten sizes studied, the first nine columns being the code for the specific subpool arrangement evaluated and the tenth column containing the corresponding storage allocation. Thus, the first row had the smallest requirements, the second row the next smallest, and so on. Table 7 contains a portion of the output for sizes 10,...,19. Looking at this output, if the j<u>th</u> column contained a relatively large number of zeros toward the top of the output, this suggested that the j<u>th</u> and j+1<u>st</u> sizes should not be pooled together; similarly, a large number of ones toward the top of the j<u>th</u> column of the output suggested that the j<u>th</u> and j+1<u>st</u> sizes should be pooled together. Based on this computation and subsequent analyses, a simple rule of thumb for subpool construction was postulated.

Rounding up of Request Sizes

The general idea of rounding up the size of a storage request is not new. It has, for example, been studied via simulation by Randell [7]. He investigated the special case of rounding up request sizes to the nearest multiple of a given quantum of space, e.g. to the nearest power of 2. His conclusion was that rounding up in general is not desirable in that it offers no gain in storage utilization.

The proposed rule for subpool construction by rounding up requests requires sampled data on the number extant for each size over a reasonable and representative period of time, data similar to that in table 5. Each size that typically has a large number extant will be called a mode of the distribution of the number extant. The proposed rule of rounding then is to round a given size up to the nearest mode. This is called "modal roundup". In table 5, the modes are sizes 4, 5, 3, 17, 10 and 1 in decreasing order of the number extant.

This rule contains a negative statement in that rounding up modes is to be avoided. Intuitively this seems reasonable; if a size with a large number of requests typically extant is rounded up, wastage will, in all likelihood, increase. If one holds to the modal round-up rule, then one can see the unattractiveness of rounding up to the nearest multiple of a given quantum Q as reported by Randell. One will almost surely round up many modes in the process, thereby increasing the storage waste.

Beyond what is implied in modal round-up, there appears to be no generality that can be stated about rounding up request sizes except, perhaps, that there appears to be some insensitivity of storage usage to the rounding up of many of the rarely occurring smaller sizes.

Finally, it should be noted that the gains in space allocation by good subpooling over no subpooling are nowhere as dramatic as the gains in time by special chaining of certain sizes over no special treatment of any sizes.

ON-LINE EXPERIMENTATION

The Experiment

The next step in this investigation was the ultimate validation of the emulation findings concerning dynamic chains and subpooling for the most frequently observed sizes. This was an on-line experiment comparing the original free-storage algorithm's performance with the performance of an algorithm of the third class, with dynamically growing chains for certain subpooled sizes.

The specific new algorithm chosen for the experiment provided special service for the 13 sizes most frequently used in our system. These sizes were 1, 2, 3, 4, 5, 6, 7, 8, 9, 10, 17, 18 and 29; they accounted for approximately 98% of all the calls to free storage. The subpooling employed for the thirteen sizes was: 1, 2-3, 4, 5, 6-7-8, 9-10, 17-18, 29. This arrangement satisfies the modal round-up rule by not rounding up the modes in table 4, with the exception that 17's were rounded up to 18. The desirability of this exception is readily apparent from table 7, which clearly shows that 17's and 18's are grouped together in most of the better arrangements.

Since those sizes not assigned to subpools by the new algorithm are handled according to the rules governing the original algorithm, the latter algorithm can be obtained as a special case by not specifying any sizes for subpool handling. The ease of changing algorithms by parameter settings thus facilitates comparative evaluation via designed experiments carried out in an operational environment. An experiment of this nature was conducted shortly after incorporation of the new algorithm. At 5:00 am each morning during a two-week period, a CP-67 system containing either the new or old FREE-FRET algorithm was loaded according to a specified schedule and allowed to run for 24 hours. If a system-abnormal termination occurred during the day, the same system would be reloaded. The experiment was designed to consist of only eight days, namely Monday through Thursday of each week. The reason for this was twofold. First, weekend data were excluded due to low system usage during such periods. Furthermore, in order to have a statistically balanced experiment with respect to sequencing of algorithms within a week, and day-by-day comparisons between weeks, an even number of days in each week was required. The design of the experiment is shown in table 8.

This design provides protection against possible linear or quadratic time trends in usage during the course of the experiment, and allows comparisons between algorithms, among days of the week, and between weeks. During the experimental period, data on forty different variables were recorded at five-minute intervals by means of a software monitoring program. These included cumulative counts of FREE and FRET calls broken down by subpool class, free-storage occupancy statistics, and accumulated supervisor time spent in FREE and FRET.

These measurements were provided specifically to compare the two FREE/FRET algorithms. Additionally, data normally collected on the system for measurement and analysis of system performance were recorded during the experimental period. These data included counts of various types of paging and I/O activity, numbers of active and inactive users, and subdivisions of CPU time into several different states.

Experimental Results

The principal results of the experiment are summarized in table 9, which shows the average supervisor time per FREE and per FRET call during each day of the experiment. The data are presented according to the layout of table 8, with results corresponding to the new algorithm italicized for ease of comparison. It is readily apparent that the average CP time per call to FREE or FRET is reduced by a factor of 7 or 8 to 1 by the new algorithm. These differences are so large that formal statistical analyses and tests of significance are scarcely required to verify the conclusions.

Finally, the question of how the new FREE/FRET algorithm affects the overall system performance naturally arises. A simple answer to this question, provided by direct system measurement during the experiment, is that CP time spent in FREE/FRET has been reduced from an average of 14.6% of total CP time to an average of 2.4%. A more extensive answer to the question may be gleaned from tables 10 and 11, which have been abstracted from a paper by Bard [8]. Table 10 shows the reductions in CP time for various types of I/O operations, as estimated by fitted regression models, while table 11 shows how throughput rates have been increased by the introduction of the new algorithm. As a result of the study, the improved algorithm was incorporated into CP-67. The salient feature of the study was the demonstration that significant improvement in system performance can be effected by designing resource allocation algorithms to take advantage of observed demand and utilization patterns.

Acknowledgement. The authors gratefully acknowledge R. Adair, J. Ravin and J. Seymour for their suggestions and assistance in this research. In particular, J. Ravin wrote the data reduction and emulation programs and carried the major burden of the detailed validation of the emulation and the sample data used to drive it, and J. Seymour designed and programmed the FREE/FRET algorithm incorporated in CP, together with the software measurements required in the on-line experiment.

REFERENCES

1. D.T. Ross, "The AED free storage package", Comm. ACM 10, 181-192 (1967).
2. D.E. Knuth, "Dynamic storage allocation", The art of computer programming 1, Addison-Wesley, 1968.

3. J.A. Campbell, "A note on an optimal-fit method for dynamic allocation of storage", Comp. J. 14, 7-9 (1971).
4. R.A. Meyer and L.H. Seawright, "A virtual machine time-sharing system", IBM Syst. J. 9, 199-218 (1970).
5. This algorithm was incorporated into CP-67, version 3.
6. W.S. Humphrey, Jr., Switching circuits with computer applications, McGraw-Hill, 1958, p. 109.
7. B. Randell, "A note on storage fragmentation and program segmentation", Comm. ACM 12, 365-372 (1969).
8. Y. Bard, "Performance criteria and measurement for a time-sharing system", IBM Syst. J. 10, 193-216 (1971).

Table 1 Block size request patterns

Order	Size	Frequency	Cumulative frequency	Proportion
1	4	121106	121106	0.248
2	5	106780	227886	0.467
3	29	75637	303523	0.623
4	1	54453	357976	0.734
5	8	54331	412307	0.846
6	10	19938	432245	0.887
7	3	18074	450319	0.924
8	9	10018	460337	0.944
9	18	9188	469525	0.963
10	17	4548	474073	0.972
11	7	2530	476603	0.978
12	23	1530	478133	0.981
13	6	1486	479619	0.984
14	2	1285	480904	0.986
15	12	979	481883	0.988
16	21	918	482801	0.990
17	11	657	483458	0.992
18	104	400	483858	0.993
19	146	386	484244	0.993
20	27	310	484554	0.994
.	.	.	.	.
.	.	.	.	.
.	.	.	.	.
71	77	1	487504	1.000
72	84	1	487505	1.000
73	98	1	487506	1.000
74	103	1	487507	1.000
75	109	1	487408	1.000
76	112	1	487509	1.000
77	124	1	487510	1.000
78	193	1	487511	1.000

Table 2 Cumulative proportions of requests

Block sizes	Days			
	1	2	3	4
4	0.248	0.258	0.471	0.300
5	0.467	0.464	0.753	0.540
29	0.623	0.673	0.782	0.650
1	0.734	0.744	0.822	0.724
8	0.846	0.851	0.893	0.799
10	0.887	0.886	0.907	0.836
3	0.924	0.919	0.933	0.877
9	0.944	0.940	0.944	0.889
18	0.963	0.953	0.949	0.902
17	0.972	0.960	0.955	0.910
7	0.978	0.975	0.965	0.974
23	0.981	0.977	0.966	0.975
6	0.984	0.984	0.969	0.986
2	0.986	0.986	0.970	0.988
12	0.988	0.987	0.972	0.989

Table 3 Frequency of requests by size and duration

Size	*Duration in seconds* 0 < 5	5 < 10	10 < 30	30 < 60	60+	*Proportion less than 5 seconds*
65	0	0	1	0	21	0.000
17	1284	617	694	194	189	0.431
31	36	5	8	4	27	0.450
101	18	0	8	5	2	0.545
146	161	17	27	14	37	0.629
16	103	2	3	1	25	0.769
10	11761	1734	621	178	152	0.814
18	4395	862	23	0	0	0.832
3	13528	1672	488	83	135	0.850
4	106147	5510	4275	947	532	0.904
33	51	1	0	1	3	0.911
104	296	22	4	1	0	0.916
12	255	21	1	1	0	0.917
15	138	3	3	0	6	0.920
2	915	25	30	6	0	0.938
9	4543	212	31	3	0	0.949
14	209	10	0	0	0	0.954
5	90248	1678	1246	318	369	0.962
13	258	5	3	0	2	0.963
11	1032	39	0	0	0	0.964
6	4138	97	49	7	0	0.964
8	29028	176	3	1	0	0.994
1	28884	48	36	1	0	0.997
7	25065	63	2	1	0	0.997
29	43029	0	0	0	0	1.000
	367419	12821	7556	1767	1505	0.940

Note:
Any sizes not shown, and for which there are more than five observations, had all blocks returned in less than five seconds.

Table 4 Number of blocks extant by size at 10-second intervals

Size	*Time in ten-second intervals*																			
	541	542	543	544	545	546	547	548	549	550	551	552	553	554	555	556	557	558	559	560
1	12	11	11	11	11	11	11	11	11	11	13	12	11	11	11	11	12	12	13	12
2	1	1	1	2	2	1	0	0	1	1	1	1	0	0	0	1	0	0	0	0
3	32	31	30	33	33	32	26	26	33	28	32	36	32	38	38	40	37	39	43	44
4	123	122	121	123	122	123	120	115	117	116	124	131	109	132	126	115	113	131	133	137
5	93	93	93	92	93	94	92	94	85	86	88	97	89	88	89	88	92	90	92	99
6	0	0	0	0	0	0	0	1	1	0	2	0	0	0	0	0	0	1	1	0
7	1	0	1	0	1	1	0	1	0	1	1	2	1	0	0	1	3	1	1	0
8	0	0	1	1	1	1	0	0	1	0	1	0	1	1	1	0	1	0	0	1
9	0	0	0	0	1	0	0	1	0	0	0	2	0	0	0	0	1	0	0	2
10	16	17	17	17	17	16	17	17	15	17	16	16	15	15	14	13	15	17	18	15
11	1	0	0	0	0	0	0	0	0	0	0	0	0	1	0	0	0	0	0	0
12	0	0	0	0	0	0	0	0	0	0	1	0	0	0	0	0	0	0	0	0
15	3	3	3	3	3	3	3	3	3	3	3	3	3	3	3	3	3	3	3	3
16	13	13	13	13	13	13	13	13	12	12	12	13	13	13	13	13	13	13	13	14
17	27	28	28	27	29	24	26	28	22	24	25	25	25	22	21	22	22	25	26	25
18	2	1	1	2	2	3	2	2	3	3	3	3	2	4	4	3	3	3	3	4
28	0	0	0	0	0	0	0	0	0	0	0	0	0	0	0	0	0	0	0	0
29	0	0	0	0	0	0	0	0	0	0	1	0	0	0	3	0	1	1	0	0
31	13	13	13	13	13	13	13	13	12	12	13	13	13	13	13	13	13	14	14	14
33	3	3	3	3	3	3	3	3	3	3	3	3	3	3	3	3	3	3	3	3
65	10	10	10	10	10	10	10	10	9	9	9	10	10	10	10	10	10	10	10	11
69	0	0	0	0	0	0	0	0	0	0	0	0	0	0	0	0	0	0	1	0
101	1	1	1	1	1	1	0	0	0	0	0	0	0	0	0	0	0	0	1	0
104	0	0	0	0	0	0	0	0	0	0	1	0	0	0	0	0	0	1	1	0
129	3	3	3	3	3	3	3	3	3	3	3	3	3	3	3	3	3	3	3	3
146	5	7	6	6	6	7	5	5	6	4	3	5	5	5	6	6	6	6	7	6
TOTALS	359	357	356	360	364	359	344	346	337	333	355	375	335	362	358	345	351	373	386	393

Table 5 Maximum number of blocks extant during 10-second intervals for 14 most frequently requested sizes

Size	*Time in ten-second intervals*																			
	541	542	543	544	545	546	547	548	549	550	551	552	553	554	555	556	557	558	559	560
1	13	14	13	13	13	13	13	13	12	14	15	15	15	13	14	14	13	13	14	14
2	1	1	1	2	2	2	1	1	2	2	2	3	2	2	1	1	1	0	1	2
3	36	37	34	37	35	36	33	30	36	37	35	44	39	40	39	40	40	41	44	48
4	126	125	125	126	125	124	124	123	124	120	126	135	133	136	135	128	117	137	136	140
5	95	97	96	96	97	97	94	95	95	89	93	99	97	94	94	93	93	94	97	107
6	2	2	1	1	0	0	1	2	2	2	2	4	1	1	1	1	1	1	2	3
7	5	4	3	3	1	2	2	2	2	2	3	3	2	3	2	2	4	5	4	4
8	1	1	2	2	3	2	2	2	3	2	1	2	3	2	2	2	1	3	1	3
9	1	1	1	1	2	1	1	1	1	1	1	3	2	1	1	2	2	3	2	4
10	18	19	18	18	17	18	19	18	19	18	20	17	19	17	17	18	15	18	20	19
12	1	0	1	0	1	0	1	0	0	0	2	1	0	0	0	0	0	0	0	0
17	27	28	29	29	29	29	26	28	28	26	26	25	27	25	24	24	23	25	26	27
18	4	4	3	3	3	5	5	3	3	4	4	6	4	5	5	5	4	4	5	5
29	4	3	1	2	5	1	2	2	2	2	4	3	3	4	6	5	6	4	5	5

Table 6 Average depths of FREE and FRET searches during blocks 3601 – 3800, and number of pages used after 3800 blocks

Algorithms	*Day* 1	2	3	4	5
A	19.14 13.45 12	24.61 21.23 12	31.64 22.21 15	27.35 20.92 12	48.28 40.69 17
B	1.23 1.04 12	4.95 4.12 12	9.88 7.76 15	4.44 3.35 13	13.53 10.36 17
C	1.18 1.04 12	2.87 2.47 13	8.61 3.98 16	3.42 2.19 15	5.02 3.90 18
D	1.01 1.00 12	1.34 1.26 14	2.48 2.25 17	1.10 1.04 15	2.63 1.74 18
E	1.01 1.00 12	1.24 1.18 14	1.38 1.27 16	1.05 1.04 15	2.64 1.47 18

Sizes allocated according to distribution of numbers of blocks extant observed on Day 1.

Algorithms
A – Control
B – 50% points for top 5 sizes
C – 95% points for top 5 sizes
D – 95% points for top 15 sizes
E – 95% points for top 15 sizes, with grouping

Table 7 Storage utilization for different subpool arrangements

	Sizes									*Usage*
	11	12	13	14	15	16	17	18	19	
	0	0	0	1	1	1	1	1	0	1389
	0	1	1	0	1	1	1	1	0	1389
	1	0	0	1	1	1	1	1	0	1389
	0	0	0	1	1	0	0	1	0	1391
	0	1	1	1	1	1	1	1	0	1391
	0	1	0	1	1	1	1	1	0	1391
	1	0	0	1	1	0	0	1	0	1391
A	0	1	1	1	1	1	0	1	0	1393
R	0	1	0	1	1	0	0	1	0	1393
R	0	0	1	1	1	0	0	1	0	1394
A	1	0	1	1	1	0	0	1	0	1394
N	0	0	0	1	1	0	1	1	0	1395
G	0	0	0	1	1	1	0	1	0	1395
E	0	0	1	1	1	1	1	1	0	1395
M	1	0	1	1	1	1	1	1	0	1395
E	1	0	0	1	1	1	0	1	0	1395
N	1	0	0	1	1	0	1	1	0	1395
T	0	0	0	0	1	1	1	1	0	1397
S	0	1	0	1	1	1	0	1	0	1397
	0	1	0	1	1	0	1	1	0	1397
	1	0	0	0	1	1	1	1	0	1397
	0	0	1	1	1	0	1	1	0	1398
	0	0	1	0	1	1	1	1	0	1398
	1	0	1	0	1	1	1	1	0	1398
	1	0	1	1	1	0	1	1	0	1398
	0	0	0	1	0	1	1	1	0	1399
	0	0	1	1	1	1	0	1	0	1399
	0	1	0	0	1	1	1	1	0	1399
	1	0	1	1	1	1	0	1	0	1399
	1	0	0	1	0	1	1	1	0	1399
	0	0	1	1	0	1	1	1	0	1401
	0	1	0	1	0	1	1	1	0	1401
	1	0	1	1	0	1	1	1	0	1401
	0	1	1	1	1	0	0	1	0	1402
	0	0	0	1	0	0	0	1	0	1404
	1	0	0	1	0	0	0	1	0	1404
	0	0	0	1	0	1	1	0	1	1405
	1	0	0	1	0	1	1	0	1	1406
	0	0	1	1	0	0	0	1	0	1406
	0	1	1	1	1	0	1	1	0	1406
	0	1	0	1	0	0	0	1	0	1406
	1	0	1	1	0	0	0	1	0	1406
	0	0	0	1	0	1	0	1	0	1407
										.
										.
										.
	0	0	0	0	0	0	0	0	0	1483 (no pooling)

Table 8 Experimental design

	Monday	*Tuesday*	*Wednesday*	*Thursday*
Week 1	old	new	new	old
Week 2	new	old	old	new

Table 9 Experimental results

	CP time per call (microseconds)			
	Monday	*Tuesday*	*Wednesday*	*Thursday*
Week 1				
FREE	325	*50*	*45*	426
FRET	221	*38*	*34*	287
Week 2				
FREE	*37*	346	313	*45*
FRET	*27*	236	217	*33*

Note: Italicized numbers correspond to the new algorithm.

Table 10 Estimated CP time per I/O (milliseconds)*

	VSIO	*Page I/O*	*Spool I/O*
Old	9.7	3.0	5.7
New	7.9	2.0	4.6

Abstracted from Table 11. Reference 8

Table 11 Average throughputs during test period*

	Old	*New*
Percentage of problem state time	19.8	22.0
Percentage of CP state time	21.3	18.1
VSIO per sec	11.7	12.4
VMIO per sec	8.5	8.5
SPOOL I/O per sec	1.1	1.2
PAGE I/O per sec	6.7	7.0

*Taken from Table 12. Reference 8

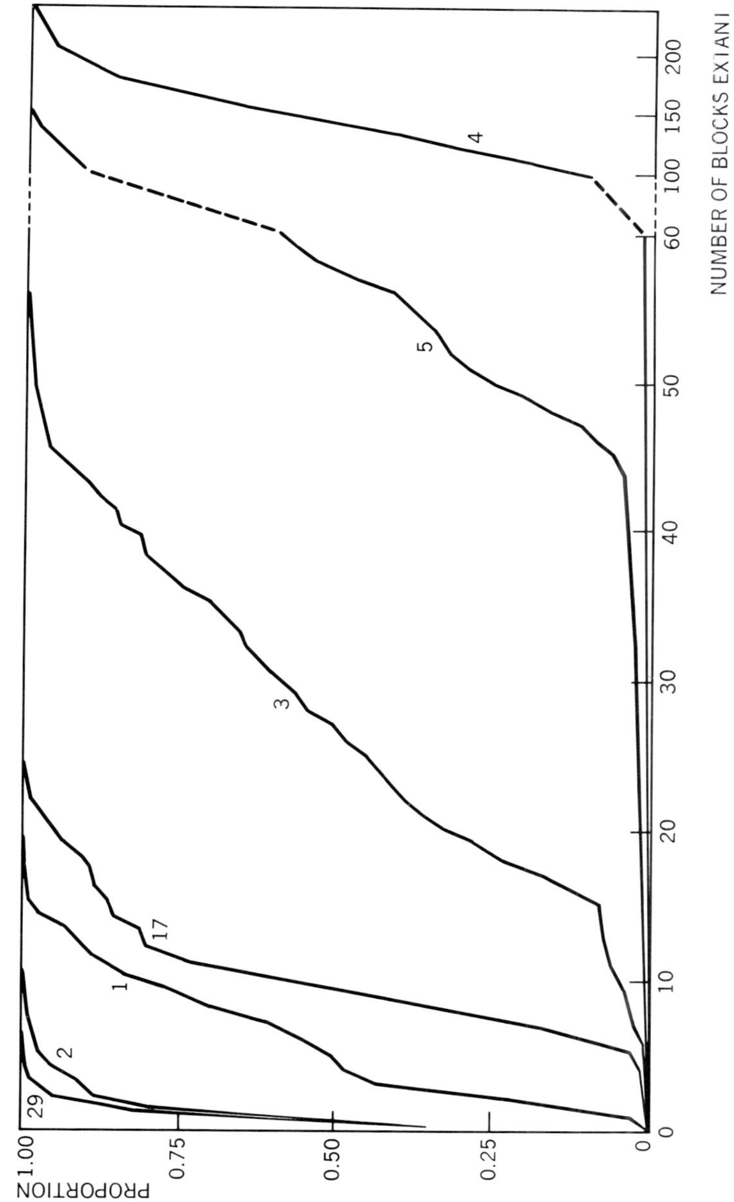

Fig. 1. Cumulative distributions of numbers of blocks extant for selected block sizes.

LOCALITY: WORKING SET, REQUEST STRING, DISTANCE FUNCTION AND REPLACEMENT STACK

D. Hatfield and J. Gerald

In the environments of computer systems which supply virtual memories by means of dynamic address translation, the number of real memory exceptions can be a critical operating parameter [1]. A real memory exception occurs whenever a virtual memory block is expected to be but cannot be found in a physical memory buffer. There can be exceptions (also called "faults") in a high-speed cache, in what is called main memory or in any level of a memory hierarchy. For a program to perform well in a memory hierarchy system it should not overload the system with memory exceptions. What constitutes an overload will vary with the speeds of the memories in the different levels of the hierarchy, but in general it is desirable to minimize the exception rate at all levels, especially if this can be done at little cost to the users of the system.

Hopefully, programs will stay localized in virtual memory space for significant periods of time. A reasonable period of time would be the time necessary to perform the I/O to bring into physical memory a virtual memory block which caused an exception. In a multiprogramming environment this time could be used to process one user's memory exception while another user is running without an exception. More localized programs should cause fewer memory exceptions.

Localization is hard to define. Most crudely, a program is more localized if it spends more time within smaller spans of space. Initially, in a project aimed at localizing programs [2], we divided programs into relocatable sectors and compared the sector-to-sector communications within virtual memory blocks (called "pages" in the working environment of that experiment) to the sector-to-sector communications from one block (page) to another [2]. We used an algorithm similar to those used in attacking backboard wiring problems to pack the relocatable sectors into pages so as to minimize the communications from one page to another. This partitioning of the communications matrix and consequent rearrangement of program sectors was found to result in a noticeable reduction of the program's page exception rate for a wide range of replacement algorithms and available real memory constraints [2].

Playing with the process of rearranging sectors, we found that it was helpful to use displays of virtual memory use, which could be updated rapidly (milliseconds) whenever a new arrangement was determined. In addition, it has been the case so far that human pattern recognition can improve (sometimes significantly) on the arrangement made by the partitioning program [2]. When working with these memory usage displays it was more natural to use for an arrangement figure of merit not the ratio of communications within pages to total communiations but the average number of pages used over some instruction period (see figure 1).

This is a figure which is easy to calculate from the information in the display itself. The display is generated by strobing the program at a fixed time interval and recording all areas in virtual memory used during that interval. A memory resolution is specified to the program which generates the display data; we have used resolutions as fine as 8 bytes, but have found that 32 or 64 byte-blocks are an acceptable lower limit. The strobe interval is in time measured in instructions rather than execution cycles simply because such an interval is easy to measure. For each display strobe (referred to hereafter as a "line") and any sector arrangement we can calculate what virtual memory addresses were used and therefore what pages were used. So now we are dealing with a working set measure. The working set is the specific set of pages (for cache machines the equivalent physical memory block is called a slot) used during a specific time interval. The working set size is the size of this set, the number of pages irrespective of their names. When a virtual memory area is displayed it is accompanied by the average working set size over the number of lines displayed and the average number of pages which were new from one line interval to the next; the latter number gives a measure of the fluctuation of the working set in time (figures 2, 3). The average working set size and average new page size are calculated assuming that the time measure for the working set is the time interval for the strobe used when generating the display file. If we wish to look at working sets calculated over a larger time interval, the page sets used during two or more consecutive strobe lines may be combined before averaging.

Our experience has been that once the relocatable sectors have been initially packed, the average working set size is a better measure of expected program performance than the ratio of communications within pages to communications between pages [2]. But the working set definition of program locality has one drawback: the time period for calculating the number of pages used is arbitrary, and it is hard to find a natural period which does not change as program execution continues. As the time interval is made smaller the number of pages used will stay constant or decrease; its limit is the number of pages used by a single instruction. The upper limit on working set size is the total number of pages used by the program. Somewhere between these limits is a measure of the page space needed for the program to execute with an "acceptable" number of exceptions. Unfortunately, this number is not obtainable from any working set calculations unless the time period for calculating the working set size becomes vanishingly small.

Our first practical difficulty woth working sets (beyond finding that occasionally reducing the working set associated with the display file did not imply a reduction in page exceptions) arose during a study of locality and page size. The results indicated that the more localized was the program, the greater the probability that a larger page size would result in less I/O transferred in page exceptions, given the same real memory constraints [3]. A typical experiment was to find that 2048-byte pages would give more than twice the number of page exceptions generated with 4096-byte pages, even though both 2048-byte pages associated with a 4096-byte page were not always used [3]. A simplified example is shown in figure 4, where with each 4096-byte page i we associate the 2048-byte pages 2i-1 and 2i. The initial state of the replacement stack is shown at the left, and page requests causing exceptions are underlined. In this case the program when divided up into smaller pages produced three times the page exceptions or, since the small pages are half the length of the large pages, 1-1/2 the I/) tranfers. Now if a second program is run which looks in sequence at both halves of the 4096-byte pages it will perform as shown in figure 5 with the 2048-byte pages, giving two-thirds the exceptions of the first program. But, assuming that four page references are associated with intervals for calculating a working set (and examples could be given for any time interval including one that spanned an entire program), the working set of the second program, which gives one-third fewer page exceptions, is larger. An even simpler example is shown graphically in figure 6, where Δt is the time for calculating the working set. The second case has a working set two-thirds the size of the first, but for nine or less available pages of real memory the second case implies 100% of the page references produce exceptions for LRU, FIFO and almost all other currently used replacement algorithms, whereas the first case implies for n real-memory page spaces ((15-n)/15)% of the page requests produce exceptions. Unless ten real page frames are always available, it would be preferable to use the first case, with the bigger working set. Comparing the page request strings (figure 7) in the two cases shows more clearly that this is so. In the second case between any two successive references to any particular page there occur references to nine other pages. In the first case there can be any number from one to fourteen references to other distinct pages, with the average number of distinct intervening pages a little greater than seven. Any reference string can be used to generate a distance function, showing for each distance of i distinct pages how many times that distance intervened between two successive references to any page [4]. The two plots in figure 8 show distance on the horizontal axis and fraction of the page references associated with that distance on the vertical. The cumulative distribution functions give a measure of the page exceptions expected for any number of real page frames and LRU replacement. Obviously this measure is more expensive to obtain, but it is a better predictor of performance than the working set size. A scalar measure related to locality is the average of the cumulative of the distance. For case (2) with a working set of 15 pages this figure is 1/2, and for case (b) with a working set of 10 pages the figure is 1/10. A limit on this localization measure is 1, which would be approached by a program which made a near-infinite number of references between two pages, then replaced one of those pages by a third and made a near-infinite number of references between these two, etc. Such a program would run in two real page spaces, with an infinitesimal page exception rate.

The examples in figure 9 are pairs of request strings.

The first member of the pair always has the larger working set. The first number to the right is the number of page exceptions, the second is the locality expressed as the average of the cumulative distribution of the distance (for the cycle). Though the distance function is a true measure of performance under LRU replacement, it is not as good a discriminator for FIFO replacement, as FIFO is relatively insensitive to recent page usage, which is what the distance distribution measures. Most replacement algorithms used today in practice are closer to LRU than FIFO so that the distance distribution for the page request string is a useful measure, and certainly more accurate than working set size. The relevant distance distribution for any replacdment algorithm uses the distance in the replacement stack, which for LRU is of course the same as the distance in the request string. There is a class of algorithms, called stack algorithms, for which the order in which pages are added to the replacement stack is independent of the number of page frames in the stack. For this class (LRU is a member) a distance distribution can be calculated which is valid for all stack sizes - i.e., all available real memory constraints. Unfortunately, the replacement algorithms used on systems which page data between real memory and secondary storage are predominantly not stack algorithms. So to be precise a new distance distribution should be computed for each stack size.

Though algorithms which examine recent page usage can be handled well by the request string distance distribution, algorithms like FIFO which have a memory extending far back into the past - for FIFO the past only stops at the time the page in the bottom of the stack came into the top of the stack - are much less sensitive to the request string pattern. The difficulties this imposes on estimating program performance can be indicated by the simple example shown in figure 10. Allowing the cycle to continue indefinitely gives a program which pages at a rate R or 2R exceptions per instruction, but since the only difference between the R and 2R versions is the prologue part of the request string, working set sizes and string distance distributions will be the same in both cases. Moreover, since one can remove any page which did not cause an exception from the cycle without affecting the replacement stack, no measure that examines the request string alone could predict which version pages at R and which at the 2R rate in all cases. Not only is FIFO or any other algorithm with a long memory relatively unpredictable, in a multiprogramming environment it can make all other programs pay for one program's accessing pattern, and because of the uncertainty this introduces replacement algorithms that have a long memory should be avoided.

The distance distribution can be used to alter the placement of relocatable sectors into pages so as to improve the distance distribution by shifting it towards the origin. The method is as follows. For each distance found while processing the request string, note the page involved and the page immediately preceding it in the string. For those distances which are large but not "infinite" (as large as the total number of pages involved in the program, indicating the first occurrence of a page), examine the sectors involved in the page with the large distance and the previous page. If the sector needed in the page with the large distance were placed in the page preceding it, the second page would probably not be needed and one instance of the large distance would be removed. Assume that the sector needed is the one with the most communications to the sectors in the previous page. Then, using the communications matrix and the display file, decide what sector to remove from the previous page to make room for the new one, and where to put it. Calculate a new page request string for the new arrangement of sectors, and reexamine the distance distribution.

In addition to repackaging sectors so as to reduce the average of the cumulative of the distance distribution, one can consider the question of principles for writing programs so that the request string distance will be kept low. Minimizing the distances in the request string implies optimizing for a LRU-type algorithm. LRU, FIFO, used-bit and almost all other replacement algorithms implemented on computers today rely on the assumption of time-symmetry. The assumption is that the block of virtual memory which was important (used, brought into a physical memory, changed or whatever) longest ago in the past will not be needed until longest in the future. Programs which are strongly time-asymmetric will tend to cause more exceptions than will occur under more intelligent algorithms, algorithms which are more costly to implement and in some cases impossible to implement in real time. The problem of smaller pages giving so many more page exceptions that the system must do more real I/O transfer is an asymmetry problem. Smaller page sizes show up asymmetries in access pattern that large page sizes mask. The MIN replacement algorithm, which is asymmetric since it looks into the fiture to order the replacement stack, will never give more than twice the page exceptions for pages half as large and therefore never increases the real I/O transferred as page size is halved. There are two approaches to the problem of symmetric algorithms and asymmetric programs: either make the algorithms more asymmetric or the programs more symmetric. Asymmetrizing algorithms is difficult because one can induce fine shadings of asymmetry and it will take some time to calibrate all the variations against enough real programs to get a measure of which variations work better. Therefore we will conclude with some comments on rules of thumb for program writing so as to keep the request string distances low and reduce local asymmetry.

It is obvious that the distance between successive references to a page tends to be reduced if all intermediate references, i.e. references that are used to locate data or instructions but are not used to analyze or change them, are minimized. One implication is that sparse data should be packed; in this a list representation of index information is preferable to an array representation. If the data of interest are not sparse, the most compact addressing scheme that locates the data with the fewest intermediate references should be used. But in addition the accessing scheme should be designed in the light of the probable use pattern over the data. For example, data could be tree-structured if there will tend to be immediate processing of all branches of a node before processing branches of other nodes, but this also implies that all information about the branches out of a node should be stored together in virtual memory. On the other hand, if only the leaves of such a tree are important, and important unconditionally, they should be reached as directly as possible, either by pointers from one leaf to another or by a hashing function on their content. Frequent garbage collection is important for keeping data more localized.

If programs are combinations of inline code, branches and loops, one may consider the sequential code as unitary relocatable sectors, and the branches as communication paths between the sectors. The rebranching implied by the loops is in its effects unlike normal branching because the virtual memory areas in a loop, once brought into physical memory, are likely to stay there until the loop is complete. As the loop begins there will be an exception for each page or code or data not available initially, but most replacement algorithms will keep all information for the loop in physical memory as long as it is frequently used. So the count for the loop should not be taken as a measure of the page exceptions generated if all information needed by the loop is not in the same page, nor should a loop cause special problems if its procedure area crosses a page boundary.

In figure 11 the relative values of α and β are unimportant unless the memory sets a, b and c take individually more than the physical memory space available. Assume that after all data overlaps and equivalences are accounted for there are n virtual pages needed for a+b+c, and n-1 pages in physical memory. Moreover, one can fit a+b or a+c or b+c together into n-1 pages. What is the best choice? If a and b are together there will be one exception after statement B, and one the next time that part of a+b pushed out by b is required, for 2α page exceptions. If a and c are together there will be one exception after "DO B..." and one exception after statement B or after "DO A...", for 2α page exceptions, and similarly for b and c. Only if one of a, b or c cannot fit into available real space at once is it best to spend most effort compacting b.

This can happen in a large loop such as the main line path of a program itself. The best thing to do with a big loop is to turn it into an oscillation (see figure 6) since oscillating is the most localized way of traveling over again through the same space of procedure and data, and looping is the least. An ideal program (figure 12) which is symmetric and more spatially localized in more localized time periods would consist of oscillations within oscillations.

There are many practical instances of loops which can be turned into oscillations. If it is necessary to find the minimum element of each row of a matrix stored columnwise, alternate rows should be searched in alternate directions. When multiplying two matrices which are both stored columnwise, the elements of adjacent rows of the first matrix should be summed in alternating directions, as well as stored temporarily in a compact list while all elements in a row of the product matrix are computed.

Compilers and assemblers which search large tables should begin searching for the present entry at the point at which the previous entry was found, and go in the opposite direction. If a table has been frequency-sorted it should be zoned, and oscillating searches carried on within the zones. Since the order of procedures is in many programs necessarily fixed it cannot be oscillated through, but the paths through data involved should alternate direction if possible. A strange corollary is that programs and data should be loaded in reverse order to the expected order of their first use.

Most generally the distances are kept low when procedures and data used together in time are placed together in virtual memory space. The more modular are procedure and access and data areas, the more easy it is to structure virtual memory when the relative nearnesses in time are known. Hashing and inverted lists (putting the pointers all next to one another) are examples of modularization of access. A simple thing which a source language programmer can do for programs he will execute often is to put after each named source language statement a statement that increments an associated counter. The values in the counters at the end of a run will indicate the frequency of use of the blocks of inline code in the source program. These blocks can be reordered by frequency of use.

In conclusion, we emphasize that program performance in a virtual memory environment is a complicated matter, involving the interaction of access pattern, replacement algorithm, page size(s), the supervisor's scheduler, and hardware timing and the facts of contention for I/O devices as well. In examining program patterns alone we have found that statistics are reliable in proportion to their expense, and that corrective tools are more useful than simple measures of access patterns.

REFERENCES

1. For an excellent general discussion of virtual memory and paging problems, see P. Denning, "Virtual memory", Computing Surveys 2, no. 3, 1970.
2. D. Hatfield and J. Gerald, "Program restructuring for virtual memory", IBM Syst. J. 10, 168, 1971.
3. D. Hatfield, "Experiments on page size, program access patterns and virtual memory performance", IBM J. Res. Development 16, 58, 1972.
4. See R. Mattson et al., "Evaluation techniques for storage hierarchies", IBM Systems Journal 9, no. 2, 1970.

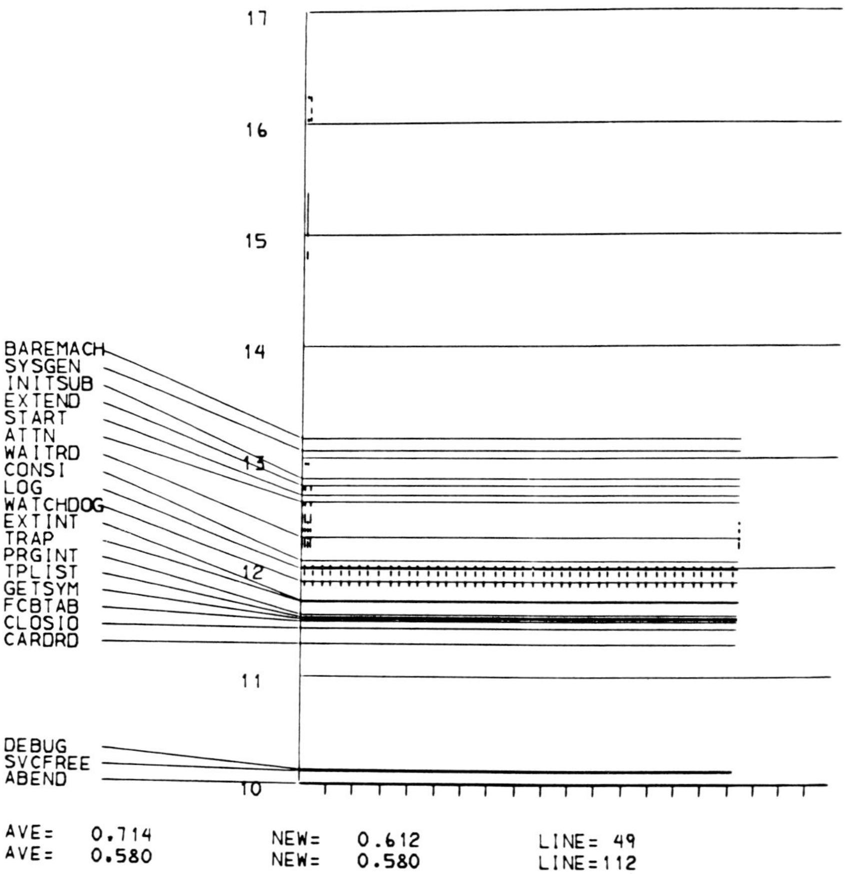
17
16
15
14
13
12
11
10
BAREMACH
SYSGEN
INITSUB
EXTEND
START
ATTN
WAITRD
CONSI
LOG
WATCHDOG
EXTINT
TRAP
PRGINT
TPLIST
GETSYM
FCBTAB
CLOSIO
CARDRD
DEBUG
SVCFREE
ABEND
AVE= 0.714
AVE= 0.580
NEW= 0.612
NEW= 0.580
LINE= 49
LINE=112

Fig. 1.

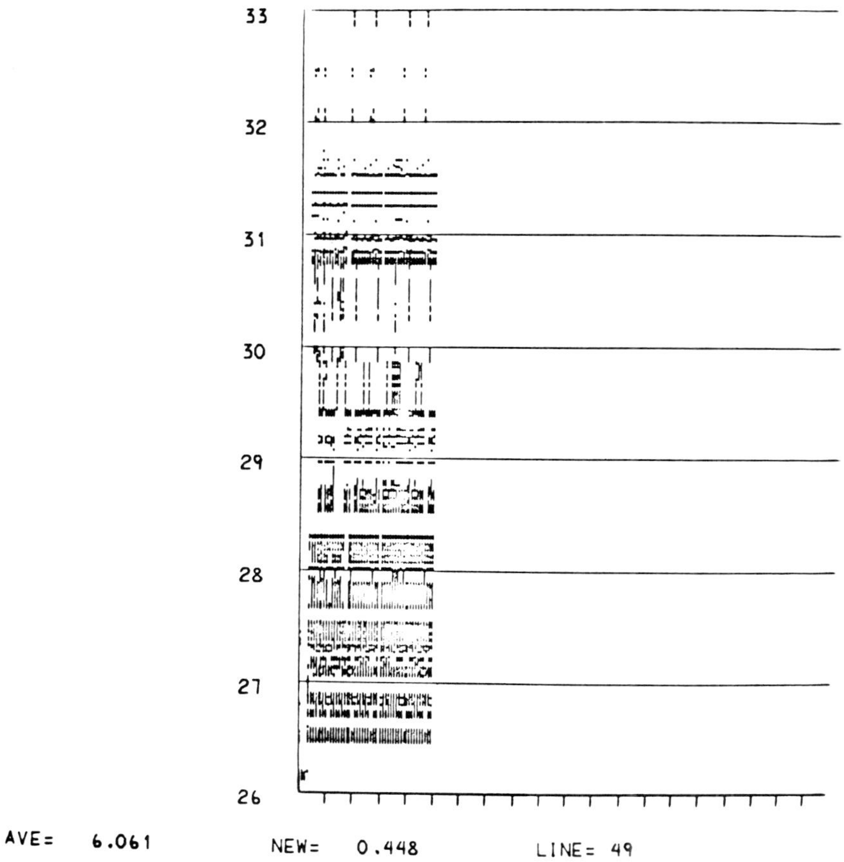

Fig. 2.

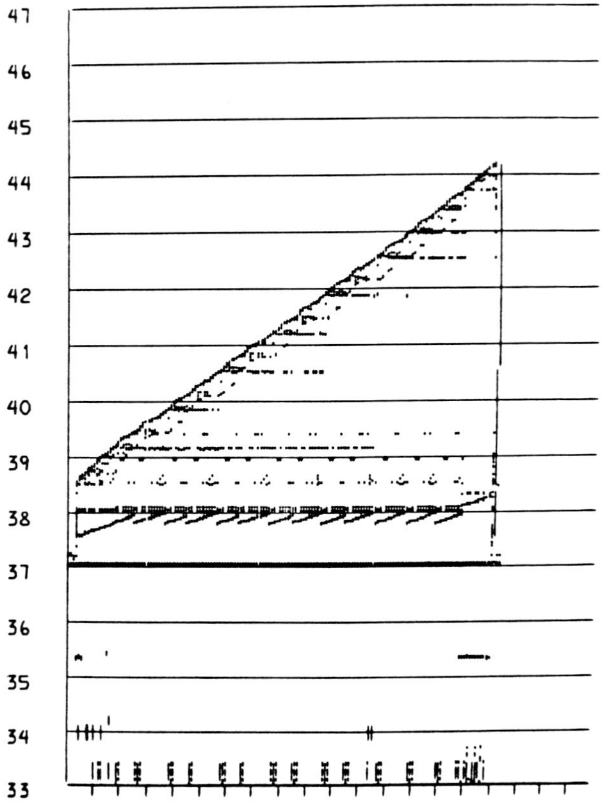

AVE=	3.530	NEW=	0.408	LINE=	49
AVE=	4.642	NEW=	0.455	LINE=	112

Fig. 3.

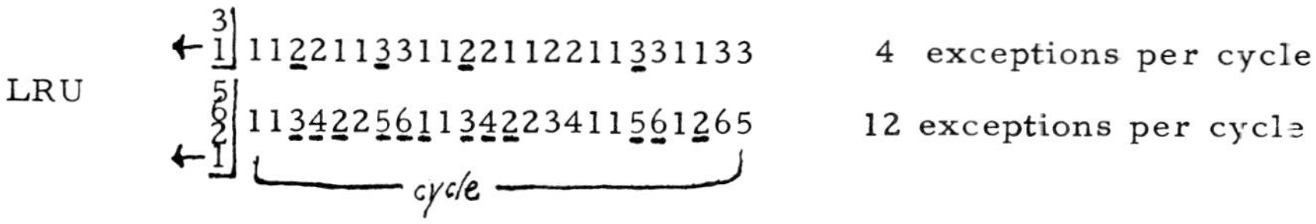

Fig. 4.

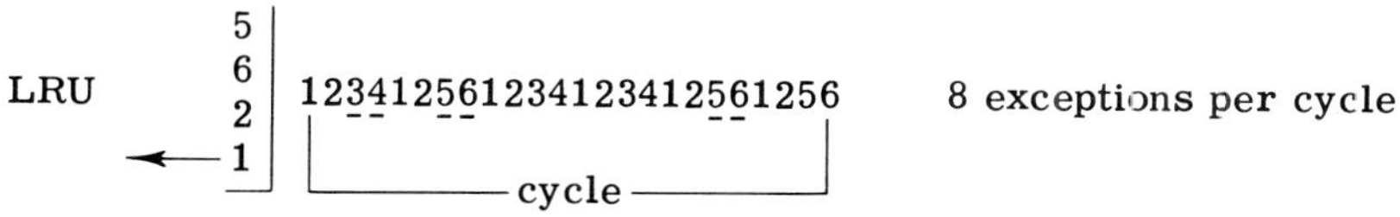

Fig. 5.

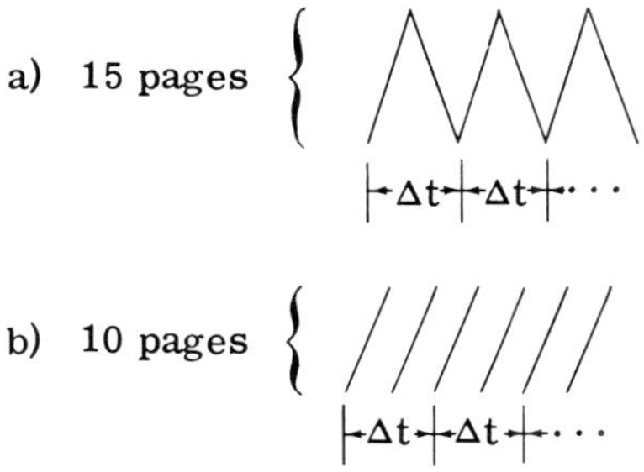

Fig. 6.

a) . . . 12345678910 11 12 12 14 15 14 13 12 11 10 98765432123456 . . .

b) . . . . 12345678910 123456789 10 123456789 10 123456789 . . .

Fig. 7.

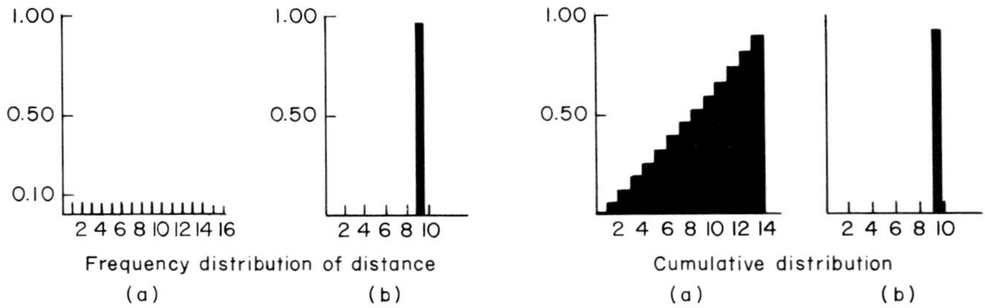

Fig. 8.

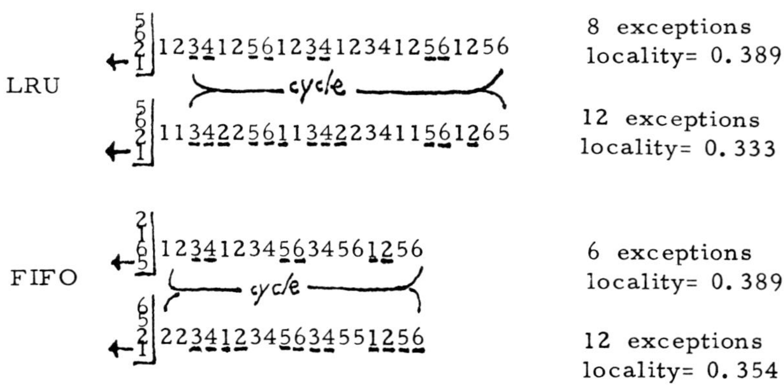

Fig. 9.

FIFO

561234563412561234123412563456341256

651244663412561234123412563456341256

cycle

Fig. 10.

```
     DO A   α times
     use memory set a
     DO B   β times
     use memory set b
B    CONTINUE
     use memory set c
A    CONTINUE
```

Fig. 11.

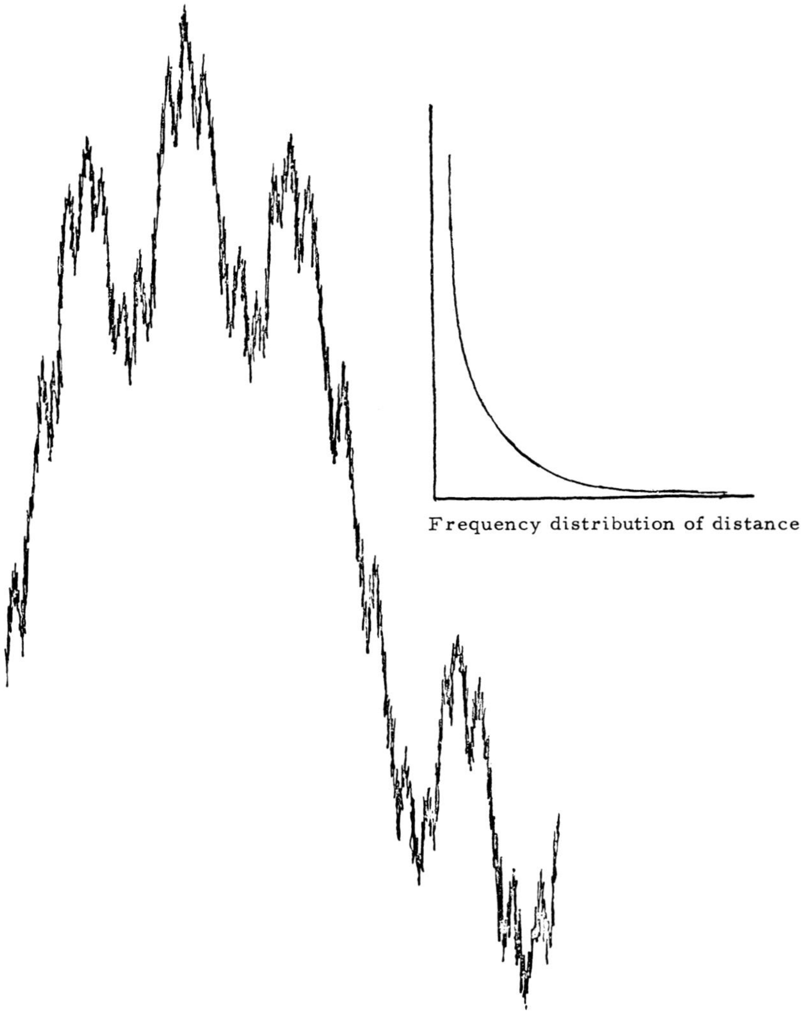

Fig. 12. Ideal program.

STATISTICAL METHODOLOGY FOR CALIBRATING A TRACE-DRIVEN SIMULATOR OF A BATCH COMPUTER SYSTEM

H. Beilner and G. Waldbaum

ABSTRACT: SOUL, a trace-driven simulator of a large batch computing system, was developed to permit controlled experiments on different hardware and software configurations for the purpose of improving system performance. SOUL's input is real data representing one day's computer workload. Its output is data describing simulated job behavior as well as a series of comparisons between the real and the simulated worlds. This paper describes the calibration methodology used for SOUL. The objective of the calibration - a "good fit" between the behavior of real and simulated jobs - was obtained by the development of a semi-automated methodology which makes use of various statistical methods.

1. INTRODUCTION

At the I.B.M. T.J. Watson Research Center, a computational simulator, SOUL (Simulation of OS/360 Under LASP - see section 2 below), has been developed to simulate the job flow through a large batch computing system. SOUL was built in order to forecast and judge effects of changes to this system's resources and operations.

This paper describes the methodology used to calibrate SOUL. Calibration of a simulator is the process of tuning certain calibration parameters and changing parts of the simulator's structure so as to yield, in some sense, a good match (as precisely defined in section 3 below) between simulator output and the desired simulator output for a selected set of input data.

A more precise definition of the calibration of SOUL can be given after examining some major model properties.

a) SOUL is a trace-driven model. During normal batch operations, data describing certain job characteristics (such as arrival time and resource requirements) are collected. These data, the so-called trace, form the main (driving) input for the simulator.

The collection of trace data is accompanied by the collection of data about the performance of the real world (i.e. data about the behavior of the jobs). The data collection was performed by the Statistical Data Gathering Package, SGP (cf. [4,14]). These real-world data are used during calibration as the desired simulator output with which the simulator output should provide a good match. In other words, we test the goodness of fit between the real and simulated worlds by comparing real job behavior with simulated job behavior. For this purpose it is necessary to define certain figures of merit which measure the divergence in the behavior of real and simulated jobs (cf. section 3).

b) SOUL is an expected-value model. Probabilistic simulators describe real-world situations by random variables. Expected-value simulators arise from probabilistic simulators by taking into account only the expected values of these random variables. As a consequence, every action in an expected-value simulator takes place under complete certainty. Thus, expected-value simulators form a subclass of deterministic simulators.

Probabilistic simulators use statistical methods to establish confidence intervals for their results by repetitions of runs with different sequences of random numbers. In contrast, deterministic simulators need not - and cannot - use statistical methods for this purpose since every repetition of a run with such a simulator leads to exactly the same output.

However, the expected-value simulator SOUL has another property which necessitates the use of statistical methods:

c) Comparisons between simulator outputs and desired simulator outputs (i.e. real-world outputs) show considerable random-like variations when parameter values are changed. A typical example of this situation is shown in figure 1.1 where two figures of merit are plotted against a certain parameter (the terms "parameter" and "figures of merit" are defined in sections 2 and 3 respectively). The character of this and similar pictures does not change if a smaller interval is examined. A definite trend always exists in the influence of the parameter on the figure of merit, but with a certain amount of "noise" superimposed on the figure of merit.

This observation leads to a mathematical formulation where figures of merit are considered not as if they have just one value per run but as if they are subject to random variations. Accordingly, we treated the problem of calibrating the model as a problem of optimization under uncertainty. As a result, we have to use statistical methods similar to those used for calibrating probabilistic simulators.

d) SOUL is an aggregate model. Aggregate simulators describe at least part of the real-world actions at some macro-level. SOUL, in particular, has been designed and built with an emphasis on keeping those simulator parts at a gross level which model the most frequent system operations. This introduces a certain amount of error into the simulation. However, it removes the well-known limitations in calibration, validation and experimentation produced by the high cost of computer time from which similar, more detailed simulators have to suffer [1,9]. SOUL is capable of simulating a 14-hour period of computer activity (involving 650 jobs) in less than 1 minute of CPU time on the same computer. As a result, we are less restrained than most previous simulation efforts from using statistical methods and iterative improvement procedures during calibration and validation.

2. THE PROJECT: A BRIEF DESCRIPTION

An IBM system/360 model 91 is used at the T.J. Watson Research Center primarily for running batch jobs. These jobs are run under the dual operating system OS/MVT-LASP*. Jobs are transmitted to the 91 through card readers, 13 remote job entry lines, and a special line which permits Time Sharing System/360 users on the research center's model 67 to submit batch jobs.

Figure 2.1 is a very general picture of the flow of a job through the system. It shows that jobs enter the system via LASP, are sent to OS (i.e. the main processor) for execution and then return to LASP for final processing and eventual deletion from the system. The state of a job which is "on main" will be referred to later as its "main phase". In an analogous manner, we shall make use of the terms "wait phase" and "system phase".

SOUL simulates this computing system. (For a detailed

*Operating System/360; Multiprocessing with a Variable Number of Tasks; Local Attached Support Processor (cf. [6,7]).

description of SOUL see [15].) Figure 2.2 depicts the overall structure of SOUL partitioned into

a) The pre-processor. The purpose of the pre-processor is to provide, via a single run, all the necessary job data for performing multiple simulations of (some period during) a single day's activities.

b) The simulator. The flow of a job through OS and LASP, as modelled by the simulator, is shown in figure 2.3. The simulators input consists of two types of data (cf. figure 2.2):

- information about the real world provided mainly by the pre-processor;
- control information provided by the user of SOUL.

Part of this control information is values for certain calibration parameters. It is this type of input which will be referred to later as "parameters".

c) The post-processor. The post-processor is the program of main interest in this paper. It is designed to

- aid in the calibration of the simulator;
- aid in improving the performance of the system being simulated.

The methods used to accomplish these design goals, especially during calibration, are described in the remainder of this paper.

3. GOODNESS OF FIT AND FIGURES OF MERIT

The main goal in calibrating a trace-driven simulator such as SOUL consists of adjusting the simulator (i.e. parameters and structure) for a certain set of input data so that the behavior of the simulated jobs matches as closely as possible the behavior of the real jobs. In the following sections (especially in section 5) we shall deal with more specific calibration goals. These will then be considered as subgoals, subordinate to the main goal.

One major problem in achieving this goal is to find a reasonable definition for the term "goodness of fit" and to create suitable "figures of merit" to measure goodness of fit.

3.1 First Attempts

Our first attempts at judging goodness of fit primarily involved the use of histograms and elementary statistics. At the conclusion of each simulation run, histograms of two kinds may be produced for each system phase. The first shows the histogram of real and simulated job phase times (figure 6.5) The second shows the histogram of job phase errors (i.e., for each job, the real time it spends in a phase minus the simulated time it spends in the same phase) (figure 6.6). Both kinds of histogram provide some visual feel for the goodness of the simulation. However, the second describes this goodness of fit somewhat better since a good distributional match does not require a good match at the job level. For example, errors of the kind where one job is x minutes too slow and another is x minutes too fast are not detected by the first type of histogram.

The histogram method was helpful during the first stage of our calibration efforts. By examining histogram shapes in relation to changes in the model, we were able to remove some of the largest inaccuracies in the simulator. However, we soon realized that this method only provides rough indications on how to calibrate the model, such as "the simulator is too slow in the main phase". Moreover, we were unable to use these histograms for making objective decisions between different versions of the simulator when neither version was consistently better for all jobs.

We had to refine our methodology to try to find more compact measures for goodness of fit which would more positively direct how to change the simulator in order to improve the calibration.

3.2 A Mathematical Model

The information we use to determine goodness of fit consists of real and simulated system phase arrival and departure times. Differences between the real and simulated worlds (i.e. the degree of goodness of fit) can be measured either by differences between real and simulated event times or by differences between real and simulated time durations. We have concentrated mostly on the second type of difference. We calculate, for each job i, its error during each phase j: e'_{ij} = real time job i is in phase j minus simulated time job i is in phase j. Two compact figures for "error in phase j" are the average error, u'_j, in phase j and the average squared erro, v'_j, in phase j:

$$u'_j = \frac{1}{n} \sum_{i=1}^{n} e'_{ij} \quad , \quad v'_j = \frac{1}{n} \sum_{i=1}^{n} (e'_{ij})^2$$

where n = total number of jobs in the run. These two figures of merit are the ones which we use to measure the goodness of fit during the calibration process. However, the average squared error has the advantage over the average error that it detects errors of the form "job a is x minutes too slow and job b is x minutes too fast". Thus, the emphasis during calibration is primarily upon improving the average squared errors.

One calibration objective could be to find minimum values for all the v_j', possibly subject to $u_j' = 0$. However, it is not clear whether it is possible to find a version of the simulator where these conditions can be satisfied simultaneously for every system phase j. Neither is it clear that there exists a minimum value for any v_j'. Indeed, in section 4.1 we show that even theoretically these conditions can be only partially fulfilled.

In section 1 we noted that the model's figures of merit (i.e. the u_j', v_j') show an underlying trend with superimposed variations when plotted against some calibration parameter. To describe this situation we created a mathematical model. We assume that:

- every error e_{ij}' has a deterministic functional relationship $e_{ij}(\bar{p})$ on the parameter vector $\bar{p} = (p_1, p_2, \ldots, p_m)$,
- the value of every e_{ij}' also depends on an additional random variable $\hat{e}_{ij}$, where the distribution of $\hat{e}_{ij}$ possibly depends on the value of $\bar{p}$.

Thus, we assume that

$$e_{ij}' = e_{ij}(\bar{p}) + \hat{e}_{ij}$$

Therefore

$$\begin{aligned} u_j' &= \frac{1}{n}\sum_{i=1}^{n} e_{ij}(\bar{p}) + \frac{1}{n}\sum_{i=1}^{n} \hat{e}_{ij} \\ &= \quad u_j(\bar{p}) \quad + \quad \hat{u}_j \end{aligned}$$

$$\begin{aligned} v_j' &= \frac{1}{n}\sum_{i=1}^{n} e_{ij}^2(\bar{p}) + \frac{2}{n}\sum_{i=1}^{n} e_{ij}(\bar{p})\cdot\hat{e}_{ij} + \frac{1}{n}\sum_{i=1}^{n} \hat{e}_{ij}^2 \\ &= \quad v_j(\bar{p}) \quad + \quad \hat{v}_{j1}(\bar{p}) \quad + \quad \hat{v}_{j2} \end{aligned}$$

where u_j', v_j', $\hat{v}_{j1}(\bar{p})$ and $\hat{v}_{j2}$ are random variables.

The trend functions $u_j(\bar{p})$ and $v_j(\bar{p})$ can be interpreted as multidimensional regression functions with respect to u_j' and v_j' while the observable values of u_j', v_j' can be thought of as lying, with a certain probability, in some band around these trend functions.*

In the calibration process described in section 4, our objective is to find an optimum parameter vector $\bar{p}_0$ with respect to the trend functions $u_j(\bar{p})$ and $v_j(\bar{p})$, given the statistical fluctuations of u_j' and v_j'. In section 4.1 and Appendices A and B we show that such a vector $\bar{p}_0$ exists.

4. METHODS FOR TUNING THE MODEL

This section describes the statistical methods which have been explored for calibrating the simulator. The aim of these methods was the subgoal of finding an optimum vector, $\bar{p}_0$, in the parameter space with respect to our figures of merit. We call this process "tuning the model".

4.1 Justification of a Subgoal

In Appendix A we show that under some assumptions (discussed in Appendix B) about the error functions $e_{ij}(\bar{p})$, there is for every phase a vector of bounds $\bar{b}_j$, where $\bar{0} < \bar{b}_j < \bar{\infty}$, such that

a) The trend function for the average squared error in this phase

$$v_j(\bar{p}) = \frac{1}{n}\sum_{i=1}^{n} e_{ij}^2(\bar{p})$$

has at least one minimum value for some parameter $\bar{p}_{j1}$ in the interval $-\bar{b}_j < \bar{p}_{j1} < \bar{b}_j$. $\bar{p}_{j1}$ can be reached by adjusting those parameters which are used in the simulation of this phase. (In Appendix B this type of parameter is said to "directly influence" the phase.)

b) The trend function $v_j(\bar{p})$ has at least one minimum

*We wish to assume that the $\hat{e}_{ij}$ have finite expected values and finite variance. However, since we restrict the functions $u_j(\bar{p})$ and $v_j(\bar{p})$ in 4.1 to a certain class of functions, we cannot assume that the $\hat{e}_{ij}$ (and thus the $\hat{u}_j, \hat{v}_{j1}(\bar{p}), \hat{v}_{j2}$) have expected values of zero. As a consequence, the bands need not be centered with respect to the trend functions.

value for some parameter vector $\bar{p}_{j2}$ in the interval $-\bar{b}_j < \bar{p}_{j2} < \bar{b}_j$, even under the restriction that the trend function for the average error in phase j has the value zero at p_{j2}, i.e.

$$u_j(\bar{p}_{j2}) = \frac{1}{n} \sum_{i=1}^{n} e_{ij}(\bar{p}_{j2}) = 0$$

$\bar{p}_{j2}$ can be reached by adjusting those parameters which are used in the simulation of this phase.

These results show that for a fixed phase it is reasonable to adjust all the parameters which are used in the simulation of this phase in order to find some minimum value for the average squared error of this phase both with or without the restraint of zero average error. However, the results do not justify a search for a set of parameter values which will simultaneously minimize the average squared error for all phases.

4.2 Calibration Parameters

Each calibration parameter has a physical-mathematical meaning. It denotes one fixed value in the range of values of a random variable which describes a particular time delay or time expansion factor. At the beginning of the calibration process, an initial value must be assigned to every calibration parameter. In order to reduce the total time required for the calibration effort, attempts were made to use initial values (for these parameters) which are realistic estimates of the expected values of the corresponding random variables.

Whenever possible, we used the trace information to estimate these expected values. As an illustration, let us consider how we derived an estimate for the parameter TIMPUNA. This parameter denotes the time delay between the printing of two outputs at the same printer. A program was written to search the trace for all cases where a printer finished printing the output for a job at a time when other jobs were waiting for their outputs to be printed. From these cases we calculated the average time gap between two activities of the same printer and assigned this number as the initial value for TIMPUNA.

If the trace did not contain any information suitable for estimating a particular parameter, we employed other means. In some cases, we made test runs with jobs having certain specific characteristics. For example, this technique was used to produce an estimate for the parameter MINSTIM denoting the minimum duration of a job step. We obtained an estimate for MINSTIM by running a number of jobs which consisted solely of dummy job steps (i.e., job steps which did not have any work to perform).

In order to obtain estimates for the remaining parameters, we even had to interview operators and observe the real system at work. For example, the parameter TIMNTOT denoting the time to mount a set-up tape was estimated from such information.

After the initial values of the calibration parameters are set, we then perform the calibration process with the aim of improving the figures of merit. This improvement is achieved by an iterative process, whereby:

- The tuning process may lead to some optimum parameter vector $\bar{p}_0$ where each element has a reasonable value (i.e. a valur which does not contradict the physical meaning of the parameter or, in other words, a value inside the range of values of the corresponding random variable).
- The tuning may reveal that one or more parameters tends to leave its reasonable range. In this case, we fix the values of these parameters at the appropriate extremes of their reasonable ranges.

At the same time, attempts are made to identify and correct the source of inaccuracy in the model. This may entail rebuilding parts of the simulator. Such structural changes are up to human judgment; their nature is found mostly without the use of statistical methods.

4.3 The Method of Good Balance: Linear Regression of Errors versus Job Characteristics

In section 3.1 we described our first attempts at calibrating the simulator. Changes in histogram shape and elementary statistics were used in a type of "guided guessing" improvement process. This process was not satisfactory. We needed to find out more precisely what the sources of model inaccuracies were. For example, we needed to know whether the deviations between the simulated and the real worlds showed up more (or less) in jobs that required large amounts of core than in jobs that required small amounts of core.

One approach could have been to assign jobs to different job classes according to their resource requirements and then to examine the simulator errors for each class. However, we use an analytic approach which is a direct evolution of this one. In particular, we perform a linear regression analysis* between different job characteristics:

CPU time consumed,
IO time consumed,
number of IO-interrupts,
core requirement,
number of job steps,
number of tape set-ups,
number of disk set-ups,

and the simulator errors for certain phases:

error of job in the wait phase,
error of job in the main phase,
error of job in the system phase.

Then t-tests are made on the significance of the regression coefficients while an analysis of variance (F-test) is made on the overall significance of the regression. The results of these tests are used very conservatively since the test assumptions are obviously violated: the sequences of job characteristics and simulator errors are not random samples (a certain degree of serial correlation exists; cf. section 5). Moreover, the normality assumptions are not necessarily valid.

Once a significant regression is observed, the significant regression coefficients are examined to determine those job characteristics which cause the largest (positive or negative) errors in the simulator. Then the model is tuned and the simulator structure is changed in order to achieve the situation where the requirement for any kind of resource, compared to requirements for other kinds, does not produce significantly larger errors. In other words, we try to find a model where errors are evenly spread out over all types of jobs. This calibration subgoal is called the "well-balanced model".

This regression method is very helpful during the calibration process. It shows whether parameter values have to be changed and if so, the direction in which they should be

---

*The necessary computations were performed using a standard subroutine package [8]. A description of the methods used in this package can be found in [3].

changed. This direction is determined directly or indirectly:

- The direction for changing a parameter can be determined directly if we can deduce the direction from the sign of some significant regression coefficient. For example, it is fairly easy to deduce the direction for changing the parameter TIMNTOT (the time to mount a set-up tape) from the sign of the regression coefficient which describes the relationship between a job's error in the wait phase and the number of tape set-ups it requires.
- The direction for changing a parameter has to be determined indirectly if it is not clear whether an increase in the value of this parameter will increase or decrease a certain regression coefficient. In this case, additional simulation runs are made to determine the relationship between the regression coefficient and the parameter.

The limitations of this method are a direct consequence of its goal. As the desired state of good balance is approached, the errors become independent of any job characteristic and the regression analysis tends to yield insignificant results.

4.4 An Optimization Method: Linear Regression of Errors versus Parameters

This section describes a method which directly addresses the goal of obtaining minimum values for the average squared errors v_j'. As pointed out in section 4.1, it is justified to look for a minimum value for only one such function. At present, we look for a minimum value only for the system phase.

Every calibration run of SOUL saves a condensed set of information about the run on disk storage. This information contains the value of the parameter vector $\bar{p}$ as well as the value of the average squared error v' for the system phase. After a certain number of runs, a multiple linear regression analysis is performed between those parameters which have been varied and v'.

In contrast to section 4.3, where we perform a regression analysis between the job characteristics (independent variables) and job phase errors (dependent variables), we perform here a regression analysis between the parameters (independent variables) and the average squared error in the system phase (dependent variable). Moreover, in section 4.3 we consider the sample of job phase errors contained in one run, while we deal here with the sample of average squared errors obtained from a series of runs.

This regression analysis is designed to show the influence of parameter values on the average squared error in order to obtain a basis for recommending parameter changes. In other words, we want to be guided to points in the parameter space where smaller values of v' are to be expected.

For this reason, the regression analysis is not performed over all previous runs but is localized around the parameter point $\bar{p}_b$ with the "best" result to date (cf. section 5). This point is considered to be the center of a rectangular subspace in the parameter space. The "local" regression takes into account every previous run where the parameter vector is situated inside this subspace.

If this regression has a significant result (F-test) then the linear regression equation formed from the significant regression coefficients (t-test) may be thought of as being a local linear approximation of the regression trend function $v(\bar{p})$ around $\bar{p}_b$. In this sense, the significant regression coefficients form a vector $\bar{r}$ which can be assumed to point in a direction close to the direction of steepest ascent for $v(\bar{p})$ and we use the equation

$$\bar{p} = \bar{p}_b - s \cdot \bar{r} \qquad (s > 0)$$

to give recommendations for further parameter vectors (for an example cf. section 6).

This optimization method is fully analogous to the well-known deterministic method of steepest descent, except that the vector of regression coefficients is used in place of the gradient.

5. A DECISION METHOD: VALIDATING IMPROVEMENT IN THE GOODNESS OF FIT

In section 4 we described the methods which help us tune the model and detect the need for structural changes to the simulator. Our objective is the best possible match between the real and the simulated worlds as measured by the average squared errors v_j'. In this process smaller values of v_j' indicate an improvement in calibration. However, since v_j' is subject to random-type variations, we have to ensure that a new value for v_j' is not only better but significantly better before we accept a new version of the simulator as superior.

To ensure the significance of an improvement, we consider the sequence of squared job errors, $(e_{i,l}')^2$, in the system phase before some model change and the sequence of squared job errors, $(e'_{i,2})^2$, in the system phase after the model change as a sequence of paired observation. The sign-test [13] and the Wilcoxon-signed-rank-test [13] are used to test the hypothesis of superiority of one model version over the other. We chose these non-parametric tests so we would not be forced to make any assumptions about the unknown error distributions.

The samples of paired observations $(e'_{i,1})^2$ and $(e'_{i,2})^2$ show a considerable degree of autocorrelation. In order not to affect the validity of the tests, we apply the tests not to the error samples but to blocks of observations (cf. [5]) according to

$$E'_{k,j} = \sum_{i=(k-1)*B+1}^{(k-1)*B+B} (e'_{i,j})^2$$

where j = 1,2, k = 1,2,..., and B = block length. B is chosen so that a run test [13] does not reject the hypothesis of randomness of the blocks at the 5% level. This has been achieved with B in the range between 25 and 30.

If the average squared error of sequence 2 is smaller than that of sequence 1, then the sign-test and the Wilcoxon-signed-rank-test are used to examine the paired blocks, $E'_{k,1}$ and $E'_{k,2}$, in order to determine whether the hypothesis that sequence 2 has a larger average squared error than sequence 1 can be rejected at a certain confidence level. If so, the new model version (i.e. the version which generated sequence 2) is accepted as superior.

6. THE CALIBRATION PROCESS IN DETAIL

The calibration process is cyclic. As shown in figure 6.1, each (simulator + post-processor) run produces:

a) information about the goodness of the run, such as
   - histograms,
   - figures of merit,
   - comparisons with previous runs,

b) information to direct changes in the simulator, such as
   - results of a regression analysis describing the state of balance of this run (cf. section 4.3),
   - results of a regression analysis, including

information about previous runs, with recommendations for parameter changes (cf. section 4.4).

Condensed information about the results of the run as well as information needed for comparisons with future runs are saved on disk.

After all this information is inspected, either the state of the calibration is found sufficient and the calibration process is concluded or the appropriate parameter and/or structural changes are made and the next calibration cycle is started. This may entail:

- changing parameter values according to the state of balance (cf. section 4.3),
- changing parameter values according to the optimization method (cf. section 4.4),
- identifying and repairing a structural inaccuracy.

At present the inspection of information and the initiation of actions (that follows this inspection) require human intervention. However, with the exception of identifying and repairing structural inaccuracies, this part of the calibration process can also be automated.

The remainder of this section illustrates the calibration process. Let us assume that we are beginning a calibration (in which case we use our initial estimates for the calibration parameters) or that a structural change was just made as a consequence of the preceding cycle (in which case we use our previous optimum parameter vector). This state marks the beginning of a tuning process where an optimum set of parameter values has to be determined.

The first tuning runs are guided by the method of good balance (cf. section 4.3). Using this method we repeatedly change the parameter values with the object of obtaining insignificant regression coefficients (for the regression between errors in the phases and the job characteristics). Figure 6.2 shows the results of the regression for the error in the system phase at some parameter vector $\bar{p}_1$. We see that:

- The regression equation is significant at the 5% level ($F = 11.59 > 2.1$).
- The error depends, at the 5% level, significantly on the CPU requirement, CPUJ, of the job ($t = 6.97 > 2.6$).
- The error has, at the 5% level, no significant dependence on any of the remaining job characteristics.

At this step, the significant dependence of the job error on the CPU requirement could not be removed by tuning. This appeared to indicate some structural inaccuracy. However, rather than wait until this inaccuracy was identified and corrected, we decided to switch to our second tuning method (cf. section 4.4) which is to perform a local regression analysis over all runs that use this version of the model structure. The results of such a regression for the average squared error in the system phase versus parameters are shown around the parameter vector $\bar{p}_2^*$ in figure 6.3. We observe that:

- The regression equation is significant at the 5% level ($F = 125.7 > 2.3$).
- The average squared error depends, at the 5% level, significantly on the parameters OV2 and OV3 ($|t| > 6.8$); at the 10% level, also on the parameter OVO ($t > 1.3$).

The values of the respective regression coefficients recommend that OVO and OV2 should be assigned smaller values and OV3 a larger value. However, since OV2 has already reached the border of its reasonability range, we change only the values of OVO and OV3. The recommendations for OV3 (made by the program) are certainly outside its reasonability range. Not knowing the exact upper bound for OV3, we decided to increase the value for this parameter, but not by the amount recommended. Two runs with (OVO,OV3) = (224, 5.4), (184, 5.6) did not yield any significant improvements. We then performed another run (corresponding to a value for $s \sim 20$) in the quadrant for OVO and OV3 that was recommended by the program: (OVO,OV3) = (0, 5.8). This point, $\bar{p}_3$, is superior to $\bar{p}_2$.

The decision information about the process in going from $\bar{p}_1$ to $\bar{p}_2$ to $\bar{p}_3$ is shown in figure 6.4:

- The run test (cf. section 5) does not reveal any indication at the 5% level that the blocked samples are not random ($7\leq\#RUN=13\leq17$; $7\leq\#RUN=11\leq17$; $7\leq\#RUN=7\leq17$).
- By comparing the paired block observations, we find that $\bar{p}_2$ is superior to $\bar{p}_1$ for 16 blocks and $\bar{p}_3$ is superior to $\bar{p}_2$ for 15 blocks. That is sufficient for the sign-test (cf. section 5) to accept, at the 10% level, the super-

*$\bar{p}_2$ was found by means of a previous local regression around the point $\bar{p}_1$. It was determined to be superior to $\bar{p}_1$ by the decision process illustrated later in this section.

iority of $\bar{p}_2$ over $\bar{p}_1$ and of $\bar{p}_3$ over $\bar{p}_2$.

- The Wilcoxon-signed-rank-test accepts both hypotheses at the 5% level.
- Some danger of a nonconvergence in the process becomes apparent since $\bar{p}_3$ is superior to $\bar{p}_1$ for only 14 blocks which is not sufficient for the sign-test to accept, at the 10% level, the superiority of $\bar{p}_3$ over $\bar{p}_1$. (The Wilcoxon-signed-rank-test rejects the hypothesis that $\bar{p}_3$ is superior to $\bar{p}_1$ at the 5% level, although it does accept this hypothesis at the 10% level.) In order to handle this situation, a record of block results of "so far best points" is held on disk and new points are compared with each of the former points. This method rejects the superiority of $\bar{p}_3$ at the 5% level and holds $\bar{p}_2$ as the best point.
- The average squared error in the system phase is reduced as follows:

$\bar{p}_1$	3400	
$\bar{p}_2$	3190	
$\bar{p}_3$	2710	(if working at the 10% level)

Figures 6.5 and 6.6 show the two types of histograms for the system phase at point $\bar{p}_3$.

The tuning phase is not continued at this point. Instead, we try to identify the source of the two detected model insufficiencies:

- Tuning of parameter values could not remove the significant dependence of the error per job in the system phase on the CPU requirement of the job.
- Parameter OV2 tends to leave its range of physical reasonability.

The source of these insufficiencies was identified, a structural change of the simulator made and a new tuning process started.

7. PLANS

7.1 Validation

Nothing has yet been said in this paper about the important problem of validating SOUL. In a very general sense, validation of a simulator can be defined as determining the domain of situations for which the simulator yields valid results once calibration has been established [10].

Validation of a trace-driven simulator such as SOUL has two different aspects:

a) A validation of the calibrated model is necessary in order to ensure that the simulator is a valid representation of the real system for more than just the particular set of traces used during calibration. This type of validation can be achieved by confirming the validity of the simulator for sets of input data which were not used during calibration.

Unfortunately, we currently have only two days' traces at our disposal. As a consequence, we can validate SOUL only in a fairly modest way by calibrating for one day's trace and confirming its validity by using the other day's trace.

However, we have to some degree validated our calibration technique. In the process of improving SOUL we did not find any situation where a significant improvement for one day was accompanied by a significant deterioration for the other day.

We hope to have at our disposal in the near future traces for a number of days. If we then detect that the present state of calibration is inadequate, we plan to re-calibrate the model by making the following changes to our methodology:

- Whereas we now use in the tuning process the average squared errors for one day's jobs, we shall use the sum of the average squared errors for a number of days.
- Whereas we now test the superiority of one model version over another by observing the samples of blocked job errors, we shall use as the basis for the statistical tests the samples of average squared errors for different days' traces.

b) It is also necessary to ensure that the simulator is a valid representation of certain unreal systems which could be real. Since the purpose of SOUL is to permit controlled experiments on different hardware and software configurations for the purpose of improving system performance, SOUL must be capable of simulating these unreal systems as well as the real system. In particular, any experimentation with SOUL should aim to find model versions where the simulator performance is better than the performance of the real world. Such models can then be used to propose improvements to the real system and to forecast the effects of these proposals.

The validity of the simulation of such an unreal system can be confirmed only by making the proposed changes to the system and then checking whether or not the simulator's forecasts are valid.

It is not reasonable to expect that a simulator is valid for every imaginable system. However, some forecasts must be shown to be valid in order to establish a sound basis for continued use of the simulator.

Despite these difficulties, simulation is recommended as the most suitable technique for performance projection [9].

7.2 Experimentation

This paper has described an effective methodology for calibrating a trace-driven simulator. We have shown that we use various statistical methods for directing the tuning of calibration parameters and for deciding between different simulator versions.

A similar methodology may be used in future experimentations with SOUL. Improving the model behavior can be performed in a directed process by adjusting our methodology in the following way:

- Whereas in calibrating we try to obtain a model which behaves, in some sense, as closely as possible to the real world, in experimenting we try to obtain a model which behaves, in some sense, as differently as possible from the real world. In this case, we can measure the distance between real and simulated worlds by job improvements (i.e. differences in job phase durations between reality and model) and try to maximize certain functions of the job improvements.
- Whereas one important part of the calibration process consists of tuning certain calibration parameters, one important part of the experimentation process consists of tuning certain (arithmetic) system parameters.

Given these changes, an experimentation process can be performed in the same way as the calibration process.

Acknowledgments. It is a pleausre to thank Drs. M.M. Lehman and D.N. Streeter for their continued encouragement and support. The work in section 4.1 where we discuss the mathematical reasonability of the tuning process resulted from a question posed by Dr. R.F. Tsao. J.E. Yang calibrated the output phase. Messrs. J.A. Cooperman, E.F. Crowding, H.W. Lynch, J.J. Lentz, J.H. Rooney and W.H. Tetzlaff were most helpful in providing us with information.

REFERENCES

1. Advanced multiprogramming analysis procedure (AMAP) service description manual, IBM Form GH20-0725.
2. G. Aumann, Reelle Funktionen, Springer, 1954.
3. W.W. Cooley and P.R. Lohnes, Multivariate procedures for the behavioral sciences, Wiley, 1962.
4. J.A. Cooperman and W.H. Tetzlaff, Analysis in an OS user environment, September 3, 1970, RC3161, IBM.
5. J.R. Emshoff and R.L. Sisson, Design and use of computer simulation models, Wiley, 1968.
6. IBM system/360 operating system: introduction, Form GC28-6534.
7. IBM system/360 attached support processor system (ASP): system programmer's manual, Form GH20-0323.
8. IBM system/360 scientific subroutine package (PL/I) program description and operations manual, Form GH20-0586.
9. H.C. Lucas, "Performance evaluation and monitoring", ACM Comp. Surveys 3, no. 3, September 1971.
10. M.H. MacDougall,"Computer system simulation: an introduction", ACM Comp. Surveys 2, no. 3, September 1970.
11. F.F. Martin, Computer modeling and simulation, Wiley, 1968.
12. T.H. Naylor, J.L. Balintly, D.S. Burdick and Kong Chu, Computer simulation techniques, Wiley, 1966.
13. S. Siegel, Nonparametric statistics for the behavioral sciences, McGraw-Hill, 1956.
14. W.H. Tetzlaff, Statistics gathering package reduction program, August 4, 1971, IBM RC3483.
15. G. Waldbaum and H. Beilner, "SOUL: a simulation of OS/360 under LASP", submitted to SJCC 72.

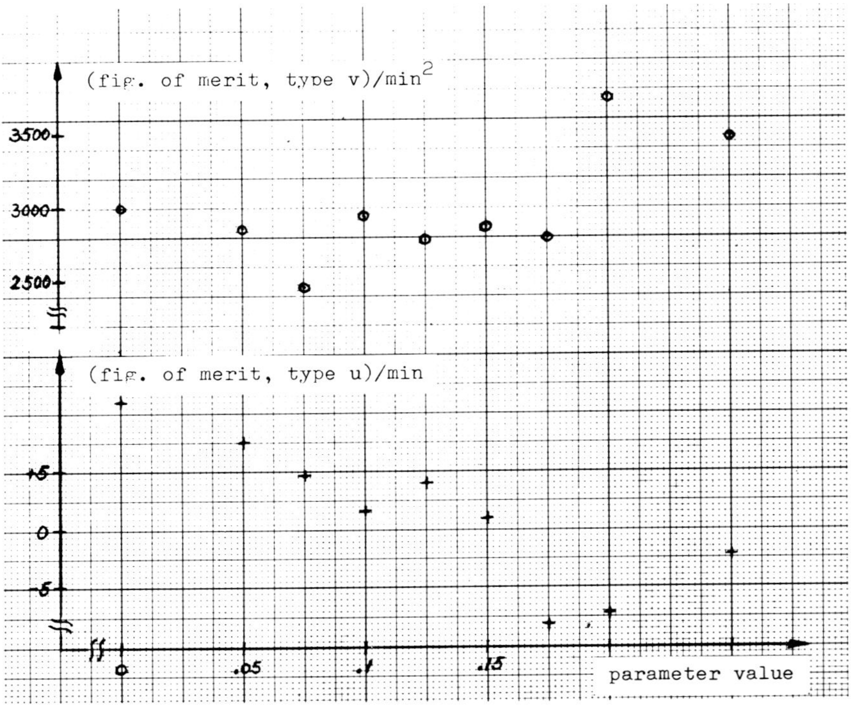

Fig. 1.1. Dependence of figures of merit on a parameter.

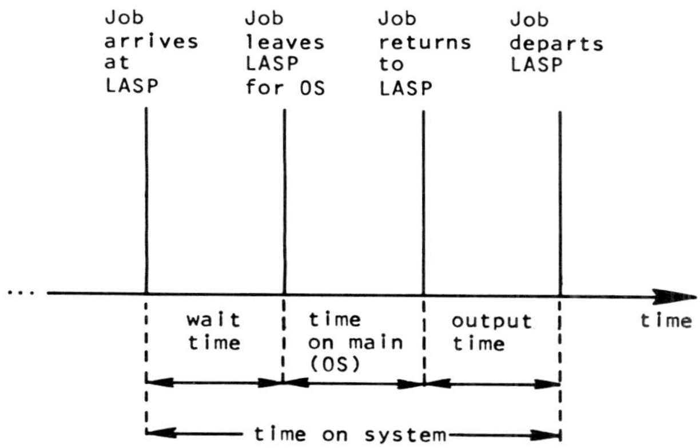

Fig. 2.1. Job flow through OS-LASP.

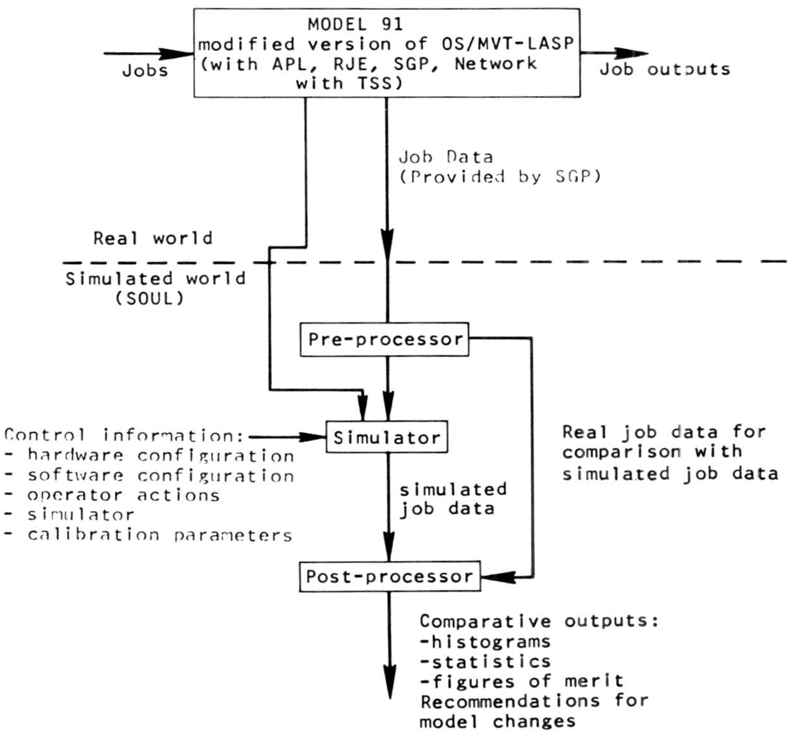

Fig. 2.2. Overall structure of SOUL.

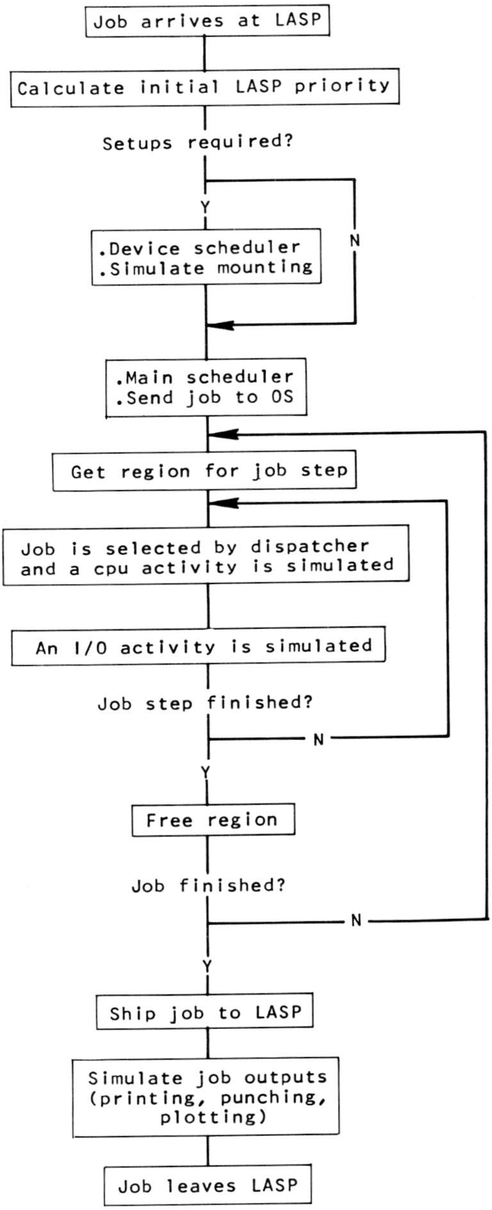

Fig. 2.3. Gross flowchart of simulator.

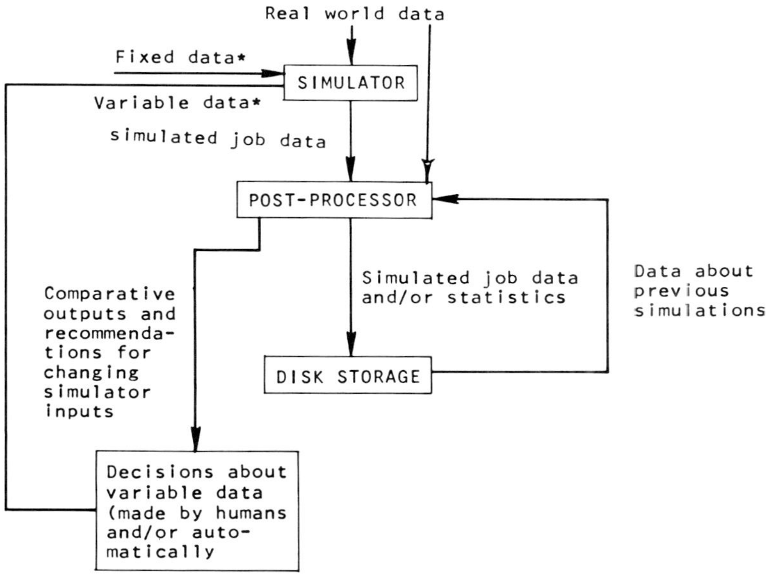

* Variable data + fixed data = control information

Variable data = { hardware configuration (when experimenting); software configuration (when experimenting); operator actions (when experimenting); calibration parameters (when calibrating) }

Fig. 6.1. Flowchart of calibration/experimentation process.

VARIABLE	MEAN	STANDARD DEVIATION	CORRELATION X VS Y	REGRESSION COEFFICIENT	STD.ERROR OF REG.COEFF.	COMPUTED T VALUE	BETA COEFF.
NJS	3.33384	2.08001	0.10858	1.14607	1.11839	1.02474	0.04362
CPUJ	0.47495	0.86007	0.29022	18.97624	2.72153	6.97263	0.29866
IOJ	1.19422	2.06901	0.10381	-1.06418	1.38028	-0.77099	-0.04029
EXCPJ	3260.97705	4152.98828	0.11207	0.00014	0.00066	0.20511	0.01031
RGNJ	259.42676	157.01662	-0.06636	-0.03016	0.01337	-2.25668	-0.08666
SETUPS	0.46494	0.62173	0.04358	5.34549	3.39000	1.57684	0.06082
DEPENDENT:							
ERRSYST	20.49178	54.64740					
INTERCEPT		13.82785					
MULTIPLE CORRELATION		0.31111					
STD. ERROR OF ESTIMATE		52.17507					

ANALYSIS OF VARIANCE FOR THE REGRESSION

SOURCE OF VARIATION	DEGREES OF FREEDOM	SUM OF SQUARES	MEAN SQUARES	F VALUE
ATTRIBUTABLE TO REGRESSION	6	1.89321E+05	3.15534E+04	11.59099
DEVIATION FROM REGRESSION	649	1.76673E+06	2.72224E+03	

Fig. 6.2. Regression analysis "error in system phase versus job characteristics" at point $\bar{p}_1$.

VARIABLE	MEAN	STANDARD DEVIATION	CORRELATION X VS Y	REGRESSION COEFFICIENT	STD.ERROR OF REG.COEFF.	COMPUTED T VALUE	BETA COEFF.
TIMSLOS	1.19232E+03	5.30937E+02	-8.22738E-02	-1.24053E+00	3.45861E+00	-0.35868	-0.01993
TIMOSSJ	4.45946E+02	1.04335E+02	1.30390E-01	1.06165E+01	2.43510E+01	0.43598	0.03064
TIMOSFJ	1.14959E+04	7.67099E+03	-5.12813E-02	7.15794E-03	1.59217E-01	0.04496	0.00152
MINSTIM	2.29595E+03	9.74115E+02	1.83943E-01	-2.38865E+00	3.01305E+00	-0.79277	-0.06436
OVO	9.55946E+01	1.49115E+02	2.27780E-01	1.31244E+01	9.26221E+00	1.41698	0.05413
OV1	4.05405E-03	1.81749E-02	-5.98244E-02	4.27733E+04	6.56178E+04	0.65186	0.02150
OV2	8.78378E-02	3.01709E-01	9.83304E-01	1.10500E+05	5.97408E+03	18.49652	0.92210
OV3	5.04054E+00	4.90158E-01	-7.33664E-01	-6.17845E+03	3.64625E+03	-1.69447	-0.08376
DEPENDENT:							
ERRSYST	1.23349E+04	3.61554E+04					

INTERCEPT	34490.14063
MULTIPLE CORRELATION	0.98637
STD. ERROR OF ESTIMATE	6746.75391

ANALYSIS OF VARIANCE FOR THE REGRESSION

SOURCE OF VARIATION	DEGREES OF FREEDOM	SUM OF SQUARES	MEAN SQUARES	F VALUE
ATTRIBUTABLE TO REGRESSION	8	4.57851E+10	5.72314E+09	125.73157
DEVIATION FROM REGRESSION	28	1.27452E+09	4.55187E+07	

RECOMMENDATION FOR PARAMETER CHANGES

	TIMNTOT	TIMNTOD	TIMSLOS	TIMOSSJ	TIMOSFJ	MINSTIM	OVO	OV1	OV2	OV3
BEST POINT	1.00E+06	1.00E+05	1.00E+03	5.00E+02	1.00E+04	3.00E+03	2.50E+02	0.00E+00	0.00E+00	5.20E+00
S = 1	1.00E+06	1.00E+05	1.00E+03	4.89E+02	1.00E+04	3.00E+03	2.37E+02	0.00E+00	0.00E+00	6.18E+03
S = 2	1.00E+06	1.00E+05	1.00E+03	4.79E+02	1.00E+04	3.00E+03	2.24E+02	0.00E+00	0.00E+00	1.24E+04
S = 5	1.00E+06	1.00E+05	1.01E+03	4.47E+02	1.00E+04	3.01E+03	1.84E+02	0.00E+00	0.00E+00	3.09E+04

Fig. 6.3. Regression analysis "average squared error in system phase versus parameters" at point $\bar{p}_2$.

```
TIMNTOT= 1.000000E+06;
TIMNTOD= 1.000000E+05;
TIMSLOS= 1.000000E+03;
TIMOSSJ= 2.500000E+02;
TIMOSFJ= 1.000000E+04;
MINSTIM= 1.000000E+03;
OV0= 0.000000E+00;
OV1= 0.000000E+00;
OV2= 0.000000E+00;
OV3= 5.000000E+00;

ERRSYST    (SQUARES)
     1   1.96E+01
     2   8.60E+02
     3   2.43E+03
     4   4.74E+03
     5   1.19E+03
     6   5.46E+03
     7   1.62E+04
     8   1.33E+03
     9   2.76E+03
    10   9.18E+02
    11   3.35E+03
    12   4.71E+03
    13   2.97E+03
    14   9.75E+03
    15   2.10E+03
    16   3.87E+03
    17   4.65E+03
    18   1.95E+03
    19   3.05E+03
    20   9.93E+02
    21   1.02E+03
    22   1.56E+02

MEAN       2. MOMENT     S          SIGMA
 3.40E+03   2.41E+07   3.54E+03   7.57E+02

#P=              11;
#N=              11;
#RUN=               13;
Z= 4.369313E-01;
```

point $\bar{p}_1$

```
TIMNTOT= 1.000000E+06;
TIMNTOD= 1.000000E+05;
TIMSLOS= 1.000000E+03;
TIMOSSJ= 5.000000E+02;
TIMOSFJ= 1.000000E+04;
MINSTIM= 3.000000E+03;
OV0= 2.500000E+02;
OV1= 0.000000E+00;
OV2= 0.000000E+00;
OV3= 5.199989E+00;

ERRSYST    (SQUARES)
     1   7.17E+01
     2   7.74E+02
     3   3.59E+03
     4   4.01E+03
     5   1.97E+03
     6   8.45E+03
     7   1.41E+04
     8   1.27E+03
     9   2.63E+03
    10   4.54E+02
    11   2.89E+03
    12   4.34E+03
    13   2.89E+03
    14   7.57E+03
    15   1.70E+03
    16   3.16E+03
    17   3.71E+03
    18   1.80E+03
    19   2.48E+03
    20   7.97E+02
    21   1.02E+03
    22   1.55E+02

MEAN       2. MOMENT     S          SIGMA
 3.19E+03   2.02E+07   3.17E+03   6.78E+02

#P=              11;
#N=              11;
#RUN=    .          11;
Z=-4.369313E-01;
```

point $\bar{p}_2$

```
TIMNTOT= 1.000000E+06;
TIMNTOD= 1.000000E+05;
TIMSLOS= 1.000000E+03;
TIMOSSJ= 5.000000E+02;
TIMOSFJ= 1.000000E+04;
MINSTIM= 3.000000E+03;
OV0= 0.000000E+00;
OV1= 0.000000E+00;
OV2= 0.000000E+00;
OV3= 5.799989E+00;

ERRSYST    (SQUARES)
     1   9.52E+01
     2   1.25E+03
     3   4.42E+03
     4   7.79E+03
     5   2.69E+03
     6   7.94E+03
     7   9.52E+03
     8   1.20E+03
     9   1.18E+03
    10   1.12E+03
    11   2.76E+03
    12   3.94E+03
    13   1.23E+03
    14   3.88E+03
    15   1.61E+03
    16   1.70E+03
    17   2.74E+03
    18   1.16E+03
    19   1.49E+03
    20   6.36E+02
    21   8.17E+02
    22   1.55E+02

MEAN       2. MOMENT     S          SIGMA
 2.71E+03   1.39E+07   2.56E+03   5.48E+02

#P=              11;
#N=              11;
#RUN=                7;
Z=-2.184657E+00;
```

point $\bar{p}_3$

Fig. 6.4. Information for the decision process $\bar{P}_1$-$\bar{P}_2$-$\bar{P}_3$.

TIME ON SYSTEM(TIME BETWEEN ARRIVAL AT LASP TO DEPART.FROM LASP)

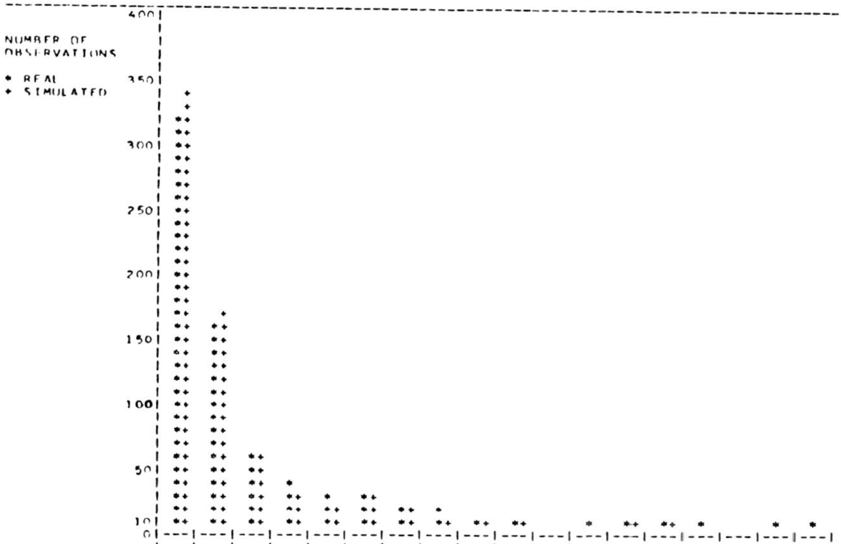

TIME/MINUTE	0–30	30–60	60–90	90–120	120–150	150–180	180–210	210–240	240–270	270–300	300–330	330–360	360–390	390–420	420–450	450–480	480–510	510–540
# REAL	316	160	55	33	22	28	15	11	5	4	0	1	1	2	1	0	1	1
# SIMULATED	338	169	57	28	13	23	12	10	2	2	0	0	1	1	0	0	0	0

NOTE:
AN INTERVAL OF THE HISTOGRAM, DENOTED BY THE LOWER BOUND LB AND THE UPPER BOUND UB, CONTAINS THE NUMBER OF ENTRIES, X, SUBJECT TO LB <= X < UB .

Fig. 6.5. Two-sample histogram at point $\bar{p}_3$.

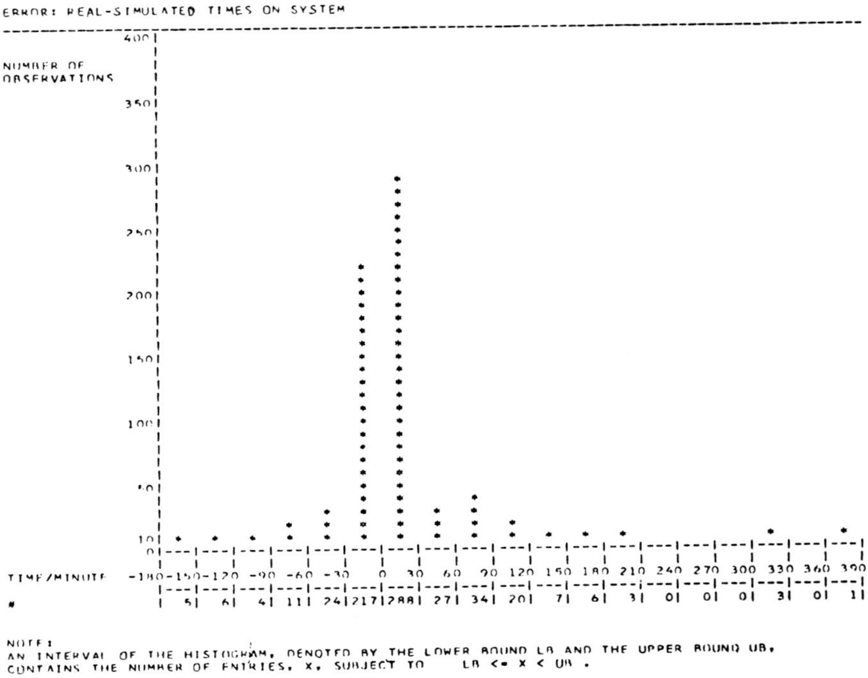

Fig. 6.6. One-sample histogram at point $\bar{p}_3$.

APPENDIX A

This appendix proves the two theorems used in section 4.1. The notation used is the same as that used in that section with the exceptions that

- all the subscripts for the phase (j) are deleted (we "fix the phase"),
- the functions u and v differ from those in section 4.1 by the factor n.

Let $\bar{p} = (p_1, p_2, \ldots, p_m)'$ denote a vector of real variables,

$I = \{1,2,\ldots,n\}$ an index set and

(1) let $e_i(\bar{p})$, $i \varepsilon I$, denote a set of real continuous single-valued functions of the vector $\bar{p}$; also, let every $e_i(\bar{p})$ be partially differentiable with respect to $\bar{p}$ and have continuous single-valued partial derivatives $\partial e_i(\bar{p})/\partial p_k$, $k=1,2,\ldots,m$; $i \varepsilon I$;

(2) Let, with respect to every variable p_k $(k=1,2,\ldots,m)$,

I_k denote a non-empty index subset $I_k \subset I$, $I_k \neq 0$,

I_k^c the complementary index subset $I_k^c = I - I_k$,

$|I_k|$, $|I_k^c|$ the number of elements in these sets and

$\bar{b}_k = (b_{k1}, b_{k2}, \ldots, b_{km})' > (0,0,\ldots,0)'$ a vector of finite positive bounds.

Let $\bar{b}$ denote the vector of maximum entries of all the $\bar{b}_k$:

$$\bar{b} = (\max_k(b_{k1}), \max_k(b_{k2}), \ldots, \max_k(b_{km}))'$$

It will be shown that, if the functions $e_i(\bar{p})$ meet the additional conditions

(3) $$|e_i(\bar{p})| < K1 = \frac{M}{|I_k^c|} \qquad i \varepsilon I_k^c,\ \bar{p} \geq \bar{b}_k,\ \bar{p} \leq -\bar{b}_k$$

(4) $$e_i(\bar{p}) < -K2 = -\frac{M}{|I_k|-1} \qquad i \varepsilon I_k,\ \bar{p} \geq \bar{b}_k$$

(5) $$e_i(\bar{p}) > K2 = \frac{M}{|I_k|-1} \qquad i \varepsilon I_k,\ \bar{p} \leq -\bar{b}_k$$

$$(6)\quad \left|\frac{\partial e_i(\bar{p})}{\partial p_k}\right| < E \qquad i\varepsilon I_k^c,\ \bar{p} \geq \bar{b}_k,\ \bar{p} \leq -\bar{b}_k,\ k=1,2,\ldots,m$$

$$(7)\quad \frac{\partial e_i(p)}{\partial p_k} < -C\cdot E \qquad i\varepsilon I_k,\ \bar{p} \geq \bar{b}_k,\ \bar{p} \leq -\bar{b}_k,\ k=1,2,\ldots,m$$

where

$$(8)\quad C = \max(1, \frac{|I_k^c|}{|I_k|-1}), \quad E > 0, \quad M > 0$$

then

A) there exists at least one vector $\bar{p}_1$ in the interval $-\bar{b} < \bar{p}_1 < \bar{b}$ so that the function

$$v(\bar{p}) = \sum_{i\varepsilon I} e_i^2(\bar{p})$$

has a minimum value $v_1 = v(\bar{p}_1)$,

B) there exists at least one vector $\bar{p}_2$ in the interval $-\bar{b} < \bar{p}_2 < \bar{b}$ so that the function

$$v(\bar{p}) = \sum_{i\varepsilon I} e_i^2(\bar{p})$$

subject to the restraint

$$u(\bar{p}) = \sum_{i\varepsilon I} e_i(\bar{p}) = 0$$

has a minimum value $v_2 = v(\bar{p}_2)$.

Proof of A. The $e_i(\bar{p})$ are, by assumption, continuous, single-valued and partially differentiable and have continuous unique partial derivatives. Thus

$$v(\bar{p}) = \sum_{i\varepsilon I} e_i^2(\bar{p})$$

is also continuous and partially differentiable and has continuous partial derivatives. $v(\bar{p})$ takes an extremum value if there is a $\bar{p}_1$ with $\partial v(\bar{p}_1)/\partial p_k = 0$, $k=1,2,\ldots,m$. The partial derivatives have the form

$$(9)\quad \frac{\partial v(\bar{p})}{\partial p_k} = \sum_{i\varepsilon I} 2e_i(\bar{p}) \frac{\partial e_i(\bar{p})}{\partial p_k}$$

$$= \sum_{i\varepsilon I_k} 2e_i(\bar{p}) \frac{\partial e_i(\bar{p})}{\partial p_k} + \sum_{i\varepsilon I_k^c} 2e_i(\bar{p}) \frac{\partial e_i(\bar{p})}{\partial p_k}$$

For any k and any $\bar{p} \geq \bar{b}_k$ we obtain from (9), by (3), (4), (6), (7),

$$\frac{\partial v(\bar{p})}{\partial p_k} > |I_k| \cdot 2 \cdot \frac{M}{|I_k|-1} \cdot C \cdot E - |I_k^c| \cdot 2 \cdot \frac{M}{|I_k^c|} \cdot E$$

$$= 2 \cdot M \cdot E \cdot (\frac{|I_k|}{|I_k|-1} \cdot C - 1)$$

and by (8)

$$\frac{\partial v(\bar{p})}{\partial p_k} > 0, \quad \bar{p} \geq \bar{b}_k, \quad k=1,2,\ldots,m$$

For any k and any $p \leq -b_k$ we obtain from (9), by (3), (5), (6), (7),

$$\frac{\partial v(\bar{p})}{\partial p_k} < - |I_k| \cdot 2 \cdot \frac{M}{|I_k|-1} \cdot C \cdot E + |I_k^c| \cdot 2 \cdot \frac{M}{|I_k^c|} \cdot E$$

$$= 2 \cdot M \cdot E \cdot (- \frac{|I_k|}{|I_k|-1} \cdot C + 1)$$

and by (8)

$$\frac{\partial v(\bar{p})}{\partial p_k} < 0, \quad \bar{p} \leq -\bar{b}_k, \quad k=1,2,\ldots,m$$

Thus the set of functions $\partial v(\bar{p})/\partial p_k$, $k=1,2,\ldots,m$, forms a system of m continuous unique functions with the properties

$$(10)\quad \frac{\partial v(\bar{p})}{\partial p_k} < 0 \quad \text{for } \bar{p} \leq -\bar{b}, \quad k=1,2,\ldots,m$$

$$(11)\quad \frac{\partial v(\bar{p})}{\partial p_k} > 0 \quad \text{for } \bar{p} \geq \bar{b}, \quad k=1,2,\ldots,m$$

This enables us to use the general mean value theorem (cf. [2]) which, under these assumptions, states that there is at least one point $\bar{p}_1$, subject to $-\bar{b} < \bar{p}_1 < \bar{b}$, with

$$\frac{\partial v(\bar{p}_1)}{\partial p_k} = 0 \quad \text{for } k=1,2,\ldots,m$$

This proves that there is at least one $\bar{p}_1$ for which $v(\bar{p}_1)$ takes an extremum value. At least one of these extrema is a minimum because of (10), (11).

<u>Proof of B.</u> Again we have to look for an extremum point $\bar{p}_2$ for which

$$\frac{\partial v(\bar{p}_2)}{\partial p_k} = 0 \quad \text{for } k=1,2,\ldots,m$$

I_k is a non-empty index subset. Let $1_k \varepsilon I_k$ denote one element (index) of I_k and $I_{k-} = I_k - \{1_k\}$.

From the restraint $u(\bar{p}) = \sum_{i\varepsilon I} e_i(\bar{p}) = 0$ we obtain

$$e_{1_k}(\bar{p}) = -\sum_{i\varepsilon I_{k-}} e_i(\bar{p}) - \sum_{i\varepsilon I_k^c} e_i(\bar{p})$$

and thus

$$v(\bar{p}) = \sum_{i\varepsilon I_{k-}} e_i^2(\bar{p}) + \sum_{i\varepsilon I_k^c} e_i^2(\bar{p})$$

$$+ [\sum_{i\varepsilon I_{k-}} e_i(\bar{p}) + \sum_{i\varepsilon I_k^c} e_i(\bar{p})]^2$$

The partial derivatives have the form

$$(12) \quad \frac{\partial v(\bar{p})}{\partial p_k} = \sum_{i\varepsilon I_{k-}} 2e_i(\bar{p}) \frac{\partial e_i(\bar{p})}{\partial p_k} + \sum_{i\varepsilon I_k^c} 2e_i(\bar{p}) \frac{\partial e_i(\bar{p})}{\partial p_k}$$

$$+ 2[\sum_{i\varepsilon I_{k-}} e_i(\bar{p}) + \sum_{i\varepsilon I_k^c} e_i(\bar{p})][\sum_{i\varepsilon I_{k-}} \frac{\partial e_i(\bar{p})}{\partial p_k} + \sum_{i\varepsilon I_k^c} \frac{\partial e_i(\bar{p})}{\partial p_k}]$$

For any $\bar{p} \geq \bar{b}_k$, by (3), (4),

$$(13) \quad \sum_{i\varepsilon I_{k-}} e_i(\bar{p}) + \sum_{i\varepsilon I_k^c} e_i(\bar{p}) < -(|I_k|-1)\cdot\frac{M}{|I_k|-1} + |I_k^c|\cdot\frac{M}{|I_k^c|}$$

$$< 0$$

For any $\bar{p} \geq \bar{b}_k$, by (6), (7),

$$(14) \quad \sum_{i \varepsilon I_{k-}} \frac{\partial e_i(\bar{p})}{\partial p_k} + \sum_{i \varepsilon I_k^c} \frac{\partial e_i(\bar{p})}{\partial p_k} < -(|I_k|-1)\cdot C\cdot E + |I_k|\cdot E$$

and by (8) < 0

We obtain from (12) by (13), (14),

$$\frac{\partial v(\bar{p})}{\partial p_k} > 2 \sum_{i \varepsilon I_{k-}} e_i(\bar{p}) \frac{\partial e_i(\bar{p})}{\partial p_k} + 2 \sum_{i \varepsilon I_k^c} e_i(\bar{p}) \frac{\partial e_i(\bar{p})}{\partial p_k}$$

and by (3), (4), (6), (7),

$$\frac{\partial v(\bar{p})}{\partial p_k} > 2\ (|I_k|-1)\cdot\frac{M}{|I_k|-1}\cdot C\cdot E - 2\cdot|I_k^c|\cdot\frac{M}{|I_k^c|}\cdot E$$

by (8) > 0

For any $\bar{p} \leq -\bar{b}_k$, by (3), (5)

$$(15) \quad \sum_{i \varepsilon I_{k-}} e_i(\bar{p}) + \sum_{i \varepsilon I_k^c} e_i(\bar{p}) > (|I_k|-1)\cdot\frac{M}{|I_k|-1} - |I_k^c|\cdot\frac{M}{|I_k^c|}$$

> 0

For any $\bar{p} \leq -\bar{b}_k$, by (6), (7)

$$(16) \quad \sum_{i \varepsilon I_{k-}} \frac{\partial e_i(\bar{p})}{\partial p_k} + \sum_{i \varepsilon I_k^c} \frac{\partial e_i(\bar{p})}{\partial p_k} < -\ (|I_k|-1)\cdot C\cdot E + |I_k^c|\cdot E$$

and by (8) < 0

We obtain from (12) by (15), (16)

$$\frac{\partial v(\bar{p})}{\partial p_k} < 2 \sum_{i \varepsilon I_{k-}} e_i(\bar{p}) \frac{\partial e_i(\bar{p})}{\partial p_k} + 2 \sum_{i \varepsilon I_k^c} e_i(\bar{p}) \frac{\partial e_i(\bar{p})}{\partial p_k}$$

and by (3), (5), (6), (7)

$$< -2\ (|I_k|-1)\cdot\frac{M}{|I_k|-1}\cdot C\cdot E + 2\cdot|I_k^c|\frac{M}{|I_k^c|}\cdot E$$

by (8) < 0

Thus, as in the proof of A, the set of functions

$\partial v(\bar{p})/\partial p_k$, k=1,2,...,m, forms a set of m continuous single-valued functions with the properties

$$(17) \quad \frac{\partial v(\bar{p})}{\partial p_k} < 0 \qquad \text{for } \bar{p} \leq -\bar{b}, \qquad k=1,2,\ldots,m$$

$$(18) \quad \frac{\partial v(\bar{p})}{\partial p_k} > 0 \qquad \text{for } \bar{p} \geq \bar{b}, \qquad k=1,2,\ldots,m$$

Again we can use the general mean value theorem to state that there is at least one point $\bar{p}_2$, subject to $-\bar{b} < \bar{p}_2 < \bar{b}$, with

$$\frac{\partial v(\bar{p}_2)}{\partial p_k} = 0 \qquad k=1,2,\ldots,m$$

We conclude that there is at least one $\bar{p}_2$ for which $v(\bar{p}_2)$ takes an extremum value under the constraint $u(\bar{p}_2) = 0$. At least one of these extrema is a minimum because of (17), (18).

APPENDIX B

Appendix B provides some arguments about the reasonability of the assumptions for the two theorems proved in Appendix A.

Each assumption (numbered as in Appendix A) is repeated in verbal form and is followed by the appropriate arguments.

(1) Assumption: The error functions $e_{ij}(\bar{p})$ are continuous single-valued real functions of $\bar{p}$. They are partially differentiable with respect to $\bar{p}$. The partial derivatives are continuous.

Since we have so far in no way restricted the functions $e_{ij}(\bar{p})$, we are free to define them as belonging to this class of functions.

(2) Assumption: There is, for any parameter p_k used in tuning a certain phase, a particular set of jobs which are "directly influenced" in this phase by p_k.

We distinguish between direct and indirect influence in order to separate two types of effects which a change in the value of a certain parameter can have on the behavior of a job:

- If a particular parameter is used in the simulation of some action of a job in a certain phase, we have a "direct influence" of this parameter on this job in this phase. For example, the parameter TIMNTOT, denoting the time to mount a set-up tape, directly influences, in the wait- and system-phases, every job which asks for a tape to be mounted, but does not directly influence any job in the main phase.
- Any effect of a parameter (on a job in a phase) which is not classified as direct will be called "indirect influence". For example, the parameter TIMNTOT certainly may affect jobs in the main phase since this parameter may cause jobs to enter the main phase in different order and the time a job spends in the main phase depends upon other jobs which are simultaneously in the main phase. However, such effects are caused only indirectly, i.e. through some jobs which are directly influenced.

Accordingly, we say that a parameter has a direct influence on a certain phase if the parameter directly influences at least one job in this phase.

In order to find a minimum value for $v_j(\bar{p})$ (i.e., for phase j) this assumption holds whenever we tune this phase by adjusting parameters which directly influence at least one job in this phase (i.e., only parameters with a direct influence upon this phase).

(3), (4), (5) Assumptions: There is, for any phase j and any parameter p_k with a direct influence on this phase, some vector of bounds $\bar{b}_{jk}$ such that

- for any $\bar{p} \geq \bar{b}_{jk}$ and any $\bar{p} \leq -\bar{b}_{jk}$ the error functions $e_{ij}(\bar{p})$ for all the jobs which are not directly influenced have absolute values not exceeding some positive bound K1,
- for any $\bar{p} \leq -\bar{b}_{jk}$ the error functions $e_{ij}(\bar{p})$ for all the jobs which are directly influenced are positive and exceed some positive bound K2, where the ratio of K2 to K1 is a constant,
- for any $\bar{p} \geq \bar{b}_{jk}$ the error functions $e_{ij}(\bar{p})$ for all the jobs which are directly influenced are negative and do not exceed the negative bound -K2.

These assumptions are easily verified. Since the calibration parameters denote time delays or time expansion factors, the larger the value of a parameter, the longer a job which is directly influenced by this parameter tends to stay in the affected phase. The error (i.e., the difference between the real and simulated times that the job stays in this phase) tends to become smaller and can even be made as negative as we want. Consequently, the error functions can be made more negative than any finite bound. The converse reasoning shows that the more negative we make a parameter, the more positive are the values of the error functions for the jobs which are directly influenced. These functions can be made larger than any finite positive bound. (For the purpose of this argument, we have to allow for negative times.)

The error functions for the jobs which are not directly influenced can certainly be estimated to have, for $\bar{p} \geq \bar{b}_{jk}$ and $\bar{p} \leq -\bar{b}_{jk}$, absolute values not exceeding some finite bound. If we assign to a certain parameter large positive or large negative values, then the actions of those jobs which are directly influenced and of those jobs which are not directly influenced become at last completely separated in time and indirect influences die out. To clarify this: if, simulating one day's activities, we decide to have a tape mounting in the simulator take 48 hours, then jobs without set-up tapes will have left the simulated system at least 24 hours before the first job with set-up tapes burdens the system.

(6), (7) Assumptions: For $\bar{p} \geq \bar{b}_{jk}$ and $\bar{p} \leq -\bar{b}_{jk}$, the partial derivatives $\partial e_{ij}(\bar{p})/\partial p_k$

- have absolute values below some positive bound E for every job which is not directly influenced,
- have negative values below some negative bound -C·E for every job which is directly influenced.

The reasonability of these assumptions can be shown with the same arguments as before. For large positive and negative parameter values any indirect influence dies out so that all the partial derivatives for jobs which are not directly influenced approximate zero, whereas all the partial derivatives for jobs which are directly influenced continue to have finite negative values.

DISCUSSION OF SESSION VI

by Akira Sekino

All the papers of this session happen to use the trace data of real computer systems. But they were applied to different problems. The paper by Margolin, Parmelee and Schatzoff discusses the new "dynamic storage allocation" of CP-67. Taking a standard orthodox approach consisting of data collection, data reduction, creation of trace data, simulation of algorithms, analysis of their performance, and on-line experiment, the authors succeeded in producing dramatic improvement of performance, measured in space and time requirements, of the storage allocation algorithm. The paper includes a nice example of statistics concerning the characteristics of storage demand on real computer systems.

The paper by Hatfield and Gerald tackled a "pagination problem", i.e. the problem of clustering relocatable program sectors into pages so as to reduce the number of page exceptions. The authors have proposed a pagination method and applied it to some CP-67 programs. The results show the reduction of page exceptions amounting to an improvement on the order of from two to one to ten to one. This apparently implies that language processors (compilers, assemblers) of virtual memory computers should be carefully paginated. But this kind of costly optimization technique can be practically applied only to those programs that are frequently used and tend to have a large working set. Probably nobody else has investigated the program behavior under paging in such detail. Therefore, I would like the authors to examine (1) the stationariness of program behavior viewed as a stochastic process, and (2) the validity of (first-order) Markov chain approximation of the sector reference pattern. Depending on the result of these examinations, the pagination method based on the inter-sector distances (the inter-sector distance from sector i to sector j is defined as the expected number of steps required before reaching sector j for the first time starting from sector i) may turn out to be better than the one proposed in the paper.

AKIRA SEKINO

The paper by Beilner and Waldbaum describes the calibration methodology of SOUL, a trace-driven simulator of a large-scale batch computer system, from a statistical viewpoint. This paper is unique in its careful construction of a statistically balanced simulator. But it fails to discuss the problem of overtuning (because the trace data intended to represent the unknown input data population often contain some irrelevant characteristics (e.g. noises), the simulator may not perform very well for the unknown input data if it is tuned too exactly to the trace data). The use of smoothed trace data rather than raw trace data often gives a solution to the overtuning problem of trace-driven simulators.

SESSION VII

SOFTWARE RELIABILITY

CHAIRMAN: L. A. BELADY

SOFTWARE RELIABILITY RESEARCH

Z. Jelinski and P. Moranda

INTRODUCTION AND SUMMARY

Software reliability study was initiated by Advanced Information Systems subdivision of McDonnell Douglas Astronautics Company, Huntington Beach, California, to conduct research into the nature of the software reliability problem including definitions, contributing factors and means for control.

Discrepancy reports which originated during the development of two large-scale real-time systems form two separate primary data sources for the reliability study. A mathematical model, descriptively entitled the De-Eutrophication Process, was developed to describe the time pattern of the occurrence of discrepancies (errors). This model has been employed to estimate the initial (or residual) error content in a software package as well as to estimate the time between discrepancies at any phase of its development. Means of predicting mission success on the basis of errors which occur during testing are described.

Problems in categorizing software anomalies are described and the special area of the genesis of discrepancies during the integration of modules is discussed. Management techniques which should reduce the number of software anomalies are described.

STATE-OF-THE-ART REVIEW

First, it was decided to look into the experience of several major systems contractors and other organizations from whom data would be readily available. This turned out to be more of a problem than was realized at first because hardly any company collects data on software failures in a format suitable for analysis. Secondly, available literature was surveyed to determine what has been done in software reliability so far. This produced a variety of approaches to software reliability starting with using the same treatment given to hardware, through automatic program proving techniques, to theoretical mathematical approaches.

From the start it became evident that software reliability, quality assurance and performance were often used interchangeably. The importance of software and its reliability as an integral part of space transportation systems or strategic defense systems could not be overemphasized. Studies were conducted to determine the best way to control development of software to ensure reliability. At the other end, attempts were made to analyze software failures by applying statistical methods and by looking at the results to determine causes of failures and to devise procedures and tests to prevent them. Lt. Buchley of the U.S. Army Computer Command [1] stated that one approach to verification of software performance is to accept the program as issued but require the manufacturers to fix bugs as they happen. The natural approach to monitoring development of software programs was to adopt the existing procedures applicable to hardware. Milton Ratynski [2] discussed the concept of adapting software to the Air Force 375 hardware acquisition techniques rather than developing a separate software management process. He stated that the concepts of uniform specifications, baseline management, change control, specification maintenance, design reviews and a test program were as pertinent to computer programs as they were to hardware acquisition. He realized, however, that there existed differences between software and hardware and items and that the modification of the current hardware-oriented 375 system management process was necessary.

S.V. Schnurbrush stated [3] that verification of the Apollo Real Time Mission Program was accomplished by assigning responsibility to the system integration group and development of such testing activities that involved the total system and assured that program requirements were realized. To make the program more comprehensive a dynamic network data generator system was developed. The generator provided trajectory and vehicle systems data, and also simulated the radar and telemetry data. Once a particular mission program was developed, the system itself became the development base for subsequent mission systems. The testing effort was then directly related to the volume of functional changes incorporated and the confidence in the base system. The high level of confidence in the Apollo 8 mission program was attributed largely to the level of development and verification completed on the base system.

Development studies of a software reliability program [4] were conducted by the Federal Electric Corporation (an ITT service associate). These called for a quality control function to be established alongside software development programs. Each stage of software development, starting with definition of requirements, system design, coding and verification, would have a software reliability function associated with it. An attempt was also made to define subsystems reliability as follows:

$$R_K = e^{-\lambda_E t}$$

where R_K = subsystem reliability, λ_E = error rate for the subsystem and t = subsystem mission time. Thus, reliability of the software system was defined as the probability that the system will perform without error during its assigned mission, within the assumptions as to the equation's form being correct.

Barry W. Boehm of the Rand Corporation [5] implied that software problems in a strategic defense system could result in escalating strategic crises, thus precipitating a dangerous confrontation, or, even worse, a software error could incapacitate a key component of our defense structure at a critical time. Donald K. Slayton [6] of NASA stated that on-board software reliability must be such that, during the entire space shuttle program, not one vehicle is lost or damaged beyond repair because of software problems.

Universities and research departments of industrial concerns have conducted a number of theoretical studies with a view to having at some time to develop an automatic tool to prove programs correct. Several approaches to the formal validation of computer programs were made by Floyd, Naur, London and Manna in their respective papers (see bibliography). Ralph L. London [7] defined program validation as a certification, by means of mathematical proof, that the program specification is met in all cases. He did not claim to have invented a general technique to be used. James C. King [8] was concerned with the same problem except that he took an abstract model for a program and, having its execution, he defined correctness of programs and proof of correctness of programs. He used simple applied predicate calculus. He then constructed a computer program capable of proving the correctness of many simple programs written in Algol-like programming language over integers. Formal validation of computer programs [9] falls into two classes: those that put the burden of the proof on the human programmer, using a comparatively simple computer tool to do the actual checking, and those which relieve the programmer of much of the effort but require a much more sophisticated theorem prover.

DEFINITIONS

From the above it is evident that 'software reliability' may be a misnomer and that the same treatment cannot be applied to software as to hardware. It is generally agreed that once a software package is completely debugged then it is 100% reliable and, unlike hardware, does not deteriorate in time. The concern is more then with software performance without failure - a term quality assurance may be more appropriate at this time. This in turn implies existence of quality control of software with associated procedures and checks and even then it does not provide assurance that the program performs in accordance with requirements. Software validation is therefore another term that comes to mind as a function whose sole purpose is to make sure that a software system which may perform without failure does in fact do the job for which it was designed. Finally, a closer look at the nature of software failures is taken at an early stage of software development in order to determine the pattern and distribution of these failures. Therefore, software production implies any program coded by a programmer and compiled on a computer - a record of software failures is made from that time throughout the debugging period. When a program is 'debugged' test and evaluation is made, usually by another group of programmers who subject the program to a number of benchmark tests. As a result of these tests more software failures are uncovered. When these are corrected and benchmarks are satisfied then the program is delivered to the user and more failures occur in field operation. These three periods - software production, test and evaluation and users' experience - have formed the basic framework for this investigation.

IMPORTANCE OF SOFTWARE RELIABILITY

The importance of software reliability in our strategic

defense systems and in our space programs cannot be overemphasized. This was stated earlier by Boehm and Slayton; however, there is another practical reason why a company whose main business is systems management must be concerned with the reliability of delivered software - namely, the good will of its customers and opportunities for future business. How can a contractor state that his software is reliable? What is a better way than to be able to measure that reliability? Why not deliver a software package with built-in confidence limits of its future performance? Why not have the ability to predict failures? This can all be done, of course, provided that there exist such data on software failures which lend themselves to statistical treatment, and therefore prediction and what is involved in it. This sounds very simple but is not, and that is what this investigation is about. As systems manager, a company is also responsible for the performance of its subcontractors, and when accepting their efforts the company would like to be able to measure them. How is this possible? It is the contention of the authors that, given a clean set of data, it is possible to develop a reasonably accurate model which would permit the development of confidence limits for any software package for which such a set of data was collected. Furthermore, some attempt at prediction of future software failures could be made. These two techniques, i.e. model development and prediction, find their place in an overall system of ensuring software reliability illustrated schematically by figure 1. If there exists an organized software quality control system, then software failure data should be available for analysis. The first step is software failure classification, since much can be deduced from the occurrence of repetitive bugs and corrective action can be taken by changing operating and quality control procedures. The next step would be model development and testing, leading to prediction and confidence limits. This, in turn, results in additional remedial actions through development of better procedures, thus closing the loop.

Needless to say, current practice seldom, if ever, follow such a system. The concern of McDonnell Douglas Astronautics Company in software reliability resulted in the simultaneous institution of several research projects. These are listed in table 1. Project 2 deserves further elaboration, since it is an attempt to use an automatic tool to ensure 100% software reliability.

PROGRAM TESTING TRANSLATOR

If it were possible to have a computer program that would provide a tool to check out any computer program by exercising all the options and all the branches through all the ranges of values, then our work would consist only of supplying parameters to such a program and the computer would do the rest. Unfortunately, this is not possible because even with our marvelous computers with a nanosecond cycle the number of combinations involved in any average-size software system exceeds 10^{23} and the processing time, therefore, reaches an astronomical amount (10^{14} seconds or 100 teraseconds) of processing time. If only some assistance were provided it would certainly improve computer program checkout. A program testing translator, such a tool, is being developed at McDonnell Douglas Company; the proposed manner of operation will be to merge the program testing translator with the program to be checked out, using the metacompiler, and then have the programmer supply ranges of values to test it. At the present time the program is designed to work with Fortran programs only. It prints out how many times each branch was executed. It counts how many times DO loops have been executed, also simple GO TO or complex GO TO on each of the branches. It does a number of similar counts on IF statements, subroutines, etc. From a programmer's point of view, at the end of this check he will get a good idea of which branches of his program are checked out for a given range of values; this at least is a big step towards ensuring reliability.

SOFTWARE RELIABILITY RESEARCH OBJECTIVES

The original research assignment was simply stated and required the researcher to itemize factors that contribute towards software unreliability, to provide means for assessing these factors and to design procedures for programming of software that will decrease the probability of software malfunctions. When these factors are reviewed they usually start with incomplete or insufficient user requirements and continue with a variety of reasons such as faulty systems design, lack of experienced programmers, lack of programming management, lack of documentation, faulty systems integration, lack of (or insufficient) user training, hardware/software interface, operating problems and many others. The assessment of some or all of these factors would, by necessity, start with the purification of data, classification of software failures, determination of parameters, application of statistical techniques, model development and testing. The results of these analyses would then end in recommendations for new procedures. Nothing could be simpler, the way the assignment sounded.

SOURCES OF DATA

The first stumbling block occurred when the search for software failure data began. It was surprising to find that there was no such thing as software failure data collection for analysis' sake. The obvious place to look was inside the company where many software systems are developed; strangely enough, nobody is interested in software failures, though everybody is concerned with software reliability. It may be a paradox, but most 'bugs' are fixed without failures being documented, and yet perfect software performance is required. After a long and fruitless search for sources of data, two sources became available, by courtesy of NASA and the U.S. Navy. NASA anomaly reports contained symptoms of problems occurring in Apollo software and U.S. Navy data contained trouble reports on one of their tactical software problems; therefore, symptoms rather than software failures were reported and thus the first task, that of classifying failures, became almost impossible.

Raw data from NASA are in the form of software anomaly reports on five groups of programs: Colossus (110 reports), Luminary (90 reports), Sundance (72 reports), Sundisk (17 reports) and Comanche (10) reports. Each anomaly report contains information which describes the error, the 'environment' or type of run being employed when the anomaly occurred, and a detailed analysis. The analysis isolates the cause of error, how it became manifest, its effect on the mission, avoidance procedures which might be used, recovery, or bypass procedures, and a statement of what is required in the program to resolve the difficulty. After subsequent patches are made the closing action taken is recorded.

Details of the NTDS are classified, so that only refined or summary data can be obtained at present. However, tab runs obtained from the Navy provide cryptic descriptions of the symptoms but complete identification of the source, time of occurrence, severity, and time to fix for any system error. The so-called 'trouble reports' have been encoded and placed in a file on magnetic discs, and cuts through the data bank are easy to obtain.

Each trouble report contains information on the following: the origin or discovery time, the name of the module which has the error, the severity of the error, the category or type of trouble (31 possible choices), the machine status with respect to mode of operation, time from run start, etc. Also kept is a record of the corrections scheduled and applied, and the station responsible for implementing corrections. The original trouble reports are generally classified and are unavailable at the present time.

Summaries of the trouble report status can be obtained on inquiry through a keyboard request to the data bank. These printouts list, by program modules, the total number of TRs charged, factored into those open and those closed out (by corrective action). The totals are further broken down by severity code (high, medium, low).

MODEL DEVELOPMENT

The homogeneous Poisson point process holds a central position in hardware reliability studies. Experience shows that for many components and systems this mathematical model is appropriate for the mid-phase of maturity, where the mid-phase is defined as the period of time between the infant phase and the wear-out phase. There is presumably a very large collection of independent failure-producing elements in the component or system, and the rate of occurrence of these failures is essentially constant in time.

Software errors or malfunctions do not, at first, seem to be of the same kinds that occur in hardware: there is no physical mechanism for generation of a software failure; the failures do not appear to be random in any active sense; and once an error has been found it is in all likelihood removed from the pool of potential failure-producers.

Another view of software failures, one which emphasizes their passive instead of their active roles, makes the hypothesis of randomness more acceptable. If it is imagined that a large utility program is resident in a computing system and is servicing a steady stream of dissimilar computing jobs, each with its own peculiar demands on the master program, the jobs can be considered as entering at arbitrary points in the master utility program, with attendant opportunities to detect deficiencies in the utility program. In this concept, that of passive deficiencies being uncovered by the variety of jobs handled by the master, the effect is that the errors are detected in a random way.

Large real-time control programs, such as those used on board Apollo flights, can be viewed in a somewhat similar way. Many errors which exist in such a software system are detected only when a peculiar combination of requirements is imposed or a novel demand is made.

Thus, while there may be no mechanism for generation of a software anomaly, there are good reasons for believing that in the larger view they can be considered as being detected in a random way; the net effect is the same as if there were a mechanism for their generation.

The assumption of a uniform failure rate is unrealistic for software failures, since once failures have been detected and corrected, the number of failures in the software package is reduced. On the other hand, over periods of time between the detection of successive errors, the assumption of uniformity may have merit.

The basic model for describing the detection of software failure is indicated in figure 2. The failure rate at any time is assumed to be proportional to the current error content of the tested program. The initial error content is then denoted by N and the proportionality constant is denoted by ϕ; the failure rate drops to $(N-1)\phi$ after the first error is detected, and so forth. A typical realization of such a process is depicted in the lower portion of the figure, where the increasing time between errors is purposely indicated by spacing. The term 'de-eutrophication' would seem to be appropriate in describing such a stochastic process.

The parameters N and ϕ can be estimated by application of the maximum likelihood principle. Let $X_1, X_2, \ldots, X_n$ denote the sample of time intervals between successive errors. Under the assumption that there is a uniform failure rate between errors, as indicated above, the density for X_i is given by

$$p(X_i) = \phi[N-(i-1)] \exp\{-\phi[N-(i-1)]X_i\}$$

and the likelihood function is

$$L(X_1,X_2,\ldots,X_n) = \prod_{i=1}^{n} \phi[N-(i-1)] \exp\{-\phi[N-(i-1)]X_i\}$$

It is more convenient to maximize the logarithm of the likelihood:

$$\log_e L = \sum_{i=1}^{n} \log\{[N-(i-1)]\phi\} - \sum_{i=1}^{n} (N-(i-1))\phi X_i$$

$$= \sum_{i=1}^{n} \log\{N-(i-1\}+n \log\phi - \sum_{i=1}^{n} (N-(i-1)\phi X_i$$

Taking partials and setting to zero, we obtain

$$\partial \log L/\partial N = \sum_{i=1}^{n} \frac{1}{N-(i-1)} - \sum_{i=1}^{n} \phi X_i = 0$$

$$\partial \log L/\partial \phi = \frac{n}{\phi} - \sum_{i=1}^{n} (N-(i-1))X_i$$

Letting $\sum X_i = T$, we can solve the second equation for ϕ to yield

$$\phi = \frac{n}{NT - \Sigma(i-1)X_i}$$

and the first equation becomes

$$\sum_{i=1}^{n} \frac{1}{N-(i-1)} = \frac{nT}{NT-\Sigma(i-1)X_i} = \frac{n}{N-\frac{1}{T}\sum_{i=1}^{n}(i-1)X_i}$$

The last equation is free of ϕ and presents the key equation to solve. The two data-derived statistics are $T = \sum X_i$ and $\sum(i-1)X_i$. Knowing these, we can solve

$$\frac{1}{N} + \frac{1}{N-1} + \ldots + \frac{1}{N-(n-1)} = \frac{n}{N-\frac{1}{T}\sum_{i=1}^{n}(i-1)X_i}$$

Solving this for N and calling the result $\hat{N}$, we can obtain the estimate for $\hat{\phi}$:

$$\hat{\phi} = \frac{n}{\hat{N}T - \sum(i-1)X_i}$$

APPLICATION

Because of the succinct and highly esoteric terminology employed in describing the NTDS software deficiencies (to a degree this same statement applies to the NASA anomaly reports), the nature of the particular error or group of software errors which cause the anomaly is very difficult or impossible to determine. Many of the trouble reports or software anomaly reports are generated whenever a system-level symptom of a deficiency is noted by operators or users, and further trace back to the exact cause in software of this symptom is always impossible for the NTDS data and sometimes very difficult for the NASA data.

In consequence the analysis was focused on the time pattern of the detection of system-level symptoms of deficiencies and not on the software errors as such.

Cursory looks at the NTDS and NASA data showed that while most of the symptoms were reported individually, as indicated by the dates of the report, there were many days in which there were many errors reported. The cryptic nature of the text describing the troubles was not conducive to identification of the dependencies between errors, but in all likelihood, even with much clearer description of the symptoms, it would require a much deeper knowledge of the systems involved than the analysts possessed to link up errors in a dependent chain.

As an expediency, only the dates on which one or more errors occurred were employed; these are denoted as system stop dates. This approach, which was forced by the nature of the data, has merit when the purpose of the analysis is examined: a means is desired of predicting the length of time the system is free of stops - the number of software anomalies uncovered at a stop is really unimportant.

With this change in view, the observables X_i of our model will represent the time between the <u>ith</u> and <u>i+lst</u> system stop; the interpretation of N and ϕ is correspondingly modified.

The NTDS software consisted of some 38 different modules which were developed by several contractors adhering to different project schedules. Each module is supposed to follow an evolution indicated by three stages: the production (or development) phase, the test phase, and the user phase. It should be stated that many of the modules did not evolve in the fashion indicated. Nonetheless, it seemed better to analyze isolated modules than the total system, since the structure of the latter is impossible to define, consisting at any time of the large number of modules in various stages of their evolution.

The modules with the largest number of trouble reports were, of course, prime candidates for analysis. Many of these were rejected because they did not have the desired chronological sequencing, which is: production errors preceding test errors which precede user errors.

Two of the larger modules, the AM module and the TW module, had the desired pattern. The trouble reports which comprised the record of system stops for the AM module were analyzed by applying the maximum likelihood procedure to the observations made during the production phase. There were 26 stops in this phase. The weighted sum of the interval lengths divided by the total time was 16.032. By using these two sample-derived numbers a formula F(N) can be obtained from the key equation. A plot of F(N) versus N is shown in figure 3. The only case which is meaningful is at N = 31.1. Thus the residual error or system stop content is estimated to be 5.1.

Five additional stops for this module were found during the test phase. With this additional set of observations a new estimate, N = 31.6, is obtained. The estimated time to the detection of the "next" error is about 239 days.

The results of this and similar analyses applied to data for two other modules are shown in figure 4.

Similar analyses were made of some of the NASA data, with results shown in figure 5. In this figure a simple extension of the results is made, to illustrate the real purpose of the analysis. Forecasts for the period of error-free performance of a mission are made, based on the indicated failure rate at the end of software testing. This period is synonymous with the term 'successful' in the present context.

REFINEMENTS

One of the tacit assumptions in the model described above is that all time periods of the same length represent equal intensities of testing. In reality this is seldom the case, and adjustments have to be made. The probability of detection of an error is related to the product of the failure rate and what will be called, for lack of a better term, the exposure rate. The latter is supposed to reflect the intensity of testing both in hours per day and in the number of different sites which are independently testing or using the same software.

Let E(t) be the exposure rate for the process. Rather than using the product $E(t)\times(N-i)\phi$ as the failure rate for the ith error, an alternate approach is suggested. The approach is to employ E(t) as a normalizer of time; more specificially, E(t) = 1 for any period of normal exposure. A normal exposure, of course, is defined in terms of the specific problem at hand: for example, the normal situation might correspond to the development, testing or use of a particular program (or component or module of this program) for a one-hour period per day by a single computer. The test or use of the same program on a particular day for a two-hour period by one computer and a three-hour period by another would correspond to an exposure which is five times the normal; correspondingly, the magnitude of E(t) would be 5 for such a day. An error-free day of this type under ordinary considerations would be counted as equivalent to five normal days.

There are two ways to proceed with the analysis in the case in which there is a variable exposure rate. In the commonly used method, the Poisson process which serves as a model is considered non-homogeneous in time; that is to say, the failure rate (or exposure rate) is a time-dependent quantity. In such a representation the probability density for the length of time between the i-1st and ith failures in the de-eutrophication process would be

$$p(X_i) = E(t_{i-1}+X_i)[N-(i-1)]\phi \exp\{-\int_{t_{i-1}}^{t_{i-1}+X_i} E(u)[N-(i-1)]\phi du\}$$

where the function E(t) is the exposure rate, $[N-(i-1)]\phi$ is the failure rate and X_i are the units of time. This representation presents an analytically very untractable model and it is necessary to employ some variant in order to proceed. One method which has been applied in the very special case of a periodically varying failure rate involves essentially averaging the probability density over one cycle; of course, this procedure is not generally applicable.

A universally applicable procedure which involves averaging can be carried out after the times of a set of failures have been recorded. Since the unknowns ϕ and N are time-independent and are not affected by the averaging, it can be carried out over the intervals between failures. At best, however, this is cumbersome.

Instead of representing the non-homogeneity in the process in terms of a variable failure rate, it can be effected by transforming the time scale in accordance with the variable exposure rate. More specifically, a new time τ is constructed through the change in variables

$$\tau = F(t) = \int_0^t E(u)du$$

where τ is a monotone increasing (or, more precisely, a non-decreasing) function of time. On intervals where $E(t) > 0$, the function $F(t)$ is strictly increasing and there is a one-to-one correspondence between the variables t and τ. On intervals where $E(t) = 0$, the function $F(t)$ is constant and all time points in the interval are associated with a single value of τ. This is as it should be, of course, since if there is no exposure on the interval there is no liability for the process to produce a failure.

The modified model then can be represented by the expression

$$[N-(i-1)]\phi\, e^{-[N-(i-1)]\phi X_i^1}$$

where

$$X_i^1 = \tau_i - \tau_{i-1} \quad , \quad \tau_i = F(t_i) = \int_0^{t_i} E(u)du$$

where all of the terms have been described above. Thus the same basic model and the operations which were carried out with it apply, with the difference that $\tau_i-\tau_{i-1}$ replaces t_i-t_{i-1}. The maximum likelihood procedure will produce the equations given in the preceding section, again with X_i^1 replacing X_i.

MODULARITY AND RELIABILITY

In addition to model development, a brief look was taken at the impact of modularity on software failures. From the U.S. Navy system, designed in modular fashion, it appeared that whereas the number of failures per module diminished progressively through the three stages - production, test and evaluation, and users' operation - failures ascribed to module integration increased significantly through test and evaluation and users' operation. In trying to determine the cause of this, at least one reason came to light: that programmers coding separate modules do not talk to each other and therefore their interpretation and execution of module specification differ, causing integration problems and subsequent failures. Although further investigation seems necessary, one solution presents itself immediately, that is: instead of horizontal control over a period of time such as preliminary design review, detailed design review, etc., some vertical control should be introduced so that communications between modules developers take place.

CONCLUSION

In this study a look was taken at software reliability, and several avenues of investigation were explored. A specific effort was made to analyze both NASA anomaly reports and U.S. Navy trouble reports. The attempt to classify software failures based on these sources of data was not successful; however, a software failure predictive model was developed and some surprising results were obtained. It is realized that predictive techniques depend on sources of good data and therefore that areas of future research should include an organized data collection of software failures which would be enforced by stringent quality control procedures. Such data should be simple to collect and should at the same time contain information which would allow the inclusion of various parameters such as programmer's experience, number of program instructions, computer run time, etc. It is felt that only then can a precise predictive model be developed, and that its application would result in a tool for measuring the 'goodness' of software; this would, in turn, result in more reliable software.

REFERENCES

1. F.J. Buchley: Verification of software programs, Computers and Automation, February 1971.
2. M.J. Ratynski: The Air Force computer program acquisition concept. SJCC, 1967.
3. J.V. Schnurbush: Verification of the Apollo real-time mission program, AIAA Aerospace Computer Systems Conference, 1969.
4. O.L., Williamson, G.G. Dorris, A.J. Ryberg, and W.E. Straight: Development studies of a software reliability program, NASA report, March 1970.
5. B.W. Boehm: Some information processing implications of Air Force space missions in the 1970's, Astronautics and Aeronautics, January 1971.
6. Donald K. Slayton: Review of shuttle integrated avionics, NASA, March 1971.
7. R.L. London: Software reliability through proving programs correct, 1971 International Symposium on Fault-Tolerant Computing (IEEE), March 1971.
8. J.C. King, Proving programs to be correct, 1971 International Symposium on Fault-Tolerant Computing (IEEE), March 1971.
9. B. Elspas, K.N. Levitt, A. Waksman, and R. Waldinger: A comparison of formal program validation techniques, 1971 International Symposium on Fault-Tolerant Computing (IEEE), March 1971.

Table 1
Related software reliability research and development

	Title	Objective
1.	Software reliability	To study factors contributing to software unreliability and to decrease software malfunctions, with emphasis on manned rating and defence systems.
2.	Program translator	To design and implement a computer testing tool that aids in program acceptance testing and thus improves software reliability.
3.	Onboard checkout software design concepts	To derive software requirements applicable to inflight maintenance operations for manned spacecraft.
4.	Quality engineering	To investigate current practices of software quality and assurance recommendations for improved procedures to ensure reliable software.
5.	Software validation	To recommend how to improve program debug and test, to reduce development time, and to develop methods for estimating confidence factors for program operation.

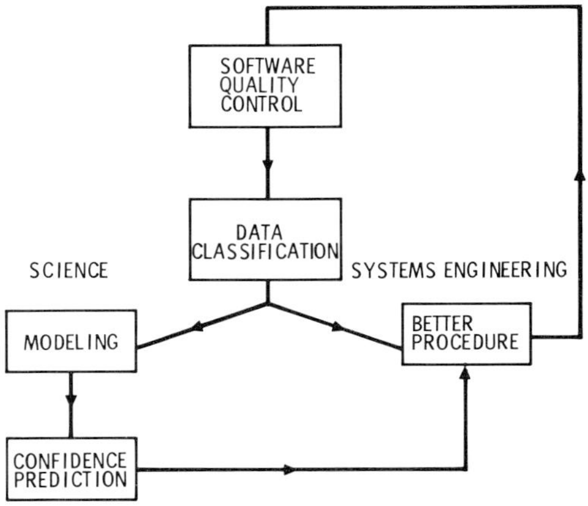

Fig. 1. Software reliability system schematic.

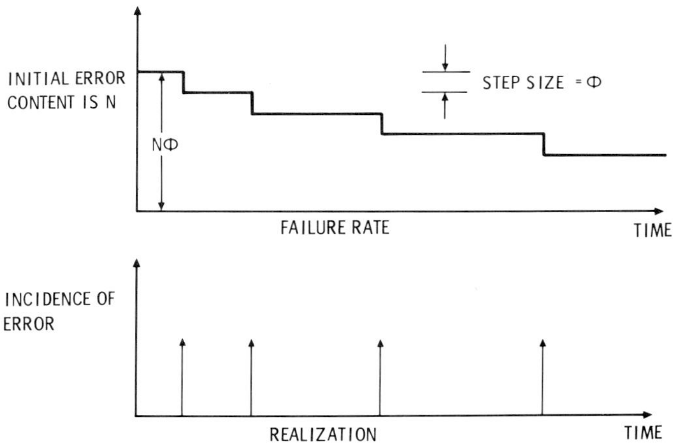

Fig. 2. De-eutrophication process.

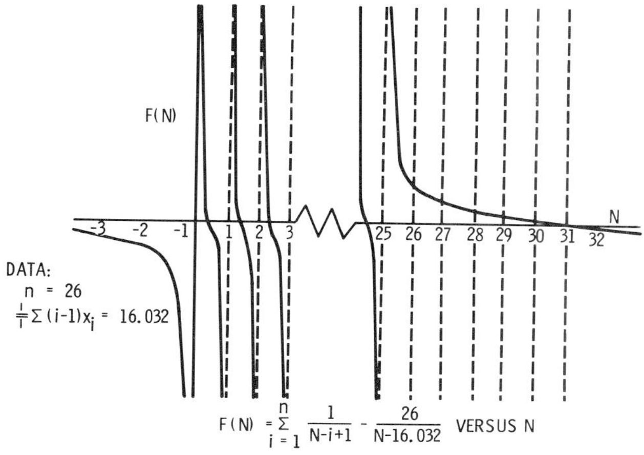

Fig. 3. Solutions to the key equation.

I. AM-MODULE
- OBSERVED 26 SYSTEM STOPS DURING PRODUCTION (CHECKOUT PHASE
- ESTIMATED INITIAL ERROR CONTENT 31.1 (5.1 ADDITIONAL)
- 5 ADDITIONAL ERRORS WERE FOUND IN THE TEST PHASE
- SECOND ESTIMATE OF ERROR CONTENT IS 31.6
- ESTIMATE OF AVERAGE TIME TO DETECTION IS 239 DAYS

II. TW-MODULE
- OBSERVED 14 SYSTEM STOPS DURING THE TEST PHASE
- ESTIMATED 17.4 ERRORS INITIALLY
- 1 ADDITIONAL ERROR FOUND IN USER PHASE
- SECOND ESTIMATE 17.1 ERRORS (2.1 REMAINING)
- ESTIMATE OF AVERAGE TIME TO NEXT ERROR IS 89 DAYS

III. EW-MODULE
- OBSERVED 12 STOPS MIXED PHASES (7/69 TO 12/70)
- ESTIMATED 14.45 ERRORS INITIALLY
- 1 ERROR FOUND IN 1971
- NEW ESTIMATE 14.54 ERRORS
- ESTIMATE OF AVERAGE TIME TO NEXT ERROR IS 207 DAYS

Fig. 4. Applications to NTDS data.

I. SUNDANCE
- OBSERVED 41 SYSTEM ANOMALIES (4/23/68 TO 1/17/69)
- ESTIMATED INITIAL ERROR CONTENT: 45.8 ERRORS
- ESTIMATED AVERAGE TIME TO NEXT ERROR IS 19.4 DAYS
- PROBABILITY OF SUCCESSFUL 8-DAY MISSION: 66.2%

II. COLOSSUS
- OBSERVED 42 SYSTEM ANOMALIES (8/27/68 TO 2/27/70)
- ESTIMATED INITIAL ERROR CONTENT: 54 ERRORS
- ESTIMATED AVERAGE TIME TO NEXT ERROR IS 10.3 DAYS
- PROBABILITY OF SUCCESSFUL 8-DAY MISSION: 46.0%

III. LUMINARY AND LUMINARY 1B
- OBSERVED 48 SYSTEM ANOMALIES (10/1/68 TO 9/30/69)
- ESTIMATED INITIAL ERROR CONTENT 200,000

Fig. 5. Applications to NASA data.

PROBABILISTIC MODELS FOR SOFTWARE RELIABILITY PREDICTION

Martin L. Shooman

Summary. With the advent of large sophisticated hardware-software systems developed in the 1960s, the problem of computer system reliability has emerged. The reliability of computer hardware can be modeled in much the same way as other devices using conventional reliability theory; however, computer software errors require a different approach. This paper discusses a newly developed probabilistic model for predicting software reliability. The model constants are calculated from error data collected from similar previous programs. The calculations result in a decreasing probability of no software errors versus operating time (reliability function). The rate at which reliability decreases is a function of the man-months of debugging time. Similarly, the mean time between operational software errors (MTBF) is obtained. The MTBF increases slowly and then more rapidly as the debugging effort (man-months) increases. The model permits estimation of software reliability before any code is written and allows later updating to improve the accuracy of the parameters when integration or operational tests begin.

INTRODUCTION

During the decade of the 1940s computers were born. In the 1950s commercial hardware and assembly programs became available. In the 1960s programming languages became popular, complex operating systems emerged, hardware became huge and sophisticated, large programs consisted of several hundred thousand words of code. In the 1970s we begin to speak of automatic programming, huge virtual memory and interconnecting networkds of computers. Along with these new developments one begins to speak of programs containing millions of words of code and performing huge and complex real-time tasks.

Along with this growth has come a realization that the largest effort in developing software is due to software integration, test, correction, retest, operational release, correction and rerelease. Actual writing of the first set of code is a small task in comparison. An even more compelling observation is that computers are increasingly being used as the heart of complex real-time systems such as air traffic control, vehicle control, space systems, and military systems. System reliability is the most important performance measure in such systems. After all, if a large batch computer crashes occasionally it probably only means one rather than three turnarounds that day. Similarly, if a time-sharing system goes down infrequently for half an hour, 50 annoyed users have an unanticipated coffee break. It is a far different story if the air traffic control handling the metropolitan New York area crashes on a stormy night under saturated conditions, and requires a several-minute system reloading delay before targets again appear on the video output terminals! In any of these situations, it is clear that a vital measure of computer system performance is both the hardware and software reliability.

The approach used in this paper is to study the debug history of previous software programs similar to the one in question. From these one determines constants for the model and makes predictions in the initial planning stages of the system reliability and required debugging effort. As the project proceeds and error data become available, these can be combined with the prior estimate to improve the accuracy of the prediction. It is important to identify at what point in the software development process we can stochastically model the error behavior and expect some sort of repeatable pattern or group of patterns. Certainly when 1-3 programmers get together to write a few thousand words of code for some program module, the various programming styles, changing specifications, early syntactical errors, and lack of formal documentation make this an inappropriate starting point. The earliest one could expect such patterns to emerge is during process integration and test. During this phase, several or many program modules are combined to form a single process (thread) which may be tens of thousands of words long. An even more suitable place to look for characteristic debugging behavior is during system integration and test when many processes are combined to form the overall program which may contain several hundred thousand or million words of code. Data are generally available at this point because, either here or earlier, configuration control has been instituted, where one group maintains the master program and only inserts changes periodically to introduce an accumulated number of written and approved change requests. The final step in testing is to place the program in simulated and/or real operation. In a compiler program this corresponds to manufacturer's release. Anyone who has worked as a systems programmer, or who has talked with one, knows that this supposedly perfect program now develops a completely new set of bugs induced by the unique characteristics of actual operation which were not included in previous tests. One would expect similar development phases (perhaps with process integration eliminated) and error patterns in smaller programs.

ERROR MODEL

Basically, the error model used in this paper assumes that the total number of errors in the program is fixed and that if we record the cumulative number of errors corrected during debugging, then the difference represents the remaining errors. The following section on reliability models will relate the probability of encountering a software bug to the number of residual bugs, the total number of instructions and the instruction processing rate.

We are looking for error behaviors during debugging which will possess some generality for both large and small programs. Thus, rather than plotting the error rate, we will plot a normalized error rate

$$\rho(\tau) = \text{errors/total number of instructions/month of debugging time} \tag{1}$$

In Table 1 are listed error data for seven large supervisory programs (operating systems) and applications programs (real-time computer software) collected by Dickson, Hesse and Kientz [1]. Since some of the manufacturers considered these data proprietary, the data are referred to by code name. In figures 1 and 2 the normalized error rate $\rho(\tau)$ is plotted as a function of τ, the number of months of debugging after release or the beginning of system test. Although several curve shapes might be fitted to these data, one characteristic is common for all curves. The normalized error rate decreases over the entire curve, or at least over the latter two-thirds or half of the curve. Initial behavior of $\rho(\tau)$ differs from example to example in figures 1 and 2. The curve shapes will be discussed further in the section on varying-manpower models.

Since we are interested in the total number of errors removed, we will define a cumulative error curve, $\varepsilon(\tau)$, which is the area under the $\rho(\tau)$ curve:

$$\varepsilon(\tau) = \int_0^\tau \rho(x)dx = \text{cumulative errors/total number of instructions} \tag{2}$$

and $\rho(\tau)$ is of course the slope of the $\varepsilon(\tau)$ curve:

$$\rho(\tau) = d\varepsilon(\tau)/d\tau \tag{3}$$

A curve of the cumulative error data for the supervisory system A of figure 1 is shown in figure 3. If similar curves for $\varepsilon(\tau)$ were drawn for the other six examples of figures 1 and 2 all would build up initially with a constant or increasing slope and then decreasing curvature, appearing to become asymptotic. The smoothing nature of integration makes all the $\varepsilon(\tau)$ curves look more alike than do the $\rho(\tau)$ curves. Both are needed for detailed study.

If we assume that the total number of errors in the program E_T is constant and that the program contains I_T instructions, then the asymptote which the $\varepsilon(\tau)$ curves approach is E_T/I_T. If we assume that all detected errors are corrected errors then by inspection of figure 3 we can write an expression for the number of residual errors:

$$\varepsilon_r(\tau) = (E_T/I_T) - \varepsilon_c(\tau) \tag{4}$$

We assume that in any sizeable program it is impossible to remove all errors, so

$$\varepsilon_c(\tau) < E_T/I_T \tag{5}$$

$$\varepsilon_r(\tau) > 0 \tag{6}$$

The error model developed above will be used in the following section to formulate a reliability model; however, it has uses in its own right. It allows us to describe in a quantitative fashion the debugging of a program. Thus, we can use this model to study some of the following points:

1. Is the ratio E_T/I_T relatively constant for similar classes of programs?
2. How do manpower deployment techniques affect the debugging process?
3. How effective are different test schemes and debugging philosophies?

RELIABILITY MODEL

In order to formulate a reliability model, we must propose a hypothesis as to how software errors occur.

We assume that operational software errors occur due to the occasional traversing of a portion of the program in which a hidden software bug is lurking. We begin by writing an expression for the probability that a bug is encountered in the time interval Δt after t successful hours of operation. This must be proportional to the probability that any randomly chosen instruction contains a bug, i.e. the fractional number of remaining bugs $\varepsilon(\tau)$, and the rate of instruction processing r_p. However, we must account for the fact that sometimes when a program is looping each repetition does not always represent new instructions. If data into the loop are changing for each repetition, then we would say that a 25-instruction loop repeated 10 times represents 250 unique instructions. On the other hand, if the loop were just an idle loop waiting for a flag to be raised indicating a new input, then only one repetition, 10 instructions would be counted as unique. We must now calculate a probability function from the above model.

From a study of basic probability and reliability theory [1,2] we learn that the probability of failure in time interval t to t + Δt, given that no failures have occurred up till time Δt, is proportional to the failure rate (hazard function) Z(t):

$$P(t<t_f\leq t+\Delta t \mid t_f>t) = Z(t)\Delta t = K\varepsilon_r(\tau)r_p\Delta t \qquad (7)$$

where $t_f \equiv$ time to failure (occurrence of a software error), $P(t<t_f\leq t+\Delta t \mid t_f>t) \equiv$ probability of failure in interval Δt, given no previous failure, r_p = unique instruction processing rate and K = an arbitrary constant. This arbitrary constant K is introduced as a "bulk" constant to account in an appropriate way for certain structural aspects of the program. Just as the definition of "unique instructions processed" tends to account roughly for the effects of real-time inputs and structural interaction, the constant K roughly "scales" the effect of an instructional error on program success. For example, if one-quarter of I_T were useful but nonvital, K = 0.75. The constant K may be estimated by carefully examining the history of detected errors to determine for each error if it is catastrophic to program success:

$$K = \frac{\text{number of catastrophic errors detected}}{\text{total number of errors detected}} \tag{8}$$

From reliability theory [2] we can show that the probability of no system failures in the interval 0 to t is the reliability function, which is related to the hazard function by

$$R(t) = \exp\left(-\int_0^t Z(x)\,dx\right) \tag{9}$$

If we substitute our expression for Z(t) from Eq. (7) into Eq. (9) and assume K, $\varepsilon_r(\tau)$ and r_p are independent of operating time, we obtain

$$R(t) = \exp(-(K\varepsilon_r(\tau)r_p)t) = \exp(-\gamma t) \tag{10}$$

Basically, the above equation states that the probability of successful operation without software bugs is an exponential function of operating time. When the system is first turned on, t = 0 and R(0) = 1. As operating time increases the reliability monotonically decreases, as shown in figure 2. We depict the reliability function for three values of debugging time, $\tau_0<\tau_1<\tau_2$.

In figure 4 the values of τ_0, τ_1 and τ_2 are chosen so that $\varepsilon_r(\tau_0) \simeq 2\varepsilon_r(\tau_1) \simeq 4\varepsilon_r(\tau_2)$. From this curve we may make various predictions about the system reliability. For example, looking along the vertical line $t = 1/\gamma$ we may state:

1. If we spend τ_0 hours of debugging then $R(1/\gamma)\simeq 0.35$.
2. If we spend τ_1 hours of debugging then $R(1/\gamma)\simeq 0.50$.
3. If we spend τ_2 hours of debugging then $R(1/\gamma)\simeq 0.75$.

A simpler way to summarize the results of the reliability model is to compute the mean time to (software) failure, MTTF, for the system [2]:

$$\text{MTTF} = \int_0^\infty R(t)\,dt \tag{11}$$

Substitution from Eq. (10) into Eq. (11) yields

$$MTTF = \int_0^\infty \exp(-\gamma t)\, dt = 1/\gamma \tag{12}$$

We may now study the behavior of the MTTF by substituting Eqs. (2), (4) and (10) into Eq. (12):

$$MTTF = 1/Kr_p[\frac{E_T}{I_T} - \int_0^\tau \rho(x)dx] \tag{13}$$

If we let $\rho(\tau)$ be modeled by a constant rate of error correction ρ_0 then

$$MTTF = 1/(Kr_p[\frac{E_T}{I_T} - \rho_0\tau]) \tag{14}$$

$$= 1/(\beta[1-\alpha\tau)] \tag{15}$$

where $\beta = (E_T/I_T)r_pK$ and $\alpha = (\rho_0 I_T)/E_T$.

If a triangular model is chosen for $\rho(\tau)$ where ρ builds up linearly to ρ_p from the origin in τ_p seconds and decays linearly to zero between τ_p and $2\tau_p$ [1,3], then we obtain

$$\begin{aligned} \rho(\tau) &= (\rho_p/\tau_p)\tau && 0\leq\tau\leq\tau_p \\ &= 2\rho_p - (\rho_p/\tau_p)\tau && \tau_p\leq\tau\leq 2\tau_p \end{aligned} \tag{16}$$

and substitution into Eq. (13) yields

$$\begin{aligned} MTTF &= 1/Kr_p[\frac{E_T}{I_T} - \frac{\rho_p}{\tau_p}\frac{\tau^2}{2}] && 0\leq\tau\leq\tau_p \\ &= 1/Kr_p[\frac{E_T}{I_T} + (\rho_p\tau_p - 2\rho_p\tau + \frac{\rho_p}{\tau_p}\frac{\tau^2}{2})] && \tau_p\leq\tau\leq 2\tau_p \end{aligned} \tag{17}$$

$$\begin{aligned} MTTF &= 1/\beta(1-\delta\tau^2) && 0\leq\tau\leq\tau_p \\ &= 1/\beta[1 + \delta(2\tau_p^2 - 4\tau_p\tau + \tau^2)] && \tau_p\leq\tau\leq 2\tau_p \end{aligned} \tag{18}$$

where $\delta = (\rho_p/2\tau_p)(I_T/E_T)$.

Lastly, we derive the expression for MTTF for an exponentially decreasing model for $\rho(\tau)$ which is developed in the following section.

$$\rho(\tau) = (k_1 E_T/I_T)(\exp[-k_1\tau]) \tag{19}$$

Substitution in Eq. (13) yields

$$\text{MTTF} = 1/Kr_p \frac{E_T}{I_T} \exp(-k_1\tau) \tag{20}$$

$$= 1/\beta\exp(-k_1\tau) \tag{21}$$

In order to compare the three models developed above, we must plot each versus a normalized time scale (see figure 5). In addition, for the triangular model we have set the area under $\rho(\tau=\tau_p) = \rho_p\tau_p = E_T/I_T$, i.e. the system is completely debugged at $\tau = 2\tau_p$. We see that all three models rise with τ; however, both the constant and triangular models only show significant improvement in the last quarter of the debugging process.

EXPONENTIAL ERROR RATE MODEL

Examination of figures 1 and 2 shows a generally decreasing trend in error rate versus debugging time (except for the first month). In addition, it seems reasonable to postulate that the number of errors debugged per month should decrease as the number of errors present decreases. If we assume that the number of errors detected per month is proportional to the number present we obtain an exponential error detection rate.

If we set the number of errors detected per month $\rho(\tau)$ proportional to the number of residual errors $\varepsilon_r(\tau)$ (cf. Eq. (4)), we get

$$\rho(\tau) = k_1[(E_T/I_T) - \varepsilon(\tau)] \tag{22}$$

where k_1 is the constant of proportionality. Combining this with Eq. (3) yields

$$\rho(\tau) = \frac{d\varepsilon(\tau)}{dt} = k_1[(E_T/I_T) - \varepsilon(\tau)] \tag{23}$$

$$\frac{d\varepsilon(\tau)}{dt} + k_1\varepsilon(\tau) = \frac{k_1 E_T}{I_T} \tag{24}$$

Summing the homogeneous and particular solutions to this first-order differential equation, we obtain

$$\varepsilon(\tau) = \frac{E_T}{I_T} + A \exp(-k_1\tau)$$

Substituting the initial condition that at $\tau = 0$, $\varepsilon(0) = 0$, we obtain $A = -E_T/I_T$. Thus, the cumulative error function becomes the familiar exponential rise to the asymptote E_T/I_T:

$$\varepsilon(\tau) = \frac{E_T}{I_T}(1 - \exp(-k_1\tau)) \tag{25}$$

Differentiating Eq. (25), we obtain an exponential error rate

$$\rho(\tau) = \frac{k_1 E_T}{I_T}\exp(-k_1\tau) \tag{26}$$

VARYING-MANPOWER MODEL

In the statistics presented in figures 1 and 2, one essential ingredient is missing, i.e. the number of people involved in testing and the extent of their efforts. Ideally, we might wish to measure this in terms of the skill of the programmers and the efficiency of their techniques. An easier and possibly satisfactory expedient would be to merely record the number of manhours or manmonths of work expended on testing and correction and normalize both ρ and ε with respect to this factor. Thus we may define

$$\rho = \text{errors/instruction/month} = \frac{\text{errors}}{\text{instruction-month}} \tag{27}$$

$$\rho' = \text{errors/instruction/manmonths} = \frac{\text{errors}}{\text{instruction-man-months}} \tag{28}$$

which yields

$$\rho = M\rho' \tag{29}$$

where $M \equiv$ average number of men working on testing and correction during a month. (A more refined model would also include some measure of computer test time expended.)

If we postulate that ρ' should be roughly constant for similar sizes and classes of programs, then we can postulate that the shape of $\rho(\tau)$ in figures 5 and 2 is dependent on the assigned manpower. Thus, we could explain a curve which initially increased and then decreased in several ways:

1. The underlying behavior of $\rho'(\tau)$ is constant, but initially only limited manpower was assigned and was gradually increased. Later in the testing, after significant progress was achieved, the manpower was reduced and assigned to other projects.

2. The underlying behavior of $\rho'(\tau)$ is exponential, but initially only limited manpower was assigned. Once the manpower built up to its peak, it was held constant.

3. A variant of explanation 2 is to postulate that initially it is reasonably easy to formulate tests which reveal a large number of errors, but it requires a large amount of time to correct these errors. Thus, with limited manpower and a few scheduled batch runs per day, the programmer may still detect a large number of errors initially but only correct a portion of these. Since in all probability only the corrected errors rather than the detected errors are scored, this explanation would yield a constant saturation error correction rate for a period of time followed by an exponentially declining tail. The exponential tail in the error correction rate would begin when the initial testing group (alone or augmented by reinforcements) reduced the total of the backlog queue plus the newly arrived errors below their saturation level. The above explanation implies that the time behavior of the number of detected and corrected errors could differ appreciably.

One may quantitatively introduce the concept of varying manpower into the exponential error rate model of the previous section in a simple manner. If in Eq. (22) we include the fact that the detected errors should also be proportional to the manpower, we may write

$$k_1 = k_2 M \tag{30}$$

where k_2 is a new constant of proportionality. Thus Eqs. (25) and (26) become

$$\varepsilon(\tau) = \frac{E_T}{I_T} (1 - \exp(-k_2 M \tau)) \tag{31}$$

$$\rho(\tau) = \frac{k_2 M E_T}{I_T} \exp(-k_2 M\tau) \tag{32}$$

From Eq. (31) we see that the $\varepsilon(\tau)$ curve has the same basic behavior and asymptote, and only the time constant is changed. The $\rho(\tau)$ curve experiences both amplitude and decay rate changes when M is changed.

Further investigation of the validity of the models proposed in this section requires a detailed set(s) of debug statistics which include the differences (if any) between error detection and correction rate and the manpower assigned as a function of time.

EVALUATION OF MODEL CONSTANTS

The models proposed in this paper shed valuable light on the anatomy of the error debugging process and illustrate how the reliability functions and MTBF change with debug time and manpower. The final goal, i.e. obtaining a numerical estimate of the software reliability of a particular program, will not be met unless one is able to evaluate the model constants.

In order to be specific we focus our discussion on the constant error rate model developed in Eqs. (14) and (15). The constant I_T, the total number of program instructions, is well known in all cases and though it may increase as the program is developed this should cause no serious difficulty.

The constant K can be satisfactorily estimated using Eq. (32) and will in most cases be set to unity.

The constant r_p could be chosen in a number of ways. If d_f is the delay time for the fastest instruction, d_s the delay time for the slowest instruction and d_a the average delay time over the entire instruction set, then

1. If new data continuously flow into the program so that each new instruction is essentially "unique", we would bound r_p by $1/d_f$ and $1/d_s$. If the instruction mix of the program was roughly uniform (unlikely) we would set $r_p = 1/d_a$. Using judgment and some study of the code, it should be possible to adjust r_p up and down from $1/d_a$ to achieve a more representative value.

2. If new data flow in at intervals or only affect portions of the code, or if the program is in an idle loop, one

would have to perform a rough statistical average over the changed and unchanged code to obtain r_p. Some researchers in the field of compiler design have recommended that compilers produce frequency counts (number of times each instruction is executed) as an important statistic for the programmer (see [4]). The availability of such information would be extremely valuable in estimating r_p.

The constant ρ_0 would initially be evaluated from previous data and would be updated as the debugging progressed. It is easy to visualize a Bayesian strategy to accomplish this mixing of prior knowledge and experiment if mathematical sophistication is warranted. The estimation of E_T depends, again, on past data. In fact we can calculate ρ_0 and E_T/I_T for the seven examples given in Table 1. These are shown in Table 2.

Although there is too much homogeneity in size and type of these programs to draw any general conclusions, it seems that E_T/I_T varies between 0.5% and 1.5% while ρ_0 varies between .05%/month and .25%/month.

If we now switch our attention to the exponential model given in Eq. (26), the determination of parameters becomes quite simple. First of all, one would be able initially to determine E_T/I_T and k_1 from previous data, for an initial estimate. As data became available, we would plot ρ versus τ on log paper and use the slope to estimate k_1. Knowing this, the intercept can be used to determine E_T/I_T. Clearly, updated estimates can be obtained as the debugging progresses. Furthermore, it is reasonable to assume that if I_T grows as the integration progresses, then E_T should grow at roughly the same rate, so that E_T/I_T will stay roughly constant. If we include the effects of varying manpower as was done in Eq. (32), we need merely insert the known value of M or, if M is varying, to plot ρ versus $M\tau$ on log paper and proceed as before.

CONCLUSIONS

It has been shown that by studying the debug error rate versus time during system integration or user test of large software programs one can model the number of residual errors and the probability of encountering one. Three models (the constant error rate, triangular error rate, and exponentially decreasing error rate) have been developed which are logically satisfying and fit some of the preliminary data in in many respects. The one corrupting factor which must be studied is the manpower deployment pattern during test. It remains to apply the model to test the accuracy of its predictions to a situation in which debug and operational data are available. It is probable that a more refined model of system structure than that given in eq. (7) will be required for accurate reliability prediction.

REFERENCES

1. J. Dickson, J. Hesse, A. Kientz and M. Shooman, "Quantitative analysis of software reliability", 1972 Annual Reliability Symposium Proceedings, IEEE, January 1972.
2. Martin L. Shooman, Probabilistic reliability: an engineering approach, McGraw-Hill, New York, 1968.
3. Martin L. Shooman, "Preliminary concepts of software reliability", unpublished notes, Polytechnic Institute of Brooklyn, spring 1971.
4. Donald E. Knuth, "An empirical study of FORTRAN programs", Computer Science Dept. Report No. CS-186, Stanford University, 1970.
5. Bell System Technical Journal, December 1970: articles on testing of TSPS No. 1 and ESS No. 1 ADF.

TABLE 1

CHANGE DATA FOR SEVEN DIFFERENT PROGRAMS

Month	Application A 240,000 Inst. Changes	Application A Changes/Inst.	Application B 240,000 Inst. Changes	Application B Changes/Inst.	Application C 240,000 Inst. Changes	Application C Changes/Inst.	Application D 240,000 Inst. Changes	Application D Changes/Inst.
1	514	2.15×10^{-3}	905	3.76 x 10-3	235	0.98×10^{-3}	331	1.38×10^{-3}
2	926	3.85	376	1.57	398	1.66	397	1.66
3	754	3.15	362	1.51	297	1.24	269	1.12
4	662	2.76	192	0.80	506	2.11	296	1.24
5	308	1.28	70	0.29	174	.72	314	1.31
6	108	0.45			55	.23	183	0.76
					60	.25	158	0.66
							368	1.54
							337	1.41
							249	1.02
							166	0.69
							108	0.45
							31	0.13

Month	Supervisory A 210,000 Inst. Changes	Supervisory A Changes/Inst.	Supervisory B 240,000 Inst. Changes	Supervisory B Changes/Inst.	Supervisory C 230,000 Inst. Changes	Supervisory C Changes/Inst.
1	110	0.52×10^{-3}	250	1.04×10^{-3}	225	0.98×10^{-3}
2	238	1.14	520	2.16	287	1.24
3	185	0.88	430	1.80	497	2.16
4	425	2.02	300	1.25	400	1.74
5	325	1.55	170	0.71	180	0.78
6	37	.18	120	0.50	50	0.22
7	5	.02	60	0.25		
8			40	0.17		

Table 2 - Computation of Model Constants from the Data of Table 1

Program	E_T/I_T	ρ_o
Supervisory A	6.14×10^{-3}	0.875×10^{-3}
Supervisory B	7.97	0.996
Supervisory C	7.48	1.25
Application A	13.20	2.20
Application B	7.70	1.54
Application C	7.00	1.00
Application D	12.90	0.995

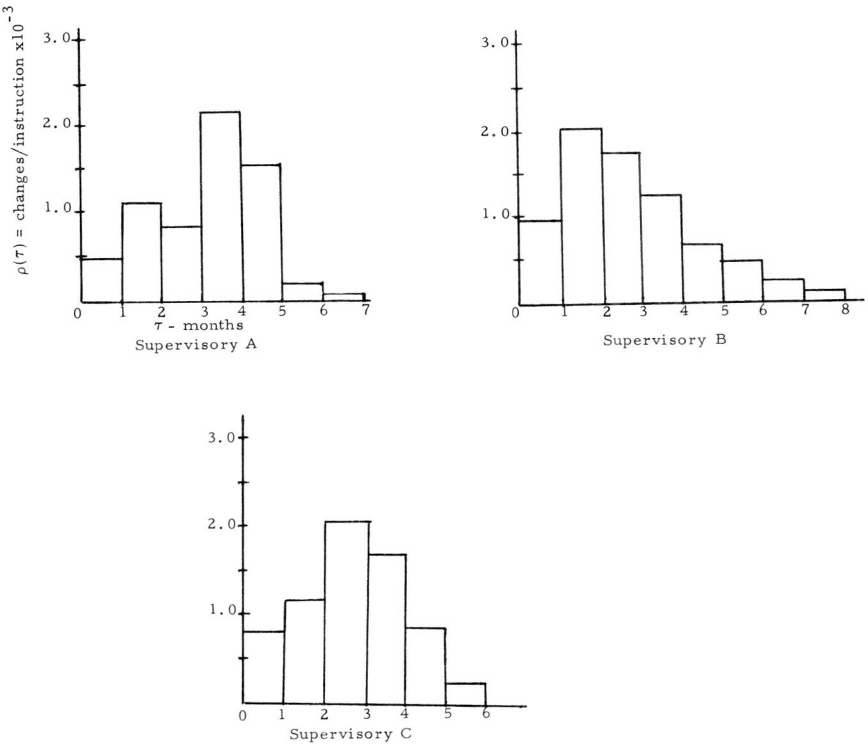

Fig. 1. Normalized error rate versus debugging time for three supervisory programs.

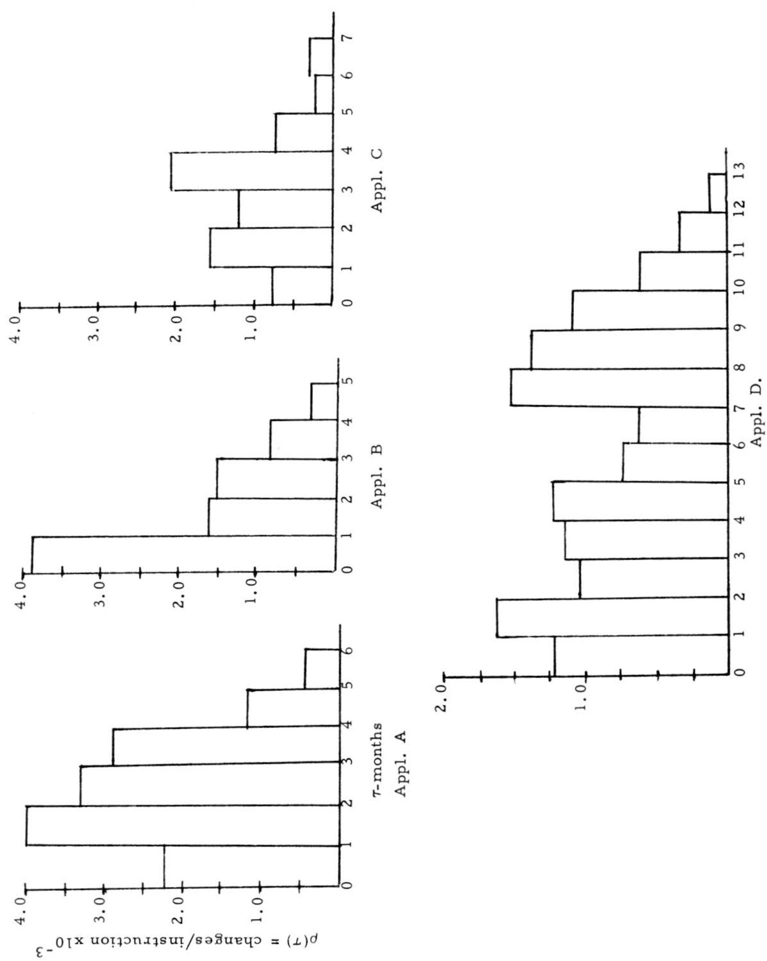

Fig. 2. Normalized error rate versus debugging time for four applications programs.

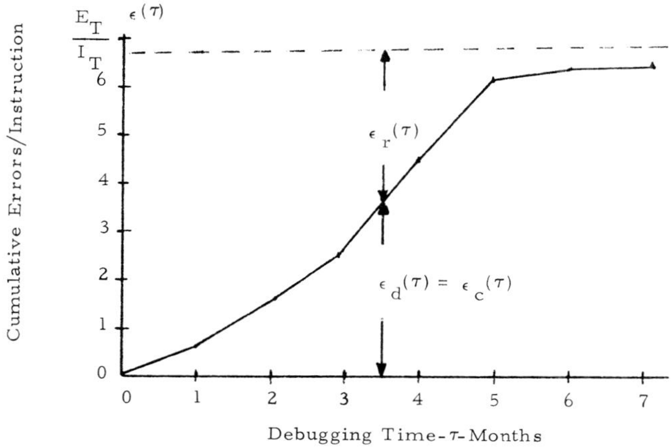

Fig. 3. Cumulative error curve for supervisory system A given in figure 1.

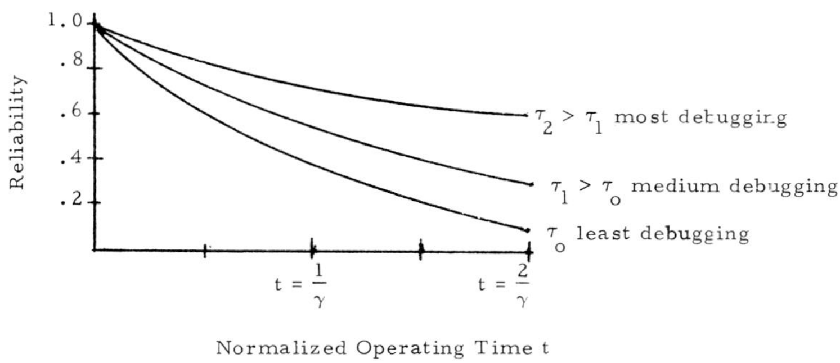

Fig. 4. Variation of reliability function R(t) with debugging time τ.

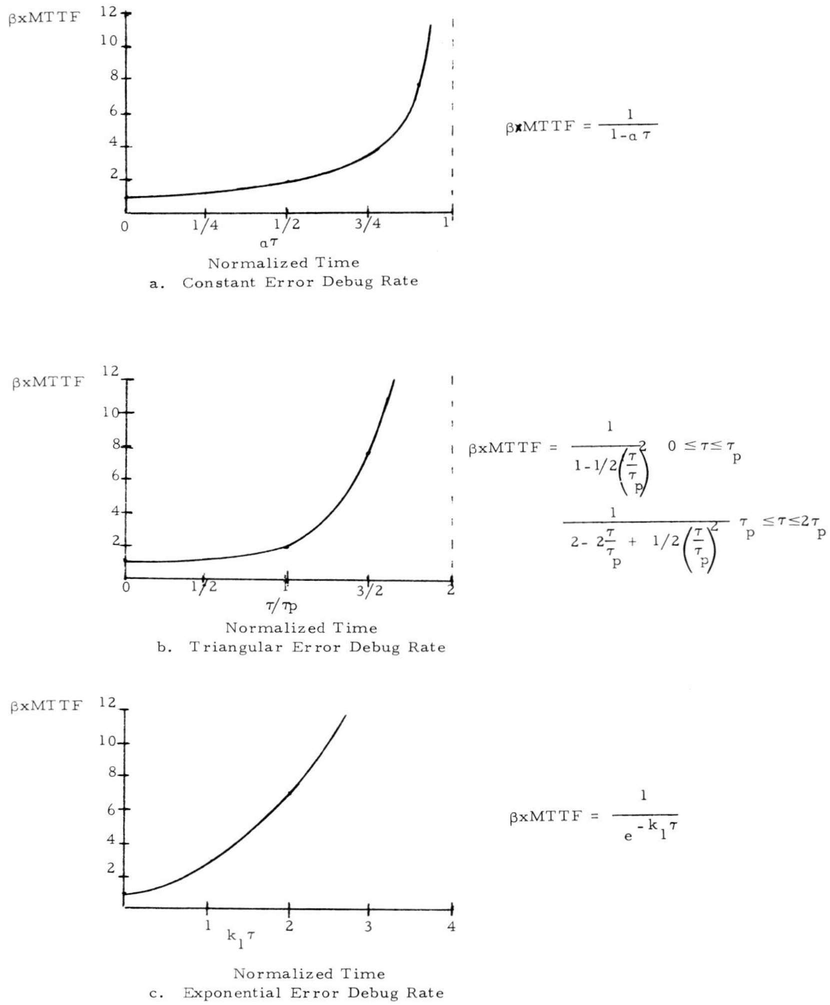

Fig. 5. Comparison of MTTF versus debugging time for three error models. (a): constant error debug rate; (b): triangular error debug rate; (c): exponential error debug rate.

AN INTRODUCTION TO GROWTH DYNAMICS

L.A. Belady and M.M. Lehman

1. PROJECT STATISTIC

The lifetime of any large programming system normally extends over a series of releases. Each such release represents an improvement on its predecessors as regards faults that have been repaired, performance weaknesses that have been removed, functional capability that has been improved and new function that has been added.

The history of a programming project is, in part, contained in the statistics (if recorded) of these releases. Thus typically there could be data available on the number of instructions and/or components (modules) added, removed and modified during the release, the number and reporting rate of previous repair work, the man-days, machine hours and dollar cost of each release, the time interval between releases and so on.

Examples of such data for a major system are given in figures 1 - 3. Figure 1 is a plot of the number of instructions in the system as a function of release number. It displays clearly the linear growth - the size of the system. In the present paper we shall not analyze the deviations from linearity, except to observe that the ripples on the data are typical of a self-stabilizing process with both positive and negative feedback loops.* That is, from a long-range point of view the rate of system growth is self-regulatory, despite the fact that many different causes control the selection of work implemented in each release, with varying budgets, increasing number of users desiring new functions or reporting faults, varying management attitudes towards system enhancement, changing release intervals and improving methodology.

In figure 2 we show as a further example a plot of the

*This observation was suggested to us by G. Prada.

net instruction growth of the system per release-interval day. In this case the points are fairly widely scattered, as would be expected for any project stretching over several years. Surprisingly, however, if these same data are smoothed, for example by averaging as in figure 3, the resultant points lie on a very smooth curve. The points of figure 3 can in fact be least-square fitted to a straight line and the logarithm of the points to the exponent of an exponential, with a chance probability of less than 0.1% and 0.01% respectively.

An extensive study of the total project statistics for this system has consistently displayed the same phenomenon. That is, in most cases, the raw data showed large variations from release to release. Smoothing or averaging, however, resulted in points that clearly lie on very smooth curves. In general these curves were of three classes - linear, exponential and bell-shaped. Linear fits appeared for measures reflecting the growth in size of the system. Exponentials appeared to fit well to points related to measures of system complexity. Data which reflected the effort (men, machine-power, cost, etc.) going into successive releases appeared to lie on a bell-shaped curve that represented an effort at first decreasing and subsequently increasing continually. This last observation corresponds to the common experience of many projects that a settling-in period during which maintenance becomes progressively easier is followed by increasing maintenance problems that ultimately cause the replacement of the system.

2. GROWTH DYNAMICS

The behavior of smoothed data, as discussed in the previous section, suggests that there is a self-regulating dynamics that underlies the growth of the system. That is, we may consider a system that comprises the body of code (constituting the programming system) together with the programmers that implement change and enhancement, their managers and executives and the user population. The dynamics of this system (the code, the programmer population, the applications, the hardware) as its internal state changes with time is governed by laws that cause it to be self-regulatory. In Simon's terms [1] we may view this total system as an artificial system and search for the laws that govern its behavior.

3. THE MACRO-MODEL

In two recent papers [2,3] the present authors studied the programming process. They showed that the effort required to take a system from the ith to the (i+1)st release could be represented by

$$W_i = W(B(i)) + C2^{2G(i)i-D(i)A(i)L(i)} \qquad (1)$$

In this expression W(B(i)) is a function of the budget allocated to the ith release. G(i) measures the lack of structure in the system. That is, a programming system perfectly structured would have G(i) equal to zero while for a system totally unstructured (every element connected to every other element) G(i) would approach infinity. The two parameters D(i) and A(i) measure the degree of knowledge that exists and is available about the internal state of the programming system. That is, D(i) represents of the completeness and accuracy of both present and archival system documentation, while A(i) measures the accessibility of that documentation. L(i), on the other hand, represents the experience and knowledge about the system of the programmers working on the system in the ith release. The greater the awareness, the less important is the existence and/or accessibility of the documentation.

We have not at this time exhaustively validated this macro-model, nor are we able to assign numerical values to the parameters. Qualitative and corroborative evidence of the general form of the model is, however, seen in the bell-shaped curves previously mentioned. Figure 4 represents an example from the same system referred to above, measuring components handled per day (smoothed data). The increased productivity indicated during the early days of the release resulted from the dominance of the first term of (1). Thus, during this period, budget and manpower allocated to the project was increasing, as were familiarity and experience with the system. At a later point in time, however, excessive rates of raw growth were attempted, documentation (as is quite usual) slipped behind, faults from earlier releases were reported as the number of users increased, new programmers were introduced and the second term began to dominate, leading to a falling-off of productivity.

Interpretation of the model also fits many of the concepts for improved programming effectiveness and quality that have evolved heuristically over the past few years. Thus, we see the macro model as a conceptual framework that explains many previously-known observations. G(i), for example, corresponds to Dijkstra's concept of structured programming [4]. Similarly, the nature of the exponent as the difference of two numbers shows why the use of high-level, self-documenting, error-avoiding programming languages makes a system more long-lived. It tends to keep the difference between the two terms of the exponent small, however large the terms may be individually.

As another example, we note that the exponent, as a difference between two numbers, would tend to grow with time even if it was initially small or even negative. However, at any time it can be made small again by spending sufficient effort (aside from the main enhancement/repair effort represented by W(B(i))) to update the documentation, to make it more accessible (increase D(i), A(i)) and/or improve the structure of the system (reduce G(i)).

This last observation suggests that we may postulate a "second law" analogous to the second law of thermodynamics. This would state that the complexity or unstructuredness - the entropy - of a programming system increases with time unless more effort specifically directed to its reduction is executed. That is, in our model G(i) grows with time. Such a postulate is in fact presented and justified in a paper currently in preparation [5].

4. THE MICRO-MODEL

We have also proposed a phenomenological micro-model of programming systems [3]. This relates the propensity of the system to spread change and faults. G(i) is in fact an abstraction of the micro-model which attempts to represent the lack of structure, the entropy, of a programming system by measuring the topology of interconnection between its elements. These interconnections relate to the sharing of objects (variables, parameters, tables, etc.) by the elements.

We do not here elaborate on this model. It will, however, be observed that it also conforms to heuristic observations. For example, it confirms that parameter-sharing through defined programming-element interfaces is to be preferred to sharing of global variables, that goto statements (which represent spurious linkages between elements) should be avoided [6]. The micro-model is also consistent with the postulated "second law" in that almost any change to a system will increase the number of elemental interconnections, by increased sharing of objects or adding branch instructions to inserted code sequences.

5. CITIES AND OTHER SYSTEMS

In a recent paper [7] Baumol discussed the results of unbalanced productivity growth on the macroeconomics of cities. The authors have suggested [5] that there is a strong link between Baumol's model that can be used to explain city decay and the decay of programs as suggested by our macro-model (1). They have further suggested that a similar phenomenon can explain the increase of bureaucracy in government, ment, the rise of middle management in large corporations and hence the decay trends in such institutions.

These topics remain to be more fully studied. The authors, however, consider it not premature to suggest that a common mechanism underlies all the above phenomena [5]. That is, they suggest that what is observed is in fact a decay trend present in all large systems. Thus, if the systems are left to follow their own dynamics they will tend to decay. Understanding the process can, however, lead to control of this decay process. Thus the authors recommend the study of this new and intrinsically statistical topic of growth dynamics to the participants in this conference.

REFERENCES

1. H.A. Simons, The science of the artificial, M.I.T. Press, 1969.
2. M.M. Lehman, "The programming process", IBM Research Division RC2722, December 1969.
3. L.A. Belady and M.M. Lehman, "Programming systems growth dynamics", IBM Research Division RC3546, September 1971.
4. E.W. Dijkstra, "Structured programming", EWD247, T.H. Eindhoven, 1970 (privately circulated).
5. L.A. Belady and M.M. Lehman, "On the macro-dynamics of programming and other systems", in preparation.
6. E.W. Dijkstra, "GOTO statement considered harmful", letter to the editor, Comm. ACM 11 (1968), 147-8.
7. W.J. Baumol, "Macro-economics of unbalanced growth: the anatomy of urban cities", Am. Econ. Rev. 55 (1967), 415-426.

```
H←CHN[B;4]
A FPLOT(H VS B)WITH(BS LIN H LSQ2 B)VS BS
```

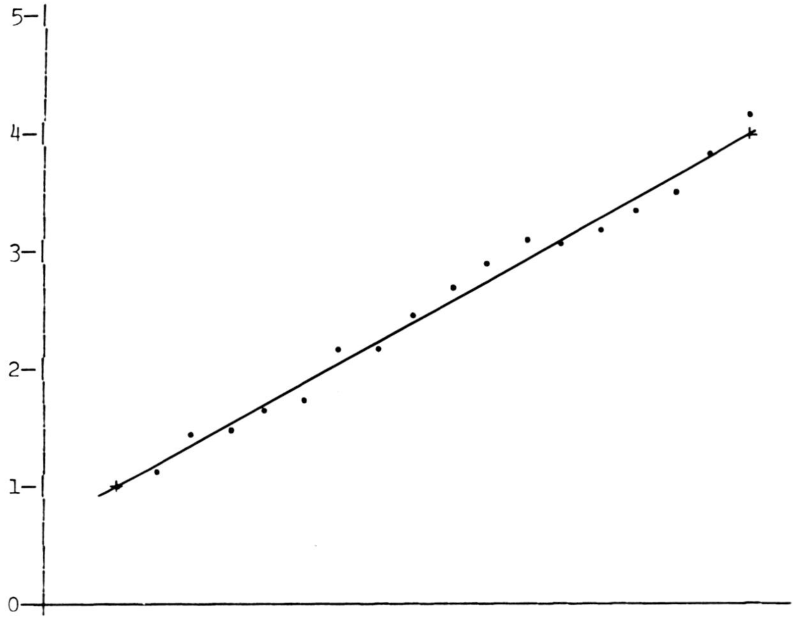

DEC. 14 1971/101

Fig. 1. Normalized component growth per release.

```
H←(CHN[B;7]-CHN[B-1;7])÷CHN[B;10]
A FPLOT H VS B
```

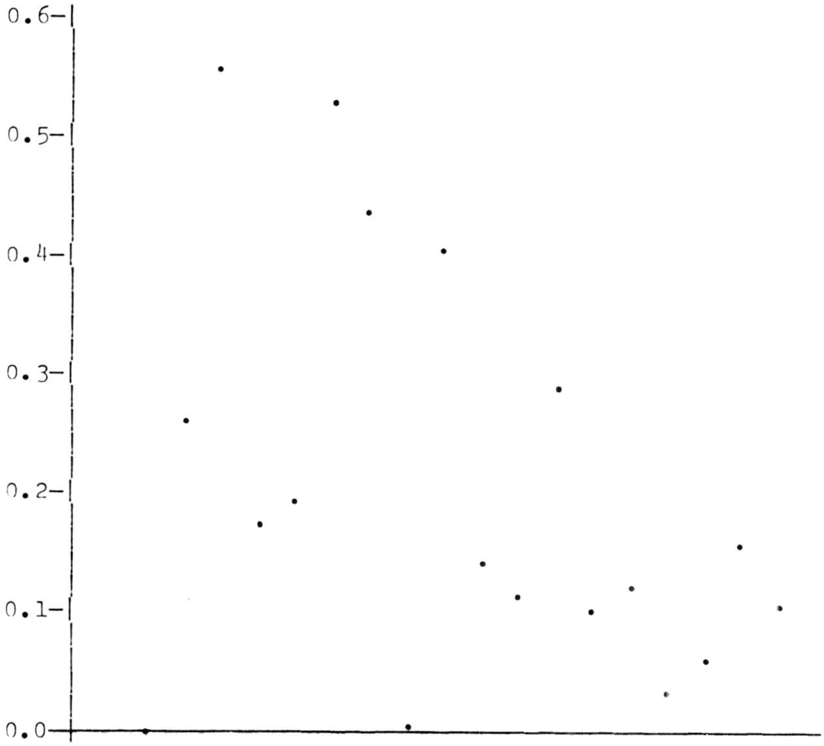

DEC. 14 1971/104

Fig. 2. Normalized net instruction growth per day.

```
H←(CHN[B;7]-CHN[B-1;7])÷CHN[B;10]
A FPLOT(C42 RDC H)VS C42 RDC B
```

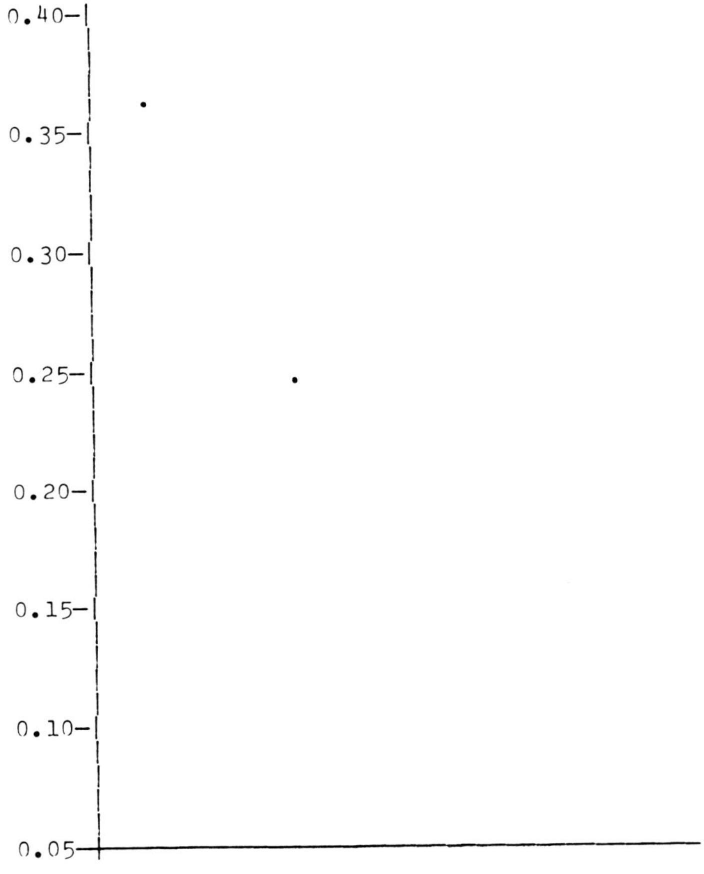

DEC. 14 1971/102

Fig. 3. Normalized net instruction growth per day, N-point average.

```
H←÷/CHN[B;5 10]
RPLOT C30
```

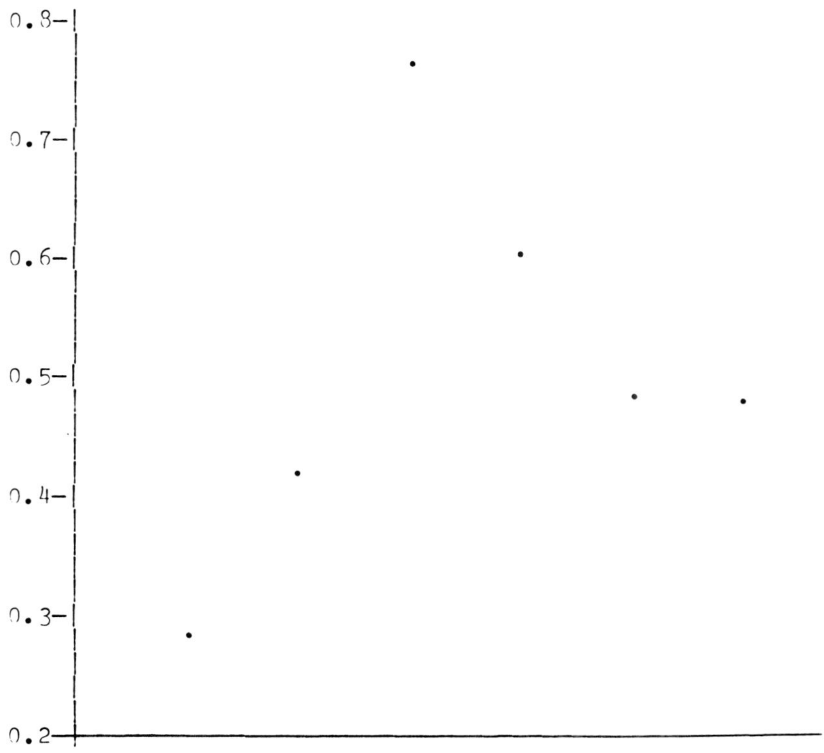

DEC. 14 1971/106

Fig. 4. Normalized components handled per release-interval day, N-point average.

DISCUSSION OF SESSION VII

by R.G. Munck

I was surprised at first that the two papers on reliability (the paper on system evolution was included at the last minute) would be included in this conference; they seemed to be more the type of thing that one would find in an IEEE conference on fault-tolerant computing. On reflection, however, reliability really is a part of systems performance. On second reflection, it's the major part, the sine qua non of a system, for there would be no point in further discussing the performance of a system which is unreliable. It is interesting that most of the papers at this conference have concerned CP-67/CMS, which is one of IBM's most reliable systems, and perhaps the first which has been reliable enough that its designers could go on to the secondary considerations of performance evaluation.

Computer charging today normally consists of charges for individual items of hardware (CPU, memory) and software (program products, programming advice). It is a pet thesis of mine that we should also charge for the reliability of a system, since that is rapidly becoming the major consideration of all involved with the system. This presents great difficulties, since we heretofore had no ways of quantifying reliability, but possibly the reliability models presented at this conference can give us a handle on doing this.

A major problem of these and similar studies is the difficulty of gathering data. The reliability of a commercial system involves data of tremendous financial import to the producing company, data which would undoubtedly be strictly proprietary. Similar data on government and military systems would undoubtedly fall under the all-enveloping shroud of government secrecy.

As a system designer, I am somewhat bothered by the models presented. From my point of view, there seem to be two types of errors, those of design and those of implementation. Design errors occur when the designers imperfectly understand the environment in which the system will run, and implementation errors occur in the mechanical processes of translation. The papers ignore this distinction, but it must be admitted that it is fuzzy in any case, and the papers don't claim to be definitive.

As a final point, one which was not stressed during the conference, I would like to present another reason why we do system performance evaluation: to find logical errors in the system. When the system is first conceived, the system model exists in the heads of the designers. This model is then translated in several ways: into machine instructions, which we somewhat loosely call "the system", and into a mathematical model, be it analytic, simulation, or Markovian, for the purposes of performance evaluation. We then "run" both models, and compare the results. Generally this comparison will point out errors in the mathematical model, since we don't understand how to translate into a mathematical model very well. Occasionally we will find that the machine model is incorrect, this being what I would call an implementation error. On rare occasions we will find that the model in our heads is wrong, this being a design error, although the comparison of models does not lead directly to this result. I would suggest that we should design systems which are easily translated into both machines and mathematics, so that a formalization of this process is possible.